Y0-BTB-223

# Contributions to Stochastics

In Honour of the 75th Birthday of Walther Eberl, Sr.

# Contributions to Stochastics

In Honour of the 75th Birthday of
Walther Eberl, Sr.

Edited by
Wolfgang Sendler

With Contributions by
M. Csörgö, S. Csörgö, M. Deistler, E. Dersch, U. Dieter
T. Drisch, E. J. Dudewicz, R. Dutter, W. Eberl jr., J. Franz
P. Gaenssler, B. Gyires, R. Hafner, J. Hartung, W. Hazod
R. Henn, L. Horváth, P. Kischka, F. Österreicher, D. Plachky
L. Rüschendorf, W. Schneemeier, W. Sendler, J. Steinebach
H. Strasser, B. K. Taneja, E. C. van der Meulen, R. Viertl
I. Vincze, B. Voet, W. Winkler

With 8 Figures

Physica-Verlag Heidelberg

Professor Dr. Wolfgang Sendler
Universität Trier
Fachbereich IV/Mathematik
Postfach 38 25
5500 Trier, FRG

0261-6944
MATH.-STAT.

ISBN 3-7908-0370-7 Physica-Verlag Heidelberg
ISBN 0-387-91302-5 Springer-Verlag New York

CIP-Kurztitelaufnahme der Deutschen Bibliothek
Contributions to Stochastics : in Honour of the 75. Birthday of Walter Eberl, Sr.
ed. by Wolfgang Sendler. With contributions by M. Csörgö ...
- Heidelberg : Physica-Verlag, 1987.
ISBN 3-7908-0370-7
ISBN 0-387-91302-5
NE: Sendler, Wolfgang [Hrsg.]: Csörgö, Miklós [Mitverf.];
Eberl, Walther: Festschrift

This work is subject to copyright. All rights are reserved, whether the whole or part of the material is concerned, specifically the rights of translation, reprinting, reuse of illustrations, recitation, broadcasting, reproduction on microfilms or in other ways, and storage in data banks. Duplication of this publication or parts thereof is only permitted under the provisions of the German Copyright Law of September 9, 1965, in its version of June 24, 1985, and a copyright fee must always be paid. Violations fall under the prosecution act of the German Copyright Law.

© Physica-Verlag Heidelberg 1987
Printed in Germany

The use of registered names, trademarks, etc. in this publication does not imply, even in the absence of a specific statement, that such names are exempt from the relevant protective laws and regulations and therefore free for general use.

Printing: Weihert-Druck GmbH, Darmstadt
Bookbinding: J. Schäffer GmbH u. Co. KG., Grünstadt
7120/7130-543210

QA 274
C661
1987
MATH

# Preface

I met Professor Walther Eberl for the first time during the early sixties due to the fact that one of my student-fellows at the University of Vienna was his son, Dr. Walther Eberl, jr. Since Professor Eberl held a position at the "Technische Hochschule" (now "Technische Universität") of Vienna, I came not in more than a loose private contact with him at that time. But closely after having received my Ph.D., I got an assistants position at his institute and was working there from 1969 until fall 1971.

During that time I profitably took part on the excellent atmosphere of that institute as well as Professor Eberl's permanent encouragement to make mathematical research. I also had opportunity to observe his cumbersome activities to enlarge the statistical institute of the Technische Universität, a finally successful fight as is shown by the present number of three chairs in stochastics.

Early in 1986 somebody out of the circle of former coworkers of Professor Eberl reminded me of his 75$^{th}$ birthday in 1987. It was soon clear that the most adequate present would be scientific work, the thing around which Professor Eberl's life is still centered. Having in mind his broad interests, we then decided the appropriate form to be a volume with contributions from various fields of stochastics in honour of him.

I want to express my warmest thanks to all contributors and Dr. Walther Eberl, jr., for his help in organizing the presented volume. Moreover, my thanks go to Professor Reinhard Viertl for having prepared Professor Eberl's curriculum vitae and his list of publications, which can be found at the end of this volume.

Wolfgang Sendler

Trier, January 1987

# Contents

# On the Distribution of the Suprema of Weighted Empirical Processes

Miklós Csörgő
Department of Mathematics and Statistics, Carleton University
Ottawa, Canada K1S 5B6

Lajos Horváth
Bolyai Institute
Szeged University
H-6720 Szeged, Hungary

Josef Steinebach
Fachbereich Mathematik
Universität Marburg
D-3550 Marburg, FRG

SUMMARY: A complete characterization of the distributions of the suprema of weighted empirical processes is given. The limit distributions are expressed in terms of weighted Brownian bridge, Wiener, Ornstein-Uhlenbeck and Poisson processes.

## 1. Introduction

Let $U_1, U_2, \ldots$ be independent uniform (0,1) random variables (r.v.'s), and let $U_{1,n} \le U_{2,n} \le \ldots \le U_{n,n}$ be the order statistics of the first $n \ge 1$ of these with empirical distribution function

$$E_n(s) = \frac{1}{n}\#\{1 \le i \le n : U_i \le s\}, \quad 0 \le s \le 1,$$

and empirical quantile function

$$U_n(s) = U_{k,n}, \quad (k-1)/n < s \le k/n \quad (k = 1,\ldots,n).$$

The uniform empirical process is

$$e_n(s) = n^{\frac{1}{2}}(E_n(s)-s), \quad 0 \le s \le 1,$$

and the uniform quantile process is

$$u_n(s) = n^{\frac{1}{2}}(s-U_n(s)), \quad 0 < s \le 1.$$

There have been many papers studying the asymptotic distribution of weighted empirical and quantile processes. For a review on earlier results we may for example refer to Chapter 5 of M. CSÖRGŐ [2]. More recently MASON [11,12], and M. CSÖRGŐ and MASON [7] established some new results in weighted sup-norm metrics for the uniform empirical and quantile processes. Concentrating on $u_n$, M. CSÖRGŐ and HORVÁTH [4] have extended these results and also obtained similar ones for the

Contributions to Stochastics.
Ed. by W. Sendler
© Physica-Verlag Heidelberg 1987

general quantile process. Also, $L_1$ and $L_2$ functionals of weighted $e_n$ and $u_n$ were studied by M. CSÖRGŐ and HORVÁTH [5]. Most of these results are based on the following approximation theorem of CSÖRGŐ et al.

Theorem A (M. CSÖRGŐ, S. CSÖRGŐ, HORVÁTH, MASON [3]). On an appropriate probability space $(\Omega, A, P)$ a sequence of Brownian bridges $\{B_n(s); 0 \le s \le 1\}$ can be so defined that for all $\lambda > 0$ we have, as $n \to \infty$,

$$\sup_{\lambda/n \le s \le 1-\lambda/n} n^{\alpha}|e_n(s)-B_n(s)|/(s(1-s))^{\frac{1}{2}-\alpha} = O_P(1) \tag{1.1}$$

for every $0 \le \alpha < \frac{1}{4}$, and

$$\sup_{\lambda/n \le s \le 1-\lambda/n} n^{\beta}|u_n(s)-B_n(s)|/(s(1-s))^{\frac{1}{2}-\beta} = O_P(1) \tag{1.2}$$

for every $0 \le \beta < \frac{1}{2}$.

Throughout in this paper we are working on the probability space of Theorem A. A simple proof of Theorem A can be found in M. CSÖRGŐ and HORVÁTH [6].

Putting our emphasis now on $e_n$, here we will study its weighted sup-norm behaviour, similarly to that of $u_n$ in [4]. However, here the question of which portion of the unit interval determines the limiting behaviour will be studied in a more refined way, providing new results also for $u_n$.

Due to the symmetry of $e_n$ around $\frac{1}{2}$, our results will be proven only on the interval $(0,\frac{1}{2}]$. We use the notation B for a Brownian bridge and W for a standard Wiener process. The weighted processes of our results are not necessarily defined at zero. We use the convention that $\sup_{0<s<a} f(s) = \lim_{t\downarrow 0} \sup_{t<s<a} f(s)$, where the latter limit is finite a.s. A similar convention is used at $\infty$. Also, $a \wedge b = \min(a,b)$, and $a \vee b = \max(a,b)$ throughout.

## 2. Results

In this paper we will work with weight functions of the form $s^{\nu}L(s)$ $(-\infty < \nu < \infty,\ s \in (0,\frac{1}{2}])$, where

$$L \text{ is slowly varying at zero,} \tag{2.1}$$

i.e., L is measurable, positive and $L(\lambda s)/L(s) \to 1$ $(s \downarrow 0)$ for all $\lambda > 0$.

The first theorem will be in terms of the integral $I(q,c) = \int_0^{\frac{1}{2}} s^{-1}\exp\{-cq^2(s)/s\}ds$.

Theorem 2.1. Let L be as in (2.1), and assume that $I(s^{\nu}L(s),c) < \infty$

*for some* $c > 0$. *Then, as* $n \to \infty$,

$$\sup_{0 \le s \le \frac{1}{2}} |e_n(s)|/(s^{\nu}L(s)) \xrightarrow{\mathcal{D}} \sup_{0 \le s \le \frac{1}{2}} |B(s)|/(s^{\nu}L(s)), \tag{2.2}$$

$$\sup_{k_n/n \le s \le \frac{1}{2}} |e_n(s)|/(s^{\nu}L(s)) \xrightarrow{\mathcal{D}} \sup_{0 \le s \le \frac{1}{2}} |B(s)|/(s^{\nu}L(s)), \tag{2.3}$$

*and*

$$\sup_{0 \le s \le k_n/n} |e_n(s)|/(s^{\nu}L(s)) \xrightarrow{\mathcal{D}} c_0 , \tag{2.4}$$

*where* $k_n \ge 0$, $k_n \to \infty$, $k_n/n \to 0$ *and* $0 \le c_0 < \infty$.

Remark 2.1. The constant $c_0$ of (2.4) is the same as that of

$$\limsup_{s \downarrow 0} |B(s)|/(s^{\nu}L(s)) = c_0 \quad \text{a.s.},$$

whose finiteness was proved in the proof of Proposition 2.1 in [4] under $I(s^{\nu}L(s),c) < \infty$ for some $c > 0$ (cf. also Theorem 3.3 in [3]).

Theorem 2.2. *Let* $L$ *be as in* (2.1) *and* $-\infty < \nu < \frac{1}{2}$. *Assume* $0 \le m_n \le k_n$ , $k_n \to \infty$, $k_n/n \to 0$, *and* $m_n/k_n \to c \in [0,1]$, *as* $n \to \infty$. *Then*

$$\left(\frac{k_n}{n}\right)^{\nu-\frac{1}{2}} L(k_n/n) \sup_{(\lambda \vee m_n)/n \le s \le k_n/n} \frac{|e_n(s)-u_n(s)|}{s^{\nu}L(s)} = o_P(1), \tag{2.5}$$

*and*

$$\left(\frac{k_n}{n}\right)^{\nu-\frac{1}{2}} L(k_n/n) \sup_{(\lambda \vee m_n)/n \le s \le k_n/n} |u_n(s)|/(s^{\nu}L(s)) \xrightarrow{\mathcal{D}} \sup_{c \le s \le 1} |W(s)|/s^{\nu}. \tag{2.6}$$

The standard deviation of a Brownian bridge $B(s)$ is $(s(1-s))^{\frac{1}{2}}$ and the thus standardized empirical process is $\{e_n(s)/(s(1-s))^{\frac{1}{2}};\ 0 < s < 1\}$. The latter is a natural enough process to be studied on its own. EICKER [9] and JAESCHKE [10] initiated such studies. Further variations are also given here.

Let

$$a(x) = (2 \log x)^{\frac{1}{2}}, \tag{2.7}$$

$$b(x) = 2 \log x + (\tfrac{1}{2}) \log\log x - (\tfrac{1}{2}) \log \pi \tag{2.8}$$

and

$$c(x) = \log((1-x)/x). \tag{2.9}$$

Define $Y$ to be the maximum of two independent r.v.'s whose distribution function is $\exp(-\exp(-x))$, $-\infty < x < \infty$. Let $\{V(t);\ -\infty < t < \infty\}$ be the Ornstein-Uhlenbeck process, i.e., a stationary Gaussian process with $EV(t) = 0$ and covariance function $EV(t)V(s) = \exp(-|t-s|)$.

Theorem 2.3. *Assume* $0 \le m_n \le k_n$, $k_n \to \infty$, $k_n/n \to 0$, *and* $m_n/k_n \to c \in [0,1]$, *as* $n \to \infty$. *Then*

$$a((\tfrac{1}{2})\log n) \sup_{0\le s\le \frac{1}{2}} |e_n(s)|/(s(1-s))^{\frac{1}{2}} - b((\tfrac{1}{2})\log n) \xrightarrow{\mathcal{D}} Y, \tag{2.10}$$

$$a((\tfrac{1}{2})\log k_n) \sup_{0\le s\le k_n/n} |e_n(s)|/(s(1-s))^{\frac{1}{2}} - b((\tfrac{1}{2})\log k_n) \xrightarrow{\mathcal{D}} Y. \tag{2.11}$$

$$a((\tfrac{1}{2})\log k_n) \sup_{k_n/n\le s\le \frac{1}{2}} |e_n(s)|/(s(1-s))^{\frac{1}{2}} - b((\tfrac{1}{2})c(k_n/n)) \xrightarrow{\mathcal{D}} Y. \tag{2.12}$$

*Also, if* $c > 0$, *then*

$$\sup_{m_n/n\le s\le k_n/n} |e_n(s)|/(s(1-s))^{\frac{1}{2}} \xrightarrow{\mathcal{D}} \sup_{0\le s\le (\frac{1}{2})\log(1/c)} |V(s)|, \tag{2.13}$$

*while if* $c = 0$, *then*

$$a((\tfrac{1}{2})\log(k_n/m_n^*)) \sup_{m_n/n\le s\le k_n/n} |e_n(s)|\,|(s(1-s))^{\frac{1}{2}} - b((\tfrac{1}{2})\log(k_n/m_n^*)) \xrightarrow{\mathcal{D}} Y, \tag{2.14}$$

*where* $m_n^* = m_n \vee 1$.

The statements (2.10)-(2.12) were first proven under some conditions on $k_n$ in [9], [10], and in the present form in [7].

Remark 2.2. When compared to (2.14), the statement of (2.13) shows that, in order to have an extreme value distribution, the length of the indicated intervals, whose end points tend to zero, cannot be too small. It is also interesting to note that ANDERSON and DARLING [1] have obtained that

$$\sup_{a\le s\le b} |e_n(s)|/(s(1-s))^{\frac{1}{2}} \xrightarrow{\mathcal{D}} \sup_{0\le s\le (\frac{1}{2})\log(\frac{b}{a}\frac{1-a}{1-b})} |V(s)|,$$

where $0 < a \le b < 1$, the same limit as in (2.13).

Corollary 2.2. *Under the conditions of Theorem 2.3 we have also*

$$\sup_{m_n/n\le s\le k_n/n} |u_n(s)|/(s(1-s))^{\frac{1}{2}} \xrightarrow{\mathcal{D}} \sup_{0\le s\le (\frac{1}{2})\log(1/c)} |V(s)|, \tag{2.15}$$

*if* $c \ne 0$, *and if* $c = 0$ *then for all* $\lambda > 0$

$$a((\tfrac{1}{2})\log(k_n/m_n^*)) \sup_{(m_n\vee\lambda)/n\le s\le k_n/n} |u_n(s)|/(s(1-s))^{\frac{1}{2}} - b((\tfrac{1}{2})\log(k_n/m_n^*)) \xrightarrow{\mathcal{D}} Y, \tag{2.16}$$

where $m_n^* = m_n \vee 1$.

The next theorems are concerned with heavy weights with $\frac{1}{2} < \nu < \infty$.

Theorem 2.4. Let $L$ be as in (2.1) and $\frac{1}{2} < \nu < \infty$. Assume $k_n \le r_n \le n/2$, $k_n \to \infty$, $k_n/n \to 0$ and $k_n/r_n \to c \in [0,1]$, as $n \to \infty$. Then

$$\left(\frac{k_n}{n}\right)^{\nu-\frac{1}{2}} L(k_n/n) \sup_{k_n/n \le s \le r_n/n} |e_n(s)|/(s^\nu L(s)) \xrightarrow{\mathcal{D}} \sup_{c \le s \le 1} |W(s)|/s^{1-\nu}.$$

The special case of $\frac{1}{2} < \nu \le 1$, $L = 1$ and $r_n = n/2$ was proved in [12].

Corollary 2.3. Under the conditions of Theorem 2.4 we have

$$\left(\frac{k_n}{n}\right)^{\nu-\frac{1}{2}} L(k_n/n) \sup_{k_n/n \le s \le r_n/n} |u_n(s)|/(s^\nu L(s)) \xrightarrow{\mathcal{D}} \sup_{c \le s \le 1} |W(s)|/s^{1-\nu}.$$

Let $\{S_i;\ i \ge 1\}$ be partial sums of independent identically distributed exponential r.v.'s with expectation 1. With $I\{A\}$ indicator function we define the Poisson process

$$N(t) = \sum_{i=1}^{\infty} i\, I\{S_i \le t < S_{i+1}\}. \tag{2.17}$$

Theorem 2.5. Let $L$ be as in (2.1) and $\frac{1}{2} < \nu < \infty$. Assume $0 \le m_n \le k_n \le n/2$, $m_n \to c_1 \in [0,\infty)$ and $k_n \to c_2 \in [c_1,\infty]$, as $n \to \infty$. Then, if $c_1 > 0$,

$$n^{\frac{1}{2}-\nu} L(1/n) \sup_{m_n/n \le s \le k_n/n} |e_n(s)|/(s^\nu L(s)) \xrightarrow{\mathcal{D}} \sup_{c_1 \le s \le c_2} |N(s)-s|/s^\nu. \tag{2.18}$$

If $c_1 = 0$ and $\frac{1}{2} < \nu < 1$, then

$$n^{\frac{1}{2}-\nu} L(1/n) \sup_{m_n/n \le s \le k_n/n} |e_n(s)|/(s^\nu L(s)) \xrightarrow{\mathcal{D}} \sup_{0 \le s \le c_2} |N(s)-s|/s^\nu, \tag{2.19}$$

and if $\nu > 1$, then

$$n^{\frac{1}{2}-\nu} L(1/n) \sup_{m_n/n \le s \le k_n/n} |e_n(s)|/(s^\nu L(s)) \xrightarrow{\mathcal{D}} \infty. \tag{2.20}$$

Also,

$$n^{\frac{1}{2}-\nu} L(1/n) \sup_{U_{1,n} \le s \le k_n/n} |e_n(s)|/(s^\nu L(s)) \xrightarrow{\mathcal{D}} \sup_{S_1 \le s \le c_2} |N(s)-s|/s^\nu. \tag{2.21}$$

If $c_1 = 0$, $\nu = 1$, and $L(s) \ge L(t)$, $s \le t$, in a neighbourhood of zero, then

$$n^{-\frac{1}{2}}L(1/n) \sup_{m_n/n \le s \le k_n/n} |e_n(s)|/(sL(s)) \xrightarrow{\mathcal{D}} \sup_{0 \le s \le c_2} |N(s)-s|/s^{\nu}. \qquad (2.22)$$

_While if_ $c_1 = 0$, $\nu = 1$, _and_ $L(s) \to 0$ _as_ $s \to 0$, _then_

$$n^{-\frac{1}{2}}L(1/n) \sup_{0<s\le k_n/n} |e_n(s)|/(sL(s)) \xrightarrow{\mathcal{D}} \infty. \qquad (2.23)$$

We should note that the special case of $\frac{1}{2} < \nu \le 1$, $L = 1$ and $m_n = 0$ was proven in [11]. An analogue of Theorem 2.5 for $u_n$ can be found in [4].

In many of our preceding results we used normalizing sequences of the form $(k_n/n)^{\nu-\frac{1}{2}}L(k_n/n)$. It is also natural to ask the question whether these sequences could be replaced by similar functions of corresponding order statistics, and also whether our weight functions of the form $s^{\nu}L(s)$ could be replaced by their random counterparts $(E_n(s))^{\nu}L(E_n(s))$ and $(U_n(s))^{\nu}L(U_n(s))$. Indeed, whenever feasible, this can always be done on account of the following theorem of SMIRNOV [13].

Theorem B (SMIRNOV [13]). _Assume_ $k_n \le n$ _and_ $k_n \to \infty$ _as_ $n \to \infty$. _Then_

$$\sup_{k_n/n \le s \le 1} |E_n(s)/s-1| \xrightarrow{P} 0$$

_and_

$$\sup_{k_n/n \le s \le 1} |U_n(s)/s-1| \xrightarrow{P} 0.$$

This result and Corollary 1.2.1 of DE HAAN [8] immediately imply

Corollary 2.4. _Let_ $L$ _be as in_ (2.1) _and_ $-\infty < \nu < \infty$. _If_ $k_n \le n$ _and_ $k_n \to \infty$ _as_ $n \to \infty$, _then_

$$\sup_{k_n/n \le s \le 1} |(E_n(s))^{\nu}L(E_n(s))/(s^{\nu}L(s))-1| \xrightarrow{P} 0$$

_and_

$$\sup_{k_n/n \le s \le 1} |(U_n(s))^{\nu}L(U_n(s))/(s^{\nu}L(s))-1| \xrightarrow{P} 0.$$

Hence by Corollary 2.4 we can use random normalizing factors, weight functions and intervals in some of our results above. For example, under the conditions of Theorem 2.2 we have, as $n \to \infty$,

$$(U_{[k_n],n})^{\nu-\frac{1}{2}}L(U_{[k_n],n}) \sup_{U_{[m_n],n} \le s \le U_{[k_n],n}} |e_n(s)|/((E_n(s))^{\nu}L(E_n(s)))$$

$$\xrightarrow{\mathcal{D}} \sup_{c \le s \le 1} |W(s)|/s^{\nu}.$$

One should be very careful with arguments like in the preceding situations. For example, the statement of (2.21) hsould be compared to

$$(U_{1,n})^{\frac{1}{2}-\nu} L(U_{1,n}) \sup_{U_{1,n}<s\le k_n/n} |e_n(s)|/((E_n(s))^{\nu}L(E_n(s)))$$

$$\xrightarrow{\mathcal{D}} S_1 \sup_{S_1<s\le c_2} |N(s)-s|/(N(s))^{\nu}.$$

3. Proofs

Proof of Theorem 2.1. First we note that finiteness of $I(s^{\nu}L(s),c)$ for some $c > 0$ implies (cf. (2.4) in [4])

$$s^{\nu}L(s)/s^{\frac{1}{2}} \longrightarrow \infty\ ,\ s \downarrow 0\ . \tag{3.1}$$

Next,

$$\sup_{0\le s\le U_{1,n}} |e_n(s)|/(s^{\nu}L(s)) = n^{\frac{1}{2}} \sup_{0\le s\le U_{1,n}} s/(s^{\nu}L(s))$$

$$\le (nU_{1,n})^{\frac{1}{2}} \sup_{0\le s\le U_{1,n}} s^{\frac{1}{2}}/(s^{\nu}L(s))$$

$$= o_P(1)$$

by (3.1). By WELLNER [14], for every $\varepsilon > 0$ there is a $0 < \lambda < 1$ such that

$$P\{\lambda s \le U_n(s) \le (1/\lambda)s,\ 1/(n+1) \le s \le n/(n+1)\} \ge 1-\varepsilon\ . \tag{3.2}$$

The latter combined with (1.1) yields that for every $\varepsilon > 0$ we have

$$\sup_{U_{1,n}\le s\le\frac{1}{2}} |e_n(s)-B_n(s)|/(s^{\nu}L(s)) \le \sup_{U_{1,n}\le s\le\varepsilon} |e_n(s)-B_n(s)|/(s^{\nu}L(s))$$

$$+ \sup_{\varepsilon\le s\le\frac{1}{2}} |e_n(s)-B_n(s)|/(s^{\nu}L(s))$$

$$= O_P(1) \sup_{0\le s\le\varepsilon} s^{\frac{1}{2}}/(s^{\nu}L(s))+o_P(1).$$

By (3.1), the latter statement results in

$$\sup_{U_{1,n}\le s\le\frac{1}{2}} |e_n(s)-B_n(s)|/(s^{\nu}L(s)) = o_P(1). \tag{3.3}$$

Hence it suffices to have that

$$\sup_{U_{1,n}\le s\le\frac{1}{2}} |B_n(s)|/(s^{\nu}L(s)) \xrightarrow{\mathcal{D}} \sup_{0\le s\le\frac{1}{2}} |B(s)|/(s^{\nu}L(s)),$$

which in turn is immediate by Remark 2.1. The proof of (2.2) is now complete. Also, the proofs of (2.3) and (2.4) follow by (3.1) and (3.3) in a similar fashion.

In the proof of Theorem 2.2 we will use the following lemma.

Lemma 3.1. _Let_ $L$ _be as in_ (2.1) _and_ $-\infty < \nu < \frac{1}{2}$. _Assume_ $0 < \lambda \le m_n^* \le k_n$, $k_n \to \infty$, $k_n/n \to 0$, _and_ $m_n^*/k_n \to c^*$, $0 \le c^* \le 1$, _as_ $n \to \infty$. _Then_

$$\left(\frac{k_n}{n}\right)^{\nu-\frac{1}{2}} L(k_n/n) \sup_{m_n^*/n \le s \le k_n/n} |B(s)|/(s^\nu L(s)) \xrightarrow{\mathcal{D}} \sup_{c^* \le s \le 1} |W(s)|/s^\nu .$$

Proof. We have

$$\{B(s);\ 0 \le s \le 1\} \overset{\mathcal{D}}{=} \{W(s)-sW(1);\ 0 \le s \le 1\}. \tag{3.4}$$

Hence, we first consider

$$\left(\frac{k_n}{n}\right)^{\nu-\frac{1}{2}} L(k_n/n)|W(1)| \sup_{m_n^*/n \le s \le k_n/n} s/(s^\nu L(s))$$

$$\le |W(1)| \left(\frac{k_n}{n}\right)^{\frac{1}{2}} \sup_{\lambda/k_n \le t \le 1} t^{1-\nu} L(k_n/n)/L(tk_n/n).$$

By (4.8) in [4] (cf. also Theorem 4.7 in [5])

$$\sup_{\lambda/k_n \le t \le 1} t^{\varepsilon} L(k_n/n)/L(tk_n/n) = O(1) \tag{3.5}$$

for every $\varepsilon > 0$. Choosing $0 < \varepsilon < 1-\nu$ in (3.5) we get

$$\left(\frac{k_n}{n}\right)^{\nu-\frac{1}{2}} L(k_n/n)|W(1)| \sup_{m_n^*/n \le s \le k_n/n} s/(s^\nu L(s)) = O_P\left(\left(\frac{k_n}{n}\right)^{\frac{1}{2}}\right) = o_P(1). \tag{3.6}$$

Therefore by (3.4) it is enough to prove

$$\left(\frac{k_n}{n}\right)^{-\frac{1}{2}} L(k_n/n) \sup_{m_n^*/k_n \le t \le 1} |W(tk_n/n)|/(t^\nu L(tk_n/n)) \xrightarrow{\mathcal{D}} \sup_{c^* \le t \le 1} |W(t)|/t . \tag{3.7}$$

Let $c^* > 0$. Then by Lemma 4.2 in [4]

$$\sup_{m_n^*/k_n \le t \le 1} \left| \frac{L(k_n/n)}{L(tk_n/n)} - 1 \right| = o(1),$$

and in the latter case of $c^* > 0$, (3.7) follows immediately.

If $c^* = 0$, we let $0 < \delta < 1$ and consider

$$\left(\frac{k_n}{n}\right)^{-\frac{1}{2}} L(k_n/n) \sup_{\lambda/k_n \le t \le \delta} |W(tk_n/n)|/(t^{\nu} L(tk_n/n))$$

$$\stackrel{\mathcal{D}}{=} O(1) \sup_{\lambda/k_n \le t \le \delta} |W(t)|/t^{\nu+\varepsilon},$$

the latter by (3.5) with $0 < \varepsilon < \frac{1}{2}-\nu$. Again by Lemma 4.2 in [4],

$$\left(\frac{k_n}{n}\right)^{-\frac{1}{2}} L(k_n/n) \sup_{\delta \le t \le 1} |W(tk_n/n)|/(t^{\nu} L(tk_n/n)) \xrightarrow{\mathcal{D}} \sup_{\delta \le t \le 1} |W(t)|/t^{\nu}.$$

Now the proof of (3.7) is complete by the law of iterated logarithm for $W$, on taking $\delta$ arbitrarily small.

Proof of Theorem 2.2. First, by (1.1) and (3.2) we get

$$\left(\frac{k_n}{n}\right)^{\nu-\frac{1}{2}} L(k_n/n) \sup_{U_{1,n} \le s \le \frac{1}{2}} |e_n(s)-B_n(s)|/(s^{\nu} L(s))$$

$$= O_P(1) \left(\frac{k_n}{n}\right)^{\nu-\frac{1}{2}} L(k_n/n) n^{-\alpha} \sup_{U_{1,n} \le s < \frac{1}{2}} \frac{s^{\frac{1}{2}-\alpha-\nu}}{L(s)}$$

$$= O_P(1) \left(\frac{k_n}{n}\right)^{\nu-\frac{1}{2}} n^{-\alpha} k_n^{\varepsilon} \sup_{U_{1,n} \le s \le \frac{1}{2}} s^{\frac{1}{2}-\alpha-\nu},$$

where the last equality is due to

$$\sup_{U_{1,n} \le s \le \frac{1}{2}} k_n^{-\varepsilon} L(k_n/n)/L(s) = O_P(1) \tag{3.8}$$

(cf. (4.5) in [4] or Theorem 4.7 in [5]). Now, assuming $\frac{1}{4} < \nu < \frac{1}{2}$ and letting $\alpha = \frac{1}{2}-\nu$, we get

$$\left(\frac{k_n}{n}\right)^{\nu-\frac{1}{2}} L(k_n/n) \sup_{U_{1,n} \le s \le \frac{1}{2}} |e_n(s)-B_n(s)|/(s^{\nu} L(s)) = o_P(1). \tag{3.9}$$

When $\nu \le \frac{1}{4}$, we let $\frac{1}{4} < \delta < \frac{1}{2}$, and note that

$$\left(\frac{k_n}{n}\right)^{\nu-\frac{1}{2}} \le \left(\frac{k_n}{n}\right)^{\delta-\frac{1}{2}} / s^{\delta-\nu}, \quad U_{1,n} \le s \le k_n/n.$$

Consequently, by (3.9) we have also

$$\left(\frac{k_n}{n}\right)^{\nu-\frac{1}{2}} L(k_n/n) \sup_{U_{1,n}\le s\le\frac{1}{2}} |e_n(s)-B_n(s)|/(s^{\nu}L(s))$$

$$\le \left(\frac{k_n}{n}\right)^{\delta-\frac{1}{2}} L(k_n/n) \sup_{U_{1,n}\le s\le\frac{1}{2}} |e_n(s)-B_n(s)|/(s^{\delta}L(s)) = o_P(1). \tag{3.10}$$

Considering now the case of $c > 0$ in the theorem, by (3.2) we first conclude that

$$\lim_{n\to\infty} P\{U_{1,n} \le m_n/n\} = 1,$$

and therefore by (3.9) and (3.10) we get

$$\left(\frac{k_n}{n}\right)^{\nu-\frac{1}{2}} L(k_n/n) \sup_{m_n/n\le s\le k_n/n} |e_n(s)-B_n(s)|/(s^{\nu}L(s)) = o_P(1).$$

Now by Lemma 3.1 with $m_n^* = m_n$ and $c^* = c$ we get the desired result.

Let now $c = 0$. First, by Lemma 4.1 in [4], and then by (3.8) with $\varepsilon = (\frac{1}{2}-\nu)/2$

$$\left(\frac{k_n}{n}\right)^{\nu-\frac{1}{2}} L(k_n/n) \sup_{0\le s\le U_{1,n}} |e_n(s)|/(s^{\nu}L(s))$$

$$= \left(\frac{k_n}{n}\right)^{\nu-\frac{1}{2}} L(k_n/n) \sup_{0\le s\le U_{1,n}} n^{\frac{1}{2}}s/(s^{\nu}L(s))$$

$$= O_P(1) \left(\frac{k_n}{n}\right)^{\nu-\frac{1}{2}} (L(k_n/n)/L(U_{1,n}))(U_{1,n})^{1-\nu} = o_P(k_n^{\nu-\frac{1}{2}+\varepsilon})$$

$$= o_P(1). \tag{3.11}$$

Thus, by (3.9)-(3.11) we get

$$\left(\frac{k_n}{n}\right)^{\nu-\frac{1}{2}} L(k_n/n) \sup_{m_n/n\le s\le k_n/n} |e_n(s)|/(s^{\nu}L(s))$$

$$= \left(\frac{k_n}{n}\right)^{\nu-\frac{1}{2}} L(k_n/n) \sup_{U_{1,n}\vee m_n/n\le s\le k_n/n} |B_n(s)|/(s^{\nu}L(s))+o_P(1).$$

By (3.2) we have that for every $\varepsilon > 0$ there is a $\lambda \in (0,1)$ such that

$$P\{\lambda/n \le U_{1,n} \le 1/(\lambda n)\} \ge 1-\varepsilon .$$

Now taking $m_n^* = \lambda/n \vee m_n/n$, or $m_n^* = 1/(\lambda n) \vee m_n/n$ in Lemma 3.1, we get

$$\left(\frac{k_n}{n}\right)^{\nu-\frac{1}{2}} L(k_n/n) \sup_{U_{1,n}\vee m_n/n\le s\le k_n/n} |B_n(s)|/(s^{\nu}L(s)) \xrightarrow{\mathcal{D}} \sup_{0\le s\le 1} |W(s)|/s^{\nu},$$

which also completes the proof of Theorem 2.2.

Proof of Corollary 2.1. It is pointed out in the proof of Theorem 2.5 in [4] that for all $\lambda > 0$ we have

$$\left(\frac{k_n}{n}\right)^{\nu-\frac{1}{2}} L(k_n/n) \sup_{\lambda/n\le s\le k_n/n} |u_n(s)-B_n(s)|/(s^{\nu}L(s)) = o_P(1) .$$

The latter statement together with (3.9) and (3.10) implies (2.5). Now Theorem 2.2 and (2.5) yield (2.6).

Proof of Theorem 2.3. Assume $c > 0$. Then we have also that $m_n \to \infty$. Hence on using (1.1) with $\alpha > 0$ we get

$$\sup_{m_n/n\le s\le k_n/n} |e_n(s)-B_n(s)|/(s(1-s))^{\frac{1}{2}} = O_P(n^{-\alpha}) \sup_{m_n/n\le s\le k_n/n} s^{-\alpha}$$

$$= O_P(m_n^{-\alpha}) = o_P(1).$$

Also,

$$\sup_{m_n/n\le s\le k_n/n} |B(s)|/(s(1-s))^{\frac{1}{2}} \stackrel{\mathcal{D}}{=} \sup_{0\le t\le (\frac{1}{2})\log(\frac{k_n}{m_n}\frac{1-m_n/n}{1-k_n/n})} |V(t)| \tag{3.12}$$

$$\xrightarrow{\mathcal{D}} \sup_{0\le s\le (\frac{1}{2})\log(1/c)} |V(t)|,$$

where $V(t)$ is the Ornstein-Uhlenbeck process of Section 2. In (3.12) we used the well known representation

$$\{V(t);\ -\infty<t<\infty\} = \{(1+e^{2t})e^{-t}B(e^{2t}/(1+e^{2t}));\ -\infty<t<\infty\} .$$

Now assume that $c = 0$. Using again (1.1) with $\alpha > 0$, we obtain

$$(\log\log k_n)^{\frac{1}{2}} \sup_{(\log k_n)/n \le s \le k_n/n} |e_n(s)-B_n(s)|/(s(1-s))^{\frac{1}{2}}$$

$$= (\log\log k_n)^{\frac{1}{2}} n^{-\alpha} O_P(1) \sup_{(\log k_n)/n \le s \le k_n/n} s^{-\alpha}$$

$$= O_P((\log\log k_n)^{\frac{1}{2}}/(\log k_n)^{\alpha}) = o_P(1).$$

Hence, whenever $m_n \ge \log k_n$, we have

$$(2 \log\log(k_n/m_n))^{\frac{1}{2}} \sup_{m_n/n \le s \le k_n/n} |e_n(s)-B_n(s)|/(s(1-s))^{\frac{1}{2}} = o_P(1).$$

Therefore Lemma 4.3 in [4] gives (2.14).

Let $m_n \le \log k_n$. Then by (2.11)

$$a((\tfrac{1}{2})\log\log k_n) \sup_{0 \le s \le (\log k_n)/n} |e_n(s)|/(s(1-s))^{\frac{1}{2}} - b((\tfrac{1}{2})\log\log k_n)$$

$$= O_P(1).$$

Therefore, by (2.11) again, we get

$$a((\tfrac{1}{2})\log k_n) \sup_{m_n/n \le s \le k_n/n} |e_n(s)|/(s(1-s))^{\frac{1}{2}} - b((\tfrac{1}{2})\log k_n) \xrightarrow{\mathcal{D}} Y.$$

Now, elementary calculations show that

$$b((\tfrac{1}{2})\log(k_n/m_n^*)) - b((\tfrac{1}{2})\log k_n) = o(1)$$

and

$$\left|\frac{a((\frac{1}{2})\log(k_n/m_n^*))}{a((\frac{1}{2})\log k_n)} - 1\right| \frac{b((\frac{1}{2})\log k_n)}{a((\frac{1}{2})\log k_n)} = o(1),$$

and hence we have also (2.14).

Proof of Corollary 2.2. The proof of (2.15) is similar to that of (2.13), only here (1.2) is used instead of (1.1). The proof of (2.16) is like that of (2.14), on using again (1.2), and also Theorem 2.6 in [4] instead of (2.11).

Proof of Theorem 2.4. In (4.35) of [4] (cf. also Theorem 4.7 in [5]) it is proved that for $\varepsilon > 0$ we have

$$(k_n/n)^{\varepsilon} \sup_{k_n/n \le s \le \frac{1}{2}} \frac{L(k_n/n)}{s^{\varepsilon} L(s)} = O(1). \tag{3.13}$$

Using (3.13) and (1.1) we get

$$\left(\frac{k_n}{n}\right)^{\nu-\frac{1}{2}} L(k_n/n) \sup_{k_n/n \le s \le \frac{1}{2}} |e_n(s)-B_n(s)|/(s^{\nu}L(s))$$

$$= O_P(1)\left(\frac{k_n}{n}\right)^{\nu-\frac{1}{2}} L(k_n/n)\, n^{-\alpha} \sup_{k_n/n \le s \le \frac{1}{2}} \frac{s^{\frac{1}{2}-\alpha}}{s^{\nu}L(s)}$$

$$= O_P\left(\frac{k_n}{n}\right)^{\nu-\varepsilon-\frac{1}{2}} n^{-\alpha} \sup_{k_n/n \le s \le \frac{1}{2}} s^{\frac{1}{2}-\alpha-\nu+\varepsilon}$$

$$= O_P(1)\, k_n^{-\alpha} = o_P(1),$$

on choosing $\varepsilon > 0$ and $\alpha > 0$ so that $\frac{1}{2}-\alpha-\nu+\varepsilon < 0$. Thus it suffices to show that

$$\left(\frac{k_n}{n}\right)^{\nu-\frac{1}{2}} L(k_n/n) \sup_{k_n/n \le s \le r_n/n} |B_n(s)|/(s^{\nu}L(s)) \xrightarrow{\mathcal{D}} \sup_{c \le s \le 1} |W(s)|/s^{1-\nu}. \tag{3.14}$$

With (3.4) in mind first we have by (3.13)

$$|W(1)| \left(\frac{k_n}{n}\right)^{\nu-\frac{1}{2}} L(k_n/n) \sup_{k_n/n \le s \le r_n/n} \frac{s^{1-\nu}}{L(s)}$$

$$= \begin{cases} O_P(1)\left(\frac{k_n}{n}\right)^{\nu-\frac{1}{2}-\varepsilon} & \text{if } \frac{1}{2} < \nu \le 1 \\ O_P(1)\left(\frac{k_n}{n}\right)^{\frac{1}{2}} & \text{if } \nu > 1 \end{cases}$$

$$= o_P(1)$$

where $\varepsilon = (\nu-\frac{1}{2})/2$.

Next,

$$\left(\frac{k_n}{n}\right)^{\nu-\frac{1}{2}} L(k_n/n) \sup_{k_n/n \le s \le r_n/n} |W(s)|/(s^{\nu}L(s))$$

$$= \left(\frac{k_n}{n}\right)^{\nu-\frac{1}{2}} L(k_n/n) \sup_{n/r_n \le s \le n/k_n} s|W(1/s)|/(s^{1-\nu}L(1/s))$$

$$\stackrel{\mathcal{D}}{=} \left(\frac{k_n}{n}\right)^{\nu-\frac{1}{2}} L(k_n/n) \sup_{n/r_n \le s \le n/k_n} |W(s)|/(s^{1-\nu}L(s))$$

$$\stackrel{\mathcal{D}}{=} L(k_n/n) \sup_{k_n/r_n \le s \le 1} |W(s)|/(s^{1-\nu}L(n/(sk_n))$$

$$\xrightarrow{\mathcal{D}} \sup_{0 \le s \le 1} |W(s)|/s^{1-\nu},$$

where the latter convergence in distribution is proven along the lines of (3.7). This completes the proof of (3.14) and also that of Theorem 2.4.

Proof of Corollary 2.3. Same as that of Theorem 2.4, only (1.2) is used instead of (1.1).

Proof of Theorem 2.5. By the well known representation of uniform-(0,1) order statistics we have

$$\{U_{i,n};\ 1 \le i \le n\} \stackrel{\mathcal{D}}{=} \{S_i/S_{n+1};\ 1 \le i \le n\} .$$

By (4.36) in [4] (cf. also Theorem 4.7 in [5]) for $\lambda > 0$ and $\varepsilon > 0$ we get

$$\sup_{\lambda/n \le s \le k_n/n} (ns)^{-\varepsilon} |L(1/n)/L(s)-1| = o(1). \tag{3.15}$$

Let $c_1 > 0$. Then by (3.15)

$$n^{\frac{1}{2}-\nu} \sup_{m_n/n \le s \le k_n/n} \left|\frac{L(1/n)}{L(s)} - 1\right| |e_n(s)|/s^{\nu}$$

$$= o(1) n^{\frac{1}{2}-\nu+\varepsilon} \sup_{m_n/n \le s \le k_n/n} |e_n(s)|/s^{\nu-\varepsilon},$$

and so it is enough to prove (2.18) with $L = 1$. Putting $s = t/S_{n+1}$, we obtain

$$n^{\frac{1}{2}-\nu} \sup_{m_n/n \le s \le k_n/n} |e_n(s)|/s^{\nu}$$

$$= n^{1-\nu} \sup_{\frac{S_{n+1}}{n} m_n \le t \le \frac{S_{n+1}}{n} k_n} \frac{\left|\frac{1}{n} \sum_{i=1}^{n} iI\left\{\frac{S_i}{S_{n+1}} \le \frac{t}{S_{n+1}} < \frac{S_{i+1}}{S_{n+1}}\right\} - \frac{t}{S_{n+1}}\right|}{(t/S_{n+1})^{\nu}}$$

$$= \left(\frac{S_{n+1}}{n}\right)^{\nu} \sup_{\frac{S_{n+1}}{n} m_n \le t \le \frac{S_{n+1}}{n} k_n} \frac{\left|\sum_{i=1}^{n} iI\left\{S_i \le t < S_{i+1}\right\} - \frac{t}{S_{n+1}}\right|}{t^{\nu}}$$

Let $\frac{1}{2} < \nu \le 1$. Then

$$\left(1 - \frac{n}{S_{n+1}}\right) \sup_{\frac{S_{n+1}}{n} m_n \le t \le \frac{S_{n+1}}{n} k_n} t^{1-\nu} = \left(1 - \frac{n}{S_{n+1}}\right) \left(\frac{S_{n+1}}{n} k_n\right)^{1-\nu}$$

$$= O_P(1) \left(\frac{k_n}{n}\right)^{1-\nu} n^{\frac{1}{2}-\nu} = o_P(1).$$

Let $\nu > 1$. Then

$$\left(1 - \frac{n}{S_{n+1}}\right) \sup_{\frac{S_{n+1}}{n} m_n \le t \le \frac{S_{n+1}}{n} k_n} t^{1-\nu} = \left(1 - \frac{n}{S_{n+1}}\right) \left(\frac{S_{n+1}}{n} m_n\right)^{1-\nu} = o_P(1).$$

Consequently, we now have to show that

$$\left(\frac{S_{n+1}}{n}\right)^{\nu} \sup_{\frac{S_{n+1}}{n} m_n \le t \le \frac{S_{n+1}}{n} k_n} \frac{\left|\sum_{i=1}^{n} iI\left\{S_i \le t < S_{i+1}\right\} - t\right|}{t^{\nu}}$$

$$\xrightarrow{\mathcal{D}} \sup_{c_1 \le t \le c_2} |N(t) - t|/t^{\nu}.$$

The latter in turn is true, because $S_{n+1}/n \xrightarrow{P} 1$, $N(\cdot)$ has jumps at $c_1$ and $c_2$ with probability zero and applying the law of iterated logarithm if $c_2 = \infty$.

In order to prove (2.21), by (3.15) and (3.2) it is enough to prove it with $L = 1$. Replacing now $m_n/n$ by $S_1/S_{n+1}$ in the proof of (2.18), (2.21) becomes immediate.

If $c_1 = 0$ and $\frac{1}{2} < \nu < 1$, then we define $A_n(\lambda) = \{m_n/n < \lambda/n < U_{1,n}\}$ and note that for every $\varepsilon > 0$ there is a $\lambda > 0$ and an $n_0$ such that, when $n \ge n_0$, $P(A_n(\lambda)) \ge 1-\varepsilon$. First by 7. of Corollary 1.2.1 in [8], for every $\mu \ne 0$ there is a slowly varying function $L^*$ such that $s^{\mu}L^*(s)$ is a strictly monotone function and

$$\lim_{s\downarrow 0} \sup_{\tau_1\le\tau\le\tau_2} \left|\frac{L^*(\tau s)}{L(s)} - 1\right| = 0, \quad 0 < \tau_1 \le \tau_2 < \infty. \tag{3.16}$$

Hence by (3.16) with $\tau_1 = \tau_2 = 1$, on the event $A_n(\lambda)$ we have

$$n^{1-\nu}L(1/n) \sup_{0\le s\le\lambda/n} s^{1-\nu}/L(s) = O(1)\lambda^{1-\nu}L(1/n)/L^*(\lambda/n)$$

$$= O(1)\lambda^{1-\nu},$$

arbitrarily small if $\lambda$ is small enough. On the other hand, by (3.16) and (3.2)

$$n^{1-\nu}L(1/n) \sup_{\lambda/n\le s<U_{1,n}} \frac{s}{s^{\nu}L(s)} = n^{1-\nu}L(1/n) \sup_{\lambda/n\le s<S_1/S_{n+1}} s^{1-\nu}/L(s)$$

$$= (1 + o_P(1))n^{1-\nu}(S_1/S_{n+1})^{1-\nu} \xrightarrow{\mathcal{D}} S_1^{1-\nu}.$$

Thus we proved that

$$n^{\frac{1}{2}-\nu}L(1/n) \sup_{U_{1,n}\ m_n/n\le s\le U_{1,n}} |e_n(s)|/(s^{\nu}L(s)) \xrightarrow{\mathcal{D}} S_1^{1-\nu} = \sup_{0\le s<S_1} \frac{|N(s)-s|}{s^{\nu}},$$

which together with (2.21) yields (2.19).

When $\nu > 1$, on the event $A_n(\lambda)$ we get by (3.16) and (3.2) again that

$$n^{1-\nu}L(1/n) \sup_{m_n/n\le s<U_{1,n}} s^{1-\nu}/L(s) = (1 + o_P(1))n^{1-\nu} \sup_{m_n/n\le s\le U_{1,n}} s^{1-\nu}$$

$$= (1+o_P(1)n^{1-\nu}(m_n/n)^{1-\nu} \to \infty .$$

As to the proof of (2.22) we have

$$n^{-\frac{1}{2}}L(1/n) \sup_{m_n/n\le s<U_{1,n}} |e_n(s)|/(sL(s)) = L(1/n) \sup_{m_n/n\le s<U_{1,n}} (1/L(s))$$

$$\xrightarrow{\mathcal{D}} 1 = \sup_{0<s<S_1} |N(s)-s|/s .$$

Also, the proof of (2.23) is immediate on account of

$$\sup_{0<s\le k_n/n} |e_n(s)|/(sL(s)) = \infty \quad \text{a.s.}$$

Acknowledgement. The research of Miklós Csörgő was supported in part by a NSERC Canada operating grant at Carleton University. The research of Lajos Horváth was done while on leave from Szeged University and visiting the Laboratory for Research in Statistics and Probability at Carleton University - University of Ottawa, also partially supported by NSERC Canada grants of M. Csörgő and D.A. Dawson and by an EMR Canada grant of M. Csörgő. The research of Josef Steinebach was done while on leave from University of Marburg and visiting the Laboratory for Research in Statistics and Probability at Carleton University - University of Ottawa, also partially supported by a NSERC Canada grant of M. Csörgő, by a NSERC Canada - Deutsche Forschungsgemeinschaft exchange grant and by a Deutsche Forschungsgemeinschaft grant.

## References

[1] T.W. ANDERSON and D.A. DARLING, Asymptotic theory of certain "goodness of fit" criteria based on stochastic processes, Ann. Math. Statist., 23 (1952), 193-212.

[2] M. CSÖRGŐ, Quantile processes with statistical applications, SIAM (Philadelphia, 1983).

[3] M. CSÖRGŐ, S. CSÖRGŐ, L. HORVÁTH and D.M. MASON, Weighted empirical and quantile processes, Ann. Probab., 14 (1986), 31-85.

[4] M. CSÖRGŐ and L. HORVÁTH, On the distributions of the supremum of weighted quantile processes, Technical Report Series of Lab. Res. Stat. Probab., Carleton University, No.55 (1985).

[5] M. CSÖRGŐ and L. HORVÁTH, On the distributions of $L_1$ and $L_2$ norms of weighted uniform empirical and quantile processes. Technical Report Series of Lab. Res. Stat. Probab., Carleton University, No.60 (1985).

[6] M. CSÖRGŐ and L. HORVÁTH, Approximations of weighted empirical and quantile processes, Statist. Probab. Lett., 4 (1986), 275-280.

[7] M. CSÖRGŐ and D.M. MASON, On the asymptotic distribution of weighted uniform empirical and quantile processes, Stochastic Process. Appl., 21 (1985), 119-132.

[8] L. DE HAAN, On regular variation and its applications to the weak convergence of sample extremes, Math. Centre Tracts 32 (Amsterdam, 1975).

[9] F. EICKER, The asymptotic distribution of the suprema of the standardized empirical process, Ann. Statist., 7 (1979), 116-138.

[10] D. JAESCHKE, The asymptotic distribution of the supremum of the standardized empirical distribution function on subintervals, Ann. Statist., 7 (1979), 108-115.

[11] D.M. MASON, The asymptotic distribution of weighted empirical functions, Stochastic Process. Appl., 15 (1983), 99-109.

[12] D.M. MASON, The asymptotic distribution of generalized Rényi statistics, Acta Sci. Math. (Szeged), 48(1985), 315-323.

[13] N.V. SMIRNOV, Limit distributions for the terms of a variational series, Amer. Math. Soc. Transl., Ser. 1, no. 67 (1949).

[14] J.A. WELLNER, Limit theorems for the ratio of the empirical distribution function to the true distribution function, Z. Wahrsch. verw. Gebiete, 45 (1978), 73-88.

# Estimating Percentile Residual Life Under Random Censorship

Sándor Csörgő
Bolyai Institute
Szeged University
H-6720 Szeged, Hungary

SUMMARY: Asymptotic confidence intervals and simultaneous confidence bands are constructed for percentile residual lifetimes under random censorship.

## 1. Introduction

Let $F(t) = P\{X \le t\}$, $t \ge 0$, be a life distribution function with support $(0,T_F)$, where $T_F = \inf\{t > 0: F(t) = 1\} \le \infty$, and corresponding quantile function $Q(s) = F^{-1}(s) = \inf\{t > 0: F(t) \ge s\}$, $0 < s < 1$, $Q(0) = 0$, $Q(1) = T_F$.
For $0 < p < 1$ consider the $(1-p)$-percentile residual lifetime

$$R(p,t) = Q(1-p(1-F(t)))-t$$

at $t > 0$, originally introduced by Haines and Singpurwalla (1974), interpreted as the $(1-p)$-percentile additional time to failure given no failure by time t. Work of Schmittlein and Morrison (1981), Arnold and Brockett (1983), Joe and Proschan (1984 a,b), Gupta and Langford (1984) and Gerchak (1984) indicates potential theoretical advantages of the median residual lifetime $R(1/2,t)$ over the more frequently used mean residual lifetime

$$E\{X-t|X>t\} = \int_t^\infty (1-F(x))dx/(1-F(t))$$

and also that values of p different from 1/2 may be of practical interest.
Given a sample $X_1,\ldots,X_n$ of X, independent random variables with distribution function F, it is natural to plug in the sample distribution and quantile functions $F_n$ and $Q_n$ for F and Q, respectively, in the

Contributions to Stochastics.
Ed. by W. Sendler
© Physica-Verlag Heidelberg 1987

defining formula of $R(\cdot,\cdot)$ above and thus to obtain a non-parametric estimator $R_n(\cdot,\cdot)$ of $R(\cdot,\cdot)$. A large-sample estimation theory of $R_n(\cdot,\cdot)$ has been worked out by M. Csörgő and S. Csörgő (1987), hereafter referred to as Cs-Cs, where emphasis is on constructing asymptotic confidence intervals for $R(p,t)$ with fixed p and t, confidence bands for $R(p,t)$ as a function of $t \geq 0$ for a fixed p, and confidence bands for $R(p,t)$ as a function of $p \in (0,1)$ for a fixed t. The methods are also illustrated there on some data of the duration of strikes in the United Kingdom.

In most practical situations when estimation of $R(\cdot,\cdot)$ may be required, however, the above sample is not complete because another sequence $Y_1,\ldots,Y_n$ of independent random variables with common distribution function $G(t) = P\{Y \leq t\}$, $t \geq 0$, concentrated on the non-negative half-line, censors on the right the X sequence so that we can only observe the pairs

$$(Z_1,\delta_1),\ldots,(Z_n,\delta_n) \ , \tag{1.1}$$

where

$$Z_k = \min(X_k,Y_k) \text{ and } \delta_k = I(Z_k = X_k) \ , \ 1 \leq k \leq n \ ,$$

and where I is the indicator function. The aim of the present paper is to work out the estimation theory of $R(\cdot,\cdot)$, parallel to that of Cs-Cs, in the case of the general random censorship model where the only assumption is that the two sequences $\{X_k\}$ and $\{Y_k\}$ are independent. In this general model, estimation is based on the celebrated Kaplan-Meier product-limit estimator of the survival function $1-F(\cdot)$. In the next section we derive asymptotic confidence intervals for the value $R(p,t)$ when p and t are fixed and then asymptotic simultaneous confidence intervals, that is, confidence bands are constructed for the function $R(.,t)$ when t is fixed.

There are serious theoretical limitations when estimating mean residual life under random censorship. One is typically forced to estimate truncated versions

$$\int_t^{T_n} (1-F(x))dx/(1-F(t)) \ ,$$

of the theoretical function above, where $T_n$ is a numerical sequence allowed to converge up to $T_F$ at a certain rate which depends heavily on a complicated relationship of the censoring distribution function G and the censored distribution function F, a relationship never known in practice (cf. Ghorai, A. Susarla, V. Susarla and Van Ryzin (1982) and S. Csörgő and Horvàth (1982)). Such limitations will not enter in the estimation process of percentile residual life functions. This fact is a practical advantage of the notion.

## 2. Confidence intervals and bands

Based upon the observations given under (1.1), where $Z_k$ is said to be uncensored or censored according to the alternative $\delta_k = 1$ or $\delta_k = 0$, consider the product-limit estimator $\hat{F}_n$ of F defined as

$$1-\hat{F}_n(t) = \begin{cases} \prod_{\{r:\, Z_{r,n} \le t\}} \left(\frac{n-r}{n-r+1}\right)^{\delta_{r,n}} & , \text{ if } t < Z_{n,n} , \\ 0 & , \text{ if } t \ge Z_{n,n} , \end{cases}$$

where $Z_{1,n} \le \ldots \le Z_{n,n}$ are the order statistics corresponding to $Z_1,\ldots,Z_n$ and $\delta_{r,n} = 1$ if $Z_{r,n}$ is uncensored and $\delta_{r,n} = 0$ if $Z_{r,n}$ is censored, $r = 1,\ldots,n$. The natural estimator of the quantile function Q is then

$$Q_n(s) = \inf\{t > 0 : \hat{F}_n(t) \ge s\}, \quad 0 < s \le 1 ,$$

$\hat{Q}_n(0) = \hat{Q}_n(0+)$, which can be written as

$$\hat{Q}_n(s) = \begin{cases} \delta_{1,n} Z_{1,n} & , \text{ if } 0 \le s \le \hat{F}_n(\delta_{1,n} Z_{1,n}) , \\ \delta_{k,n} Z_{k,n} & , \text{ if } \hat{F}_n\left(\max_{1 \le i \le k-1} \delta_{i,n} Z_{i,n}\right) < s \le \hat{F}_n(\delta_{k,n} Z_{k,n}), \ k = 2,\ldots,n, \\ Z_{n,n} & , \text{ if } \max_{1 \le k \le n} \hat{F}_n(\delta_{k,n} Z_{k,n}) < s \le 1 . \end{cases}$$

Equivalently, if we let $U_{1,n} \le \ldots \le U_{\nu_n,n}$ denote the uncensored $Z_{k,n}$ variables so that $\nu_n = \# \{1 \le k \le n : \delta_{k,n} = 1\}$ is the number of uncensored observations, we can write

$$\hat{Q}_n(s) = \begin{cases} U_{1,n} & , \text{ if } 0 \le s \le \hat{F}_n(U_{1,n}) , \\ U_{k,n} & , \text{ if } \hat{F}_n(U_{k-1,n}) < s \le \hat{F}_n(U_{k,n}), \ k = 2,\ldots,\nu_n , \\ Z_{n,n} & , \text{ if } \hat{F}_n(U_{\nu_n,n}) < s \le 1 . \end{cases}$$

Thus, while $\hat{F}_n$ jumps at the uncensored observations and $Z_{n,n}$ only, the values of $\hat{Q}_n$ are the uncensored observations and the largest observation, whether the latter is censored or not.

Now the empirical (1-p)-percentile residual life function is obviously formed as

$$\hat{R}_n(p,t) = \hat{Q}_n(1-p(1-\hat{F}_n(t)))-t \ , \ t \geq 0, \ 0 < p < 1 \ .$$

Just as in the case of lack of censoring (Cs-Cs), it follows from well-known facts that if $Q(1-p(1-F(t)))$ is the unique solution y of the inequalities $F(y-) \leq 1-p(1-F(t)) \leq F(y)$ and Q is continuous at $1-p(1-F(t))$, then $\hat{R}_n(p,t) \to R(p,t)$ almost surely as $n \to \infty$, provided also that $H(Q(1-p(1-F(t)))) < 1$. Here $H(t) = P\{Z \leq t\}$, $t \geq 0$, is the distribution function of the minimum observations for which $1-H(t) = (1-F(t))(1-G(t))$, $t \geq 0$, by the model assumption that the censored sequence $\{X_k\}$ is independent of the censoring sequence $\{Y_k\}$. Besides H we need the (sub-)distribution function

$$\tilde{H}(t) = P\{Z_k \leq t, \ \delta_k = 1\} = \int_{-\infty}^{t} (1-G)dF, \ t \geq 0 \ ,$$

of the uncensored variables and the function

$$d(s) = \int_0^{Q(s)} (1-H)^{-2} \, d\tilde{H} = \int_0^{Q(s)} (1-F)^{-2}(1-G)^{-1} \, dF, \ 0 < s < 1 \ .$$

Note that the function $(1-F(\cdot))d(F(\cdot))$, with F assumed to be continuous, is the limiting variance function of the product-limit process $\hat{\beta}_n(t) = n^{1/2} \{\hat{F}_n(\cdot)-F(\cdot)\}$ (see, e.g. in S. Csörgő and Horváth (1982)).

The proof of the first result (see (2.2) below, where $\xrightarrow{\mathcal{D}}$ denotes convergence in distribution) reveals the asymptotic structure of

$$\hat{r}_n(p,t) = n^{1/2} f(R(p,t)+t)\{\hat{R}_n(p,t)-R(p,t)\},$$

where $f = F'$ is the density function of F. This asymptotic distributional structure expressed in terms of a standard Brownian motion W will be very important when constructing confidence bands in Theorem 2.3 below. Let $N(0,v)$ denote a normal random variable with mean zero and variance $v > 0$. The first theorem is parallel to Theorem 1 of Cs-Cs.

<u>Theorem 2.1.</u> <u>Let</u> $t > 0$ <u>and</u> $0 < p < 1$ <u>be fixed so that</u> $H(R(p,t)+t) < 1$, <u>and assume that the density-quantile function</u> $f(Q(\cdot))$ <u>is positive and continuous at</u> $1-p(1-F(t))$. <u>Then, as</u> $n \to \infty$,

$$\hat{r}_n(p,t) \xrightarrow{\mathcal{D}} N\left(0, p^2(1-F(t))^2 \int_t^{t+R(p,t)} (1-H)^{-2} \, d\tilde{H}\right) \ .$$

We note that when there is no censorship (i.e. when $H = \tilde{H} = F$ and thus $d(s) = s/(1-s)$, $0 < s < 1$), the limiting variance

$$p^2(1-F(t))^2\{d(1-p(1-F(t)))-d(F(t))\}$$

here reduces to $p(1-p)(1-F(t))$, the limiting variance in Theorem 1 of Cs-Cs, that is the limiting variance in the lack of censorship.

Proof. Transforming the sequences $\{X_k\}$, $\{Y_k\}$ and $\{Z_k\}$, set $\xi_k = F(X_k)$, $\eta_k = F(Y_k)$ and $\zeta_k = F(Z_k) = \min(\xi_k, \eta_k)$, $k = 1,\ldots,n$, and form the product-limit estimator $\hat{E}_n$ of the identity function on $[0,1]$ based on $(\zeta_1, \delta_1),\ldots,(\zeta_n, \delta_n)$. Let $\hat{U}_n(s) = \inf\{u: \hat{E}_n(u) \geq s\}$, $0 \leq s \leq 1$, be the corresponding quantile function and consider the "Uniform (0,1)" product-limit and product-limit quantile processes $\hat{e}_n(s) = n^{1/2}(\hat{E}_n(s)-s)$ and $\hat{u}_n(s) = n^{1/2}(U_n(s)-s)$, $0 \leq s \leq 1$. Then, adjusting the proof of Theorem 1 in Cs-Cs to the present situation, we obtain

$$\hat{r}_n(p,t) = \hat{u}_n(1-p(1-\hat{F}_n(t)))+p\,\hat{e}_n(F(t))\{1+\varepsilon_n\}, \tag{2.1}$$

where $\varepsilon_n$ is a sequence of remainder random variables such that $\varepsilon_n \to 0$ almost surely as $n \to \infty$. Now, extending without loss of generality the underlying probability space if necessary, it follows from a special case of Theorem 3.1 of Aly, M. Csörgő and Horvàth (1985) that there exists a sequence of standard Wiener processes $W_n$ such that

$$|\hat{r}_n(p,t)-\{(1-[1-p(1-F(t))])W_n(d(1-p\{1-\hat{F}_n(t)\}))$$
$$-p\{1-F(t)\}W_n(d(f(t)))\}|$$
$$= |\hat{r}_n(p,t)-p(1-F(t))\{W_n(d(1-p\{1-\hat{F}_n(t)\}))-W_n(d(F(t)))\}| \to 0$$

almost surely as $n \to \infty$. So by standard argument and by using the fact that a standard Wiener process is strictly stationary, we obtain that

$$\hat{r}_n(p,t) \xrightarrow{\mathcal{D}} p(1-F(t))W(d(1-p\{1-F(t)\})-d(F(t))) \tag{2.2}$$

as $n \to \infty$. This proves the theorem.

The next result, the censored analogue of Theorem 3 in Cs-Cs, gives asymptotic confidence intervals for $R(p,t)$ when $p$ and $t$ are fixed. Its proof follows from (2.2) similarly to that as the proof of Theorem 3 followed from the corresponding ingredient of the proof of Theorem 1 in Cs-Cs and hence there is no need to detail it here. For estimating the limiting variance we need

$$\hat{\sigma}_n^2(p,t) = p^2(1-\hat{F}_n(t))^2\,\hat{v}_n^2(p,t)$$
$$= p^2(1-\hat{F}_n(t))^2 \int_t^{t+\hat{R}_n(p,t)} (1-H_n)^{-2}\,d\tilde{H}_n ,$$

where

$$H_n(x) = \frac{1}{n}\#\{1 \le k \le n\colon Z_k \le x\}$$

and

$$\tilde{H}_n(x) = \frac{1}{n}\#\{1 \le k \le n\colon Z_k \le x,\ \delta_k = 1\},\quad x \ge 0,$$

are the sample versions of H and $\tilde{H}$. Here the integral can be written as

$$\hat{v}_n^2(p,t) = \frac{1}{n} \sum_{\{1\le k\le n\colon\ t<Z_k\le t+\hat{R}_n(p,t)\}} \delta_k(1-H_n(Z_k))^{-2} . \tag{2.3}$$

Setting further

$$A_n^-(\lambda,p,t) = \hat{Q}_n(1-p\{1-\hat{F}_n(t)\}-\lambda\frac{\hat{\sigma}_n(p,t)}{n^{1/2}}) - t$$

and

$$A_n^+(\lambda,p,t) = \hat{Q}_n(1-p\{1-\hat{F}_n(t)\}+\lambda\frac{\hat{\sigma}_n(p,t)}{n^{1/2}}) - t ,$$

for the left and right endpoints of the interval, where $\lambda > 0$, we have the following.

<u>Theorem 2.2.</u> <u>If</u> $0 < \alpha < 1$, <u>then under the conditions of Theorem 2.1</u>

$$\lim_{n\to\infty} P\{A_n^-(\lambda_\alpha,p,t) \le R(p,t) \le A_n^+(\lambda_\alpha,p,t)\} = 1-\alpha,$$

where, with $\Phi$ denoting the standard normal distribution function, $\lambda_\alpha > 0$ is the solution of $\Phi(\lambda) = 1-\alpha/2$.

Although for many practical purposes the above confidence intervals can be completely sufficient, in other situations one may wish to have confidence bands for $R(p,\cdot)$ or $R(\cdot,t)$. For $R(\cdot,t)$ these are provided in the main result of the present paper below which is the censored analogue of Theorem 5 in Cs-Cs. The proof of this result is an extension of that of Theorem 2.2. At each step one has to ensure that the various remainder terms are uniformly negligible. The main tools for doing this are the uniform results of Aly, M. Csörgő and Horváth (1985), and since the proof of their Theorem 5.3 can be used as a guide-line, the very lenghty details are not given here except for the end of the proof of case (ii). This will explain the structure of the bands

$$B_n(\lambda,p,t) = \{\hat{Q}_n(1-p\{1-\hat{F}_n(t)\}\{1+\frac{\lambda}{n^{1/2}}\}) - t \leq R(p,t)$$

$$\leq \hat{Q}_n(1-p\{1-\hat{F}_n(t)\}\{1-\frac{\lambda}{n^{1/2}}\}) - t\}$$

where, for the time being, $\lambda$ is any positive number, and will show how the stationarity and a scale transformation of the Wiener process yield, at least in this case, asymptotically distribution-free bands.

Theorem 2.3. *Assume that* $F$ *is twice differentiable on* $(0,T_F)$ *and* $f = F' > 0$ *on* $(0,T_F)$.

(i) *If* $0 < p < 1$ *and* $T > 0$ *are fixed such that* $H(Q(1-p(1-F(T)))) < 1$, *then for any* $\lambda > 0$

$$\lim_{n\to\infty} P\{B_n(\lambda,p,t),\ 0 \leq t \leq T\} = P\{\sup_{0\leq t\leq T} |W(d(1-p\{1-F(t)\}))-W(d(F(t))| \leq \lambda\}.$$

(ii) *If* $0 < p_0 < 1$ *and* $t > 0$ *are fixed such that* $H(Q(1-p_0(1-F(t)))) < 1$, *then for any* $0 < \alpha < 1$

$$\lim_{n\to\infty} P\{B_n(\tau_\alpha \hat{v}_n(p_0,t),p,t),\ p_0 \leq p \leq 1\} = 1-\alpha\ ,$$

*where* $\hat{v}_n(p_0,t)$ *is defined in* (2.3) *and* $\tau_\alpha > 0$ *is the solution of*

$$P\{\sup_{0\leq t\leq 1} |W(t)| \leq \tau\} = 1-\alpha\ . \tag{2.4}$$

We see that the confidence band problem for $R(p,\cdot)$, with a fixed $p$, is not solved in (i) at all because the limiting distribution depends on the unknown $F$ and $G$. This limiting distribution is unknown even if we completely specify $F$ and $G$. The natural idea in such a situation is the use of the bootstrap. This is not persued here. In comparison with the corresponding case in the lack of censorship (case (i) of Theorem 5 in Cs-Cs), we note that though there the limit is distribution free, it is not known analytically and a simulation study qualitatively sketched in Section 4 of Cs-Cs should be executed to obtain percentage points.

On the other hand, the band problem for $R(\cdot,t)$, with a fixed $t > 0$, has a satisfactory solution in (ii) above. The reason for this is that $\hat{v}_n(p_0,t)$ is a consistent estimator of

$$v(p_0,t) = \left(\int_t^{t+R(p_0,t)} (1-H)^{-2}\, d\tilde{H}\right)^{1/2},$$

and the proof indicated above leads to (cf. 2.2))

$$\lim_{n\to\infty} P\{B_n(\tau_\alpha \hat{v}_n(p_o,t),p,t),\ p_o \le p \le 1\}$$

$$= P\{-\tau_\alpha p(1-F(t))v(p_o,t) \le p(1-F(t))\{W(d(1-p\{1-F(t)\}))$$

$$- W(d(F(t)))\} \le \tau_\alpha p(1-F(t))v(p_o,t),\ p_o \le p \le 1\}$$

$$= P\{\sup_{p_o\le p\le 1} |W(v^2(p,t))| \le \tau_\alpha\ v(p_o,t)\}$$

$$= P\{\sup_{0\le s\le v^2(p_o,t)} |W(s)| \le \tau_\alpha\ v(p_o,t)\}$$

since $v(1,t) \equiv 0$. The last probability in turn equals the quantity in (2.4) by a well-known scale transformation of the Wiener process. The distribution function $K(\tau)$ on the left side in (2.4) is well known as noted above and is tabulated in S. Csörgő and Horváth (1985). We have, for example $\tau_{0.2} = 1.645$, $\tau_{0.1} = 1.960$, $\tau_{0.05} = 2.242$ and $\tau_{0.01} = 2.810$. Therefore the bands in (ii) above are readily applicable. Again, this fact is in harmony with the corresponding case in the lack of censorship (case (ii) of Theorem 5 in Cs-Cs), where the obtained limiting distribution was also well known, the distribution of the absolute supremum of the Brownian bridge on [0,1]. The strike duration example in Cs-Cs suggests that bands for $R(p,t)$ in p are at least as important in practice as bands in t.

Estimation of $R(\cdot,\cdot)$ together with a number of other functions of interest in reliability and life testing in the special proportional hazards model of random censorship is considered by S. Csörgő (1988). The peculiarity of this model is that efficient estimation is based on the Abdushukurov-Cheng-Lin estimator of F rather than the Kaplan-Meier estimator.

Note: Three months after completing the first draft of the present paper I received Chung (1986). He also proved Theorem 2.1 and 2.2 above independently and somewhat earlier than I did. Furthermore, he also considers the more intricate problem of constructing confidence bands. He also proves case (i) of Theorem 2.3. However, he does not come up with asymptotically distribution-free bands for $R(\cdot,t)$ in case (ii) and hence resorts to the bootstrap in this case just as in case (i). He also applies his bootstrapped confidence bands for estimating percentile residual life on an interesting geological data set that arises from trying to measure the effect of nuclear waste disposal. It would be interesting to compare his bands and the bands proposed above on theses data.

References

ALY, E.-E., CSÖRGŐ, M. and HORVÁTH, L. (1985). Strong approximations of the quantile process of the product-limit estimator. J. Multivar. Anal. 16, 185-210.

ARNOLD, B. and BROCKETT, P.L. (1983). When does the βth percentile residual life function determine the distribution? Operat. Res. 31, 391-396.

CHUNG, C.F. (1986). Confidence bands for quantile function and percentile residual lifetime under random censorship. Ph. D. thesis, Carleton University, Ottawa, Canada.

CSÖRGŐ, M. and CSÖRGŐ, S. (1987). Estimation of percentile residual life. Operat. Res. 35, to appear.

CSÖRGŐ, S. (1988). Estimation in the proportional hazards model. Statistics 19, to appear.

CSÖRGŐ, S. and HORVÁTH, L. (1982). Statistical inference from censored samples. Alkalmaz. Mat. Lapok 8, 1-89. (in Hungarian)

CSÖRGŐ, S. and HORVÁTH, L. (1985). The baboons come down from the trees quite normally. Proc. 4th Pannonian Symp. on Mathematical Statistics, Vol. B. (W. Grossman, G. Ch. Pflug, I. Vincze and W. Wertz, eds.) Akadémiai Kiadó, Budapest, 95-106.

FELLER, W. (1966). An Introduction to Probability Theory and its Applications, Vol. 2, Wiley, New York.

GERCHAK, Y. (1984). Decreasing failure rates and related issues in the social sciences. Operat. Res. 32, 537-546.

GHORAI, J., SUSARLA, A., SUSARLA, V. and VAN RYZIN, J. (1982). Nonparametric estimation of mean residual life time with censored data. Colloquia Math. Soc. J. Bolyai 32. Nonparametric Statistical Inference (B.V. Gnedenko, M.L. Puri and I. Vincze, eds.) North-Holland, Amsterdam, 269-291.

GUPTA, R.C. and LANGFORD, E.S. (1984). On the determination of a distribution by its median residual life function: a functional equation. J. Appl. Probab. 21, 120-128.

HAINES, A.L. and SINGPURWALLA, N.D. (1974). Some contributions to the stochastic characterization of wear. Reliability and Biometry: Statistical Analysis of Life Length (F. Proschan and R.J. Serfling, eds.), SIAM, Philadelphia, 47-80

JOE, H. and PROSCHAN, F. (1984a). Percentile residual life functions. Operat. Res. 32, 668-678.

JOE, H. and PROSCHAN, F. (1984b). Comparison of two life distributions on the basis of their percentile residual life functions. Canadian J. Statist. 12, 91-97.

SCHMITTLEIN, D.C. and MORRISON, D.G. (1981). The median residual life time: a characterization theorem and an application. Operat. Res. 29, 392-399.

# An Introduction to Linear Dynamic Errors-in-Variables Models

Manfred Deistler
Institut für Ökonometrie und
Operations Research
Technische Universität Wien
A-1040 Wien, Austria

SUMMARY: In this paper we give a short survey on identification of linear systems where both inputs and outputs are possibly contaminated by noise. Such systems are called errors-in-variables- or latent variables-systems.

## 1. INTRODUCTION

The two most important branches in time series analysis are non-parametric spectral (and transfer function) estimation and identification of (finite-dimensional) linear (dynamic) systems. In the conventional approach to system identification, the classification of the observed variables into inputs and outputs is done a priori and the noise is added to the outputs or to the system equations (which makes no difference for our analysis), whereas the inputs are not contaminated by noise. This is the socalled errors-in-equations approach.

Here we consider a more general approach to noise modelling, where both inputs and outputs are possibly contaminated by noise. This is the socalled latent variables or errors-in-variables (EV) approach.

The conventional errors-in-equations approach is appropriate for a number of purposes, e.g. for prediction of the future values of the observed outputs based in their past and on the observed inputs. The main cases where the errors-in-variables setting is appropriate are:

(i) If we are interested in the"true" system generating the data, rather than in coding the data by a system, and if we cannot be sure a priori that the observed inputs are not corrupted by noise.

Contributions to Stochastics.
Ed. by W. Sendler
© Physica-Verlag Heidelberg 1987

(ii) If we want to have a symmetric treatment for all variables, since the classification into inputs and outputs or the number of equations is not known a priori

(iii) Clearly, alternatively, if we want to have a symmetric treatment of all variables, this can be done by using ARMA models, where the ARMA process is the process of all observed variables. However in many cases this leads to parameterspaces with dimension being considerably higher than for the corresponding errors-in-variables system.

(iv) If Assumption (2.9) is imposed, then errors-in-variables models are equivalent to dynamic principal components models and thus provide a decoupling between common and individual (linear) effects between the variables.

For the stable case, at least, the statistical theory of linear errors-in-equations systems has reached a certain stage of completeness [16], and there is a great number of practical applications, in particular for the single-input, single-output case. This stage by far has not been reached in errors-in-variables modelling. There is still a great number of unsolved problems, in particular those associated with the basic nonidentifiability of such systems: Whereas for errors-in-equations systems. under very general conditions (the socalled persistent exciting conditions), the transfer function is uniquely determined from the second moments of the observed processes, this is not true, in general, for errors-in-variables systems. In addition, in this case, for non-Gaussian variables, moments of order greater than two contain additional ("identifying") information about the transfer function.

For the linear static case in particular, when only two variables are observed, a rather complete analysis of the errors-in-variables model has been given by Gini [14] and Frisch [11]; see also the surveys [20] [21] [7]. The basic non identifiability alluded to above seems to have disencouraged research in the topic and to have detracted the attention of researchers for a long time, despite of its evident practical importance. In the last decade, however, there is a resurging interest in this area arising in different fields as econometrics, statistics and systems engineering [1] [2] [7] [18] [19] [27].

## 2. THE PROBLEM OF IDENTIFIABILITY

Let $\hat{z}_t$ denote the n-vector of "true" (in general unobserved) variables, which are related by a linear dynamic system of the form

$$w(z)\hat{z}_t = 0 \tag{2.1}$$

where z is a complex variable as well as the backward shift operator on $\mathbb{Z}$ (i.e. $z(\hat{z}_t) = (\hat{z}_{t-1})$) and

$$w(z) = \sum_{j=-\infty}^{\infty} W_i z^i \quad ; \quad W_i \in \mathbb{R}^{m \times n} \tag{2.2}$$

Without restriction of generality we assume that $m \leq n$ holds and that the rows of w are linearely independent in the sense that $w(e^{-i\lambda})$ has an m-th order determinant which is unequal to zero on a set of positive Lebesque measure on $[-\pi,\pi]$. The $W_i$ must be such that the infinite sum in (2.1) converges. (2.1) allows for a symmetric treatment of all components of $\hat{z}_t$ and we need not to distinguish between inputs and outputs [28]. If classification into inputs $\hat{x}_t$ and outputs $\hat{y}_t$ is known a priori then a linear dynamic system is written as

$$\hat{y}_t = k(z)\hat{x}_t \tag{2.3}$$

where

$$k(z) = \sum_{j=-\infty}^{\infty} K_j z^j \tag{2.4}$$

The special case (2.3) can be written in the form (2.1) with $\hat{z}_t = (\hat{y}_t', \hat{x}_t')'$, $w(z) = (-I, k(z))$. Clearly, even from the exact knowledge of $(\hat{z}_t)$, w(z) can be determined only up to a suitable normalization comparable to the normalization that $y_t$ is multiplied with the identity in (2.3). The observations $z_t$ are of the form

$$z_t = \hat{z}_t + u_t \tag{2.5}$$

where $u_t$ is the noise. In the errors-in-equations case, the system is of the form (2.3), and the observations are given by

$$z_t = (y_t', x_t')' = (\hat{y}_t', \hat{x}_t')' + (w_t', 0')' \tag{2.6}$$

Equation (2.1) means that $(\hat{z}_t)$ is restricted to a subspace, which is the kernel of a matrix corresponding to w(z). This subspace may also be described as the range of an other matrix by

$$\hat{z}_t = d(z) f_t \tag{2.7}$$

which together with (2.5) gives a dynamic factor model with factors $f_t$.

Throughout we assume

(i) All processes considered are (wide sense) stationary. All limits of random variables are understood in the sense of mean squares convergence.

(ii) $E\hat{z}_t = 0; \quad Eu_t = 0$

(iii) $E\hat{z}_t u_s' = 0$

(iv) The spectral density $f_u$ of the noise $(u_t)$ exists.

These assumptions are rather innocent; unless the contrary has been stated explicitely we also assume that the spectral density $\hat{f}$ of $(\hat{z}_t)$ exists. Then, for the spectral density f of $(z_t)$ we have

$$f = \hat{f} + f_u \tag{2.8}$$

However, in order to make the problem well posed, in a certain sense, some additional assumptions on the errors have to be imposed. Here, unless the contrary is stated we assume

$f_u$ is diagonal, i.e. the component processes in $(u_t)$ are mutually uncorrelated (2.9)

Assumption (2.9) together with (iii) means that all common linear dynamic effects are explained by the system (2.1), whereas the errors only contain individual effects. Clearly (2.9) is a meaningful assumption for a certain range of applications; there are, however, cases where (2.9) has to be replaced e.g. by assuming that f is the best approximation to f with corank m (in a certain norm) or by some frequency domain informations as e.g. the knowledge that the system effects correspond to low frequencies, whereas the errors are in a high frequency band.

The problem of identifiability in our context now is as follows: Given the probability law of the observed process $(z_t)$,

(i) find the maximum number, $m^*$ say, of (linearly) independent linear (dynamic) relations of the form (2.1) for $(\hat{z}_t)$.

(ii) For given $m^*$, give a characterization of the set of all $w(z)$ (and possibly in addition of all $(\hat{z}_t)$ and $(u_t)$) compatible with the probability law of $(z_t)$.

(iii) For given $m^*$ find additional conditions which guarantee uniqueness of $w(z)$ (i.e. identifiability of $w(z)$).

Note that the problem is considerably more general than the conventional problem of identifiability of linear dynamic errors-in-equations systems; in particular in our case here also the number of equations is not known a priori and has to be determined from the data.

If the external information considered are only the second moments $f$ (or equivalently, if we restrict ourselves to the case of the Gaussian observations) then the problem can be formulated as follows: for given $f$, consider all possible decompositions (2.8) into $\hat{f}$ and $f_u$, where $\hat{f}$ is a ($\lambda$-a.e) singular and $f_u$ a diagonal spectral density. As easily can be shown to every such decomposition there corresponds an errors-in-variables system (2.1) (2.5) and every errors-in-variables system gives rise to such a decomposition. Note, that for every $f$ such a decomposition exists; take for instance

$$f = U \Lambda U^* = U(\Lambda - \lambda_1 I)U^* + \lambda_1 I$$

where $U$ is unitary, $\Lambda = \text{diag}\{\lambda_1,\dots,\lambda_n\}$ is the matrix of the eigenvalues of $f$ and $\lambda_1$ is the smallest eigenvalue. The previous questions now translate to:

(i) What is the minimum rank, $n-m^*$ say, of $\hat{f}$ among all decompositions (2.8) for given $f$? Thereby we say that $f$ has rank $r$, if there is a set of positive Lebesgue measure on $[-\pi,\pi]$ such that $\hat{f}(\lambda)$ has a nonvanishing minor of size $r$ there, whereas all size $r+1$ minors are zero $\lambda$-a.e.

(ii) For given m*, characterize the set of all $\hat{f}$ of rank n-m*, compatible with given f

(iii) Under which additional assumption is $\hat{f}$ unique.

Up to now, there is no solution to this general problem. In the following we will discuss solutions for special cases.

## 3. THE STATIC CASE

3.1 The case n = 2: In this simplest case we have

$$\hat{y}_t = a\,\hat{x}_t \tag{3.1}$$

$$y_t = \hat{y}_t + w_t \quad ; \quad x_t = \hat{x}_t + v_t \tag{3.2}$$

where $a \in \mathbb{R}$, and where $(\hat{x}_t)$, $(w_t)$, $(v_t)$ are white noise processes. For a given scatter diagram this corresponds to fitting lines by least squares, where the distance of a point to the line is defined by any norm having unit balls corresponding to ellipses with achses parallel to the coordinate system.

In our stochastic framework the problem can be formulated analogously to (2.8) as follows. Let

$$\Sigma = (\sigma_{ij}), \quad \hat{\Sigma} = (\hat{\sigma}_{ij}), \quad \Sigma_u = \text{diag}\,\{\sigma_{uu}, \sigma_{vv}\}$$

denote the variance covariance matrices of $(x_t, y_t)'$, $(\hat{x}_t, \hat{y}_t)'$ and of $(v_t, w_t)'$ respectively. Then the decomposition (2.8) is of the form

$$\Sigma = \hat{\Sigma} + \Sigma_u \tag{3.3}$$

Under some minor additional assumptions we have ([14], [11])

Theorem 3.1: For the static, n=2, case (3.1), (3.2), the set of all slope parameters a compatible with given $\Sigma$ is the intervall sign $\sigma_{12}\cdot$ $[|\sigma_{12}/\sigma_{11}^{-1}|, |\sigma_{22}/\sigma_{12}^{-1}|]$.

Thus, already in this simple situation we have a basic non-identifiability for a. Note that the sign of a is uniquely given as the sign of $\sigma_{12}$ and that the endpoints of the intervall correspond to the two regression coefficients $\sigma_{12}|\sigma_{11}^{-1}$ and $\sigma_{12}|\sigma_{22}^{-1}$.

If e.g. the variance ratio $\sigma_{vv}|\sigma_{ww}$ (or $\sigma_{vv}$ or $\sigma_{ww}$) is a priori known (or can be measured) then a is uniquely determined.

In the non-Gaussian case identifiability for a may be obtained [12] [13] [20] [22] [28]; e.g. under some slight additional assumptions, if $\hat{x}_t$ is known to be non-Gaussian and $(v_t, w_t)$ is Gaussian, then a is identifiable.

3.2. The case of general n: Here the static model is of the form

$$A\hat{z}_t = 0 \; ; \quad z_t = \hat{z}_t + u_t \; ; \quad A \in \mathbb{R}^{m \times n} \tag{3.4}$$

and $(\hat{z}_t)$ and $(u_t)$ is white noise. This case has been investigated in detail by Kalman [18]. The main result is:

Theorem 3.2.: Under some additional assumptions we have

(i) $m^* = 1$ (i.e. there is only one linear relation) holds if and only if $\Sigma^{-1}$ is a positive matrix (i.e. all its entries are strictly positive) or if $\Sigma^{-1}$ can be transformed into a positive matrix by a suitable change of the signs of the variables

(ii) If $m^* = 1$, then the set of all vectors A compatible with given $\Sigma$ is the positive cone spanned by the rows of $\Sigma^{-1}$.

A complete characterization of the cases where $m^* > 1$ holds, ist still missing.
Aigain, in the non-Gaussian case, we may have identifiability [8].

## 4. THE DYNAMIC CASE

4.1. The case n = 2: Consider (2.3), (2.5) where $\hat{x}_t$ and $\hat{y}_t$ are scalar. Let $f = (f_{ij})$. The following theorem was given in [5]:

Theorem 4.1: For the danymic, n= 2 case, the set of all transfer functions k satisfying

$$|f_{21}(\lambda)| \cdot f_{11}^{-1}(\lambda) \leq |k(e^{-1\lambda})| \leq f_{22}(\lambda) \cdot |f_{12}(\lambda)|^{-1} \tag{4.1}$$

and

$$\arg (w(e^{-i\lambda})) = \arg (f_{21}(\lambda)) \tag{4.2}$$

is the set of all transfer functions compatible with given f.

Note that this theorem is the extension of Theorem 3.1. to the complex and frequency-dependent case. Again $\arg (w(e^{-i\lambda}))$, the phase of the transfer function, which is the static real case corresponds to sign a is uniquely determined, whereas the moduli of $w(e^{-i\lambda})$ in a $\lambda$-dependent intervall, whose endpoints correspond to the filter transfer functions (and thus to the dynamic regressions) $f_{21} \cdot f_{11}^{-1}$ and $f_{12} \cdot f_{22}^{-1}$.

Of course, in general, the class of f-equivalent transfer functions given by (4.1), (4.2), contains also noncausal transfer functions. If in addition k is a priori known to be causal (i.e. in (2.4)) we have $K_j = 0$, for $j < 0$) then the f-equivalence class is considerably smaller. This is discussed in detail in [4] [9] [10].

Analogously to the static case, for non-Gaussian observations, k may be identifiable [3] [9] [10] [17].

4.2. The case n = 3: Let $f = (f_{ij})$ and $s = f^{-1} = (s_{ij})$. For this case, the following dynamic analogon of Theorem 3.2 has been shown [6].

Theorem 4.2: The following statements are equivalent

(i) $m^* = 2$

(ii) $f_{12} \cdot f_{23} \cdot f_{31} > 0$ and $f_{ii} \geq |f_{ij}| \, |f_{ik}| \, |f_{jk}|^{-1}$ ($\lambda$-a.e) for $i \neq j \neq k$

(iii) $s_{12} s_{23} s_{31} < 0$ ($\lambda$-a.e)

For general n, results have been given in [15] [21] [23] [23] but up to now, there is no extension of Theorem 3.2 to this case. Again, in the non-Gaussian case, we may have identifiability.

References

[ 1] Aigner,D.J. and A.S.Goldberger (Eds.), (1977): Latent Variables in Socio-Economic Models. North Holland P.C., Amsterdam

[ 2] Aigner,D.J., C.Hsiao, A.Kapteyn and T.Wansbeek (1984): Latent Variable Models in Econometrics. In: Griliches, Z. and M.D. Intriligator (Eds.) Handbook of Econometrics. North Holland P.C., Amsterdam

[ 3] Akaike,H. (1966): On the Use of Non-Gaussian Process in the Identification of a Linear Dynamic System. Annals of the Institute of Statistical Mathematics 18, 269 - 276

[ 4] Anderson,B.D.O. (1985): Identification of scalar errors-in-variables models with dynamics, Automatica, 21, 709 - 716

[ 5] Anderson,B.D.O. and M.Deistler (1984): Identifiability in Dynamic Errors-in-Variables Models, Journal of Time Series Analysis, 5, 1 - 13

[ 6] Anderson,B.D.O. and M.Deistler (1986): Dynamic Errors-in-Variables Systems with Three Variables. Submitted for publication

[ 7] Anderson,T.W. (1984): Estimating Linear Statistical Relationships, Annals of Statistics, 12, 1 - 45

[ 8] Bekker,P.A. (1986): Comment on Identification in the Linear Errors in Variables Model, Econometrica, Vol. 54, 1, 215 - 217

[ 9] Deistler,M. (1986): Linear Dynamic Errors-in-Variables Models. In: J. Gani and M.Priestley (Eds.), Essays in Time Series and Allied Processes, Papers in Honour of E.J.Hannan, Applied Probability Trust, Sheffield.

[10] Deistler,M. (1986): Linear Errors-in-Variables Models. In: S. Bittanti (Ed.), Time Series and Linear Systems, Lecture Notes in Control and Information Sciences, Springer-Verlag, Berlin

[11] Frisch,R. (1934): Statistical Confluence Analysis by Means of Complete Regression Systems. Publication No. 5, University of Oslo, Economic Institute

[12] Geary,R.C. (1942). Inherent Relations between Random Variables. Proceedings of the Royal Irish Academy, Sec. A, 47, 63 - 76

[13] Geary,R.C. (1943): Relations between Statistics: The General and the Sampling Problem When the Samples are Large. Proceedings of the Royal Irish Academy. Sec. A, 49, 177 - 196

[14] Gini,C. (1921): Sull'interpolazione di una retta quando i valori della variable indipendente sono affetti da errori accidentali. Metron 1, 63 - 82

[15] Green,M. and B.D.O.Anderson (1985): Identification of Multivariable Errors-in-Variables Models with Dynamics. IEEE Trans. Auto Control, AC-31, 467 - 471

[16] Hannan,E.J. and L.Kavalieris (1984): Multivariate Linear Time Series Models. Advances in Applied Probability 16, 492 - 651

[17] Hinich,M.J. (1983): Estimating the Gain of a Linear Filter from Noisy Data. In: D.R.Brillinger and P.R. Krishnaiah (Eds.), Handbook of Statistics, Vol 3. North Holland, Amsterdam

[18] Kalman,R.E. (1982): System Identification from Noisy Data. In: A. Bednarek and L.Cesari (Eds.), Dynamic Systems II, a University of Florida International Symposium. Academic Press, New York

[19] Kalman,R.E. (1983): Identifiability and Modeling in Econometrics. In: Krishnaiah,P.R. (Ed.), Developments in Statistics, Vol.4, Academic Press, New York

[20] Madansky,A. (1959): The Fitting of Straight Lines when Both Variables are Subject to Error. Journal of the American Statistical Association 54, 173 - 205

[21] Maravall,A. (1979): Identification in Dynamic Shock-Error Models. Lecture Notes in Economics and Mathematical Systems, Springer-Verlag, Berlin

[22] Moran, P.A.P. (1971): Estimating Structural and Functional Relationships. Journal of Multivariable Analysis 1, 232 - 255

[23] Picci,G. and S.Pinzoni (1986): Dynamic Factor Analysis Models for Stationary Processes. IMA J. Math. Control and Information, to appear

[24] Picci,G. and S.Pinzoni (1986): A New Class of Dynamic Models for Stationary Time Series. In: S.Bittanti (ed.), Time Series and Linear Systems, Lecture Notes in Control and Information Sciences, Springer-Verlag, Berlin

[25] Reiersøl,O. (1941): Confluence Analysis by Means of Lag Moments and other Methods of Confluence Analysis, Econometrica 9, 1 - 24

[26] Reiersøl,O. (1950): Identifiability of a Linear Relation Between Variables which are Subject to Error, Econometrica 18, 375 - 389

[27] Schneeweiß,H. und H.J. Mittag (1986): Lineare Modelle mit fehlerbehafteten Daten. Physica Verlag, Würzburg

[28] Willems,J.C. (1986): From time series to linear system. To appear in Automatica

I want to express my deep gratitude to Professor Walther Eberl. He had a decisive influence on my academic career. He encouraged me to write my Ph.D. Thesis under his supervision in a time when at Austrian universities econometrics was regarded as a rather odd subject; due to his initiative the chair of Econometrics at the "Technische Hochschule Wien" was created as the first chair of Econometrics in Austria.

# On Improved Versions of General Erdös-Rényi Laws and Large Deviation Asymptotics

E. Dersch and J. Steinebach
Fachbereich Mathematik
Universität Marburg
D-3550 Marburg, FRG

SUMMARY. Somme rather general versions of the (so-called) Erdös-Rényi [14] law of large numbers have been derived by S. Csörgö [7] and Steinebach [21] for sequences of random variables satisfying a fist order large deviation theorem together with certain independence and stationarity properties. In view of recent second order large deviation asymptotics due to Dersch [13], corresponding refinements of the forementioned general Erdös-Rényi laws are discussed in this paper. The latter results can be viewed as convergence rate improvements of their earlier counterparts, typically providing the best rates, but not necessarily the best constants. Some specific examples are presented to demonstrate the applicability of our general approach.

## 1. INTRODUCTION

The Erdös and Rényi [14] (so-called) new law of large numbers is concerned with an a.s. asymptotic for the maximum moving averages of an i.i.d. sequence, taken over all subintervals of logarithmic length in a large interval of indices:

Theorem A (Erdös-Rényi). Let $\{X_i\}_{i=1,2,..}$ be an i.i.d. sequence with partial sums $S_0=0$, $S_k=X_1+...+X_k$. Suppose $M(t) = E\exp(tX_1) < \infty$ for all $t \in (0,t_1)$. Then, for each $a \in \{M'(t)/M(t) : t \in (0,t_1)\}$ and $c = c(a)$ such that $\exp(-1/c) = \rho(a) = \inf_t M(t)\exp(-ta)$, we have, with $k = k(n) = [c \log n]$,

$$\lim_{n\to\infty} \max_{0 \le i \le n-k} (S_{i+k}-S_i)/k = a \quad \text{a.s.} \tag{1.1}$$

The main essence of Theorem A is that it provides a strong noninvariance principle, since the functional dependency $\rho=\rho(a)$, which can a.s. be obtained from (1.1), via

---

AMS 198o subject classifications. Primary 6oF15; secondary 6oF1o.
Key words and phrases. General Erdös-Rényi laws, large deviations, strong limit theorems, increments of stochastic processes, strong non-invariance.

Contributions to Stochastics.
Ed. by W. Sendler
© Physica-Verlag Heidelberg 1987

$M=M(t)$ also determines the underlying distribution. Earlier, Shepp [18] obtained a related theorem under the same assumptions by showing that

$$\lim_{n\to\infty} \max_{1\leq i\leq n} (S_{i+k(i)}-S_i)/k(n) = a \qquad \text{a.s.} \qquad (1.2)$$

On observing that the Erdős-Rényi theorem is essentially based upon a first-order large deviation theorem of Chernoff [5] type and certain independence and stationarity properties of the process under consideration, some extensions of this phenomenon become available e.g. for weighted sums and non-i.i.d. sequences (Book [1], [2]), for sample quantiles (Book and Truax [3]), and also for queueing models, cumulative processes in renewal theory and linear stochastic processes (Steinebach [19], [20],[23]). Some rather general versions of Erdős-Rényi type laws, covering a considerable class of special examples, have been presented by S. Csörgő [7] and Steinebach [21].

In [6], M. Csörgő and Steinebach initiated certain refinements of the original Erdős-Rényi law by showing that

$$\max_{0\leq i\leq n-k} (S_{i+k}-S_i)/ka = 1 + o(k^{-1/2}) \qquad \text{a.s.,} \qquad (1.3)$$

and thus presenting a first convergence rate statement for (1.1). Analogous results for renewal processes have also been derived (cf. Retka [16], Steinebach [22]), based on certain improvements of the large deviation estimates involved.

The exact convergence rate in (1.3) and some related theorems have been obtained by Deheuvels, Devroye and Lynch [10] and Deheuvels and Devroye [11]. Their refinement of (1.3) e.g. says that

$$\max_{0\leq i\leq n-k} (S_{i+k}-S_i)/ka = 1 + O(k^{-1}\log k) \qquad \text{a.s.}, \qquad (1.4)$$

and it also includes the best constants. Similar results were proved for the moving quantiles (Deheuvels and Steinebach [12]) and renewal processes (Steinebach [25]), the latter results probably containing the best rates, but not necessarily the best constants. For further extensions and generalizations of the original Shepp-Erdős-Rényi law into different directions we refer to de Acosta and Kuelbs [8], Deheuvels [9], Révész [17], Steinebach [24], and the work mentioned therein.

The forementioned exact convergence rate statements are mainly based upon Petrov's [15] uniform improvements of the Chernoff [5] large deviation asymptotics, together with a precise handling of the dependencies between overlapping blocks. Recently, Dersch [13] presented the following generalization of the Petrov theorem (cf. also Chaganty and Sethuraman [4] for a local analogue), which served as a motivation for our main result below:

Theorem 1 Let $\{T_n\}_{n=1,2,..}$ be a sequence of random variables with

(i) $M_n(s) = E\exp(s\,T_n) < \infty$ for $s \in (0,s_1)$

(ii) $|\frac{1}{n}\log M_n(s)|$ and $(\frac{1}{n}(\log M_n)''(s))^{-1}$ are locally uniformly bounded in $(s_0,s_1)$, $0 \leq s_0 < s_1$.

(iii) For $s \in (s_0,s_1)$ there exists a $\delta_s > 0$, such that $M_n(z) \neq 0$ for all $z \in U_s := \{z \in \mathbb{C} : |z-s| < \delta_s\}$ and all $n \in \mathbb{N}$.

(iv) The $T_n$ are either all nonlattice or all lattice random variables with $P(T_n \in x_n + d_n\mathbb{Z}) = 1$ and $d_n n^{1/2} \to \infty$.

Furthermore, for $K \subset (s_0,s_1)$ compact, there exists an $\varepsilon_K > 0$, such that, for $b_1,b_2 > 0$:

$$\sup_{\substack{b_1 \leq |t| \leq b_2 \\ s \in K}} \left|\frac{M_n(s+it)}{M_n(s)}\right| = O(n^{-(1/2+\varepsilon_K)}),\quad \text{if } T_n \text{ is nonlattice, respectively}$$

$$\sup_{\substack{b_1 \leq |t| \leq \pi/d_n \\ s \in K}} \left|\frac{M_n(s+it)}{M_n(s)}\right| = o(n^{-1/2}d_n),\quad \text{if } P(T_n \in x_n + d_n\mathbb{Z}) = 1.$$

Let $K \subset (s_0,s_1)$ be compact. Then

$$P(T_n \geq n\,a_n) = (2\pi n)^{-1/2}\,\rho_n^n(a_n)\,B_n(a_n)\,(1+o(1))\quad (n \to \infty) \tag{1.5}$$

uniformly over all sequences $\{a_n\}_{n=1,2,..}$ with $a_n \in A_{n,K} = \{\frac{1}{n}(\log M_n)'(s) : s \in K\}$, where

$\log B_n(a_n) = O(1)$ uniformly over $\{a_n\}$,

$\rho_n(a) = \inf_s \exp(\frac{1}{n}\log M_n(s) - sa)$.

Corollary 1 Instead of (ii) assume

(ii') $|\log M_n(s) - nK(s)| = O(1)$ locally uniformly over $s \in (s_0,s_1)$, where $K:(s_0,s_1) \to \mathbb{R}$ is a strictly convex function.

Let $I \subset A := \{K'(s) : s \in (s_0,s_1)\}$ be compact. Then we have uniformly for $a \in I$:

$$P(T_n \geq na) = (2\pi n)^{-1/2}\,\rho^n(a)\,B_n(a)\,(1+o(1))\quad (n \to \infty) \tag{1.6}$$

with

$\log B_n(a) = O(1)$ uniformly over $a \in I$, and

$\rho(a) = \inf_s \exp(K(s)-sa) = \exp(K(h)-ha)$

where $h = h(a)$ denotes the unique solution of $K'(h) = a$.

The generality of the improved large deviation estimates (1.5) and (1.6) of Theorem 1 and Corollary 1 justifies a continuation of the general discussion of Erdős-Rényi laws started by S. Csörgő [7] and Steinebach [21]. In Section 2, we state and prove a corresponding result. In Section 3, some specific examples are discussed in order to illustrate the applicability of our general approach.

## 2. A GENERAL IMPROVED ERDŐS-RÉNYI TYPE LAW.

Motivated by the results of Deheuvels, Devroye and Lynch [10], Deheuvels and Devroye [11], Deheuvels and Steinebach [12] and Steinebach [25] and in view of Theorem 1 (Dersch [13]) we prove the following improved version of a general Erdős-Rényi type law:

Theorem 2. Let $\{T_{i,k}\}_{\substack{i=0,1,\dots\\ k=1,2,\dots}}$ be a double sequence of real-valued random variables.

a) Suppose there is a uniform large deviation upper bound, i.e.

$$P(T_{i,k} \geq k(a+z_k)) \leq A_1\, k^{p_1}\, \rho_1^k(a) \exp(-q_1\, k z_k)$$

holds uniformly over $i$ and all sequences $\{z_k\}$ with $k z_k = b_1 \log k$, $(p_1+1)/q_1 < b_1 < (p_1+1)/q_1 + \varepsilon_1$ for some $\varepsilon_1 > 0$, where $A_1, q_1 > 0$ and $p_1$ are constants, $0 < \rho_1(a) < 1$.

Then, for $c_1 = c_1(a)$ such that $\exp(-1/c_1) = \rho_1(a)$, and $k_1 = k_1(n) = [c_1 \log n]$, we have

$$\limsup_{n\to\infty} \max_{0 \leq i \leq n-k_1} (T_{i,k_1} - k_1 a)/\log k_1 \leq (p_1+1)/q_1 \quad \text{a.s.} \tag{2.1}$$

b) Suppose there is a sequence $\nu_{0,n} < \nu_{1,n} < \cdots$ of random indices such that, for each $n = 1,2,\dots$ and $k = k(n) = [c \log n]$:

b1) $\{T_{\nu_{i,n};k}\}_{i=0,1,\dots}$ is an independent sequence;

b2) $P(T_{\nu_{i,n};k} \geq k(a+z_k)) \geq A_0\, k^{p_0}\, \rho_0^k(a) \exp(-q_0\, k z_k)$

holds uniformly over $i$ and all sequences $\{z_k\}$ with $k z_k = b_0 \log k$, $(p_0-1)/q_0 > b_0 > (p_0-1)/q_0 - \varepsilon_0$ for some $\varepsilon_0 > 0$, where $A_0, q_0 > 0$ and $p_0$ are constants, $0 < \rho_0(a) < 1$;

b3) $\sum_{n=1}^{\infty} P(N(n) \leq \delta n/\log n) < \infty$ for some $0 < \delta < 1$ ,

where $N(n) = \#\{\nu_{i,n} : \nu_{i,n} \leq n - k(n)\}$ .

Then, for $c_0 = c_0(a)$ such that $\exp(-1/c_0) = \rho_0(a)$, and $k_0 = k_0(n) = [c_0 \log n]$, we have

$$\liminf_{n\to\infty} \max_{o\leq i\leq n-k_0} (T_{i,k_0} - k_0 a)/\log k_0 \geq (p_0 - 1)/q_0 \quad \text{a.s.} \tag{2.2}$$

Remark 1. a) If the assumptions of Theorem 2 can be satisfied with $\rho_0(a) = \rho_1(a) = \rho(a)$, a combination of assertions (2.1) and (2.2), setting $c = c_0 = c_1$ and $k = [c \log n]$ , yields

$$\max_{o\leq i\leq n-k} T_{i,k}/ka = 1 + O(k^{-1} \log k), \tag{2.3}$$

and thus provides the desired convergence rate statement in the general version of an Erdős-Rényi type law.

b) If the large deviation upper bound of Theorem 2a) also holds uniformly over $i$ and all sequences $\{z_k\}$ with $kz_k = b_1' \log k$, $p_1'/q_1' < b_1' < p_1'/q_1' + \varepsilon_1' < 0$ , for some $\varepsilon_1' > 0$, where $p_1', q_1'$ replace $p_1, q_1$, then

$$\liminf_{n\to\infty} \max_{o\leq i\leq n-k_1} (T_{i,k_1} - k_1 a)/\log k_1 \leq p_1'/q_1' \quad \text{a.s.} \tag{2.4}$$

This is immediate from the proof of part a) and the fact that $P(A_n) \to 1$ (as $n \to \infty$ ) always yields $P(A_n \text{ i.o.}) = 1$. Under (2.4), it is obvious that the rate of (2.3) is best possible.

Proof of Theorem 2. a) For sake of notational convenience we drop the index 1 in $A_1, p_1, q_1, \rho_1, c_1, k_1$ and set

$$U_n = \max_{o\leq i\leq n-k} T_{i,k} . \tag{2.5}$$

For $j$ sufficiently large, let $n_j = \max\{n : k = [c \log n] = j\}$ .
Then, for $n_{j-1} < n \leq n_j$, we have $k = k(n) = j$, $\exp((j-1)/c) < n \leq \exp(j/c)$ and $U_n \leq U_{n_j}$. Hence it is enough to prove (2.1) for the subsequence $\{n_j\}$. Now, by our assumptions on the large deviation probabilities of $\{T_{i,k}\}$ and our choice of $c = c(a)$,

$$P(U_{n_j} \geq ka + b \log k) \leq n_j\ A\ k^p\ \rho^k(a)\ \exp(-bq \log k)$$
$$= O(j^{p-bq}) \quad .$$

If $b$ is such that $p - bq < -1$, i.e. $b > (p+1)/q$, the above probabilities are summable in $j$, which by the Borel-Cantelli lemma yields the desired conclusion.
b) We also drop the index 0 in $A_0, p_0, q_0, \rho_0, c_0, k_0$ and use $U_n$ as introduced in (2.5). Similar to part a) of the proof, set $m_j = \min\{n : k = [c \log n] = j\}$, $j$ large. Then, for $m_j \leq n < m_{j+1}$, we have $k = k(n) = j, \exp((j-1)/c) < n \leq \exp(j/c)$ and $U_n \geq U_{m_j}$.
Hence it is enough to prove (2.2) for the subsequence $\{m_j\}$. Now, with $M = M_j = [\delta m_j / \log m_j]$,

$$P(U_{m_j} < ka + b \log k)$$
$$\leq P(\max_{i=0,\dots,M} T_{\nu_{i,m_j};k} < ka + b \log k) + P(N(m_j) \leq M).$$

By assumption b3), $\Sigma P(N(m_j) \leq M) < \infty$. By b1) and b2), the first probability on the right-hand side can be estimated from above by

$$\exp\{-M\ A\ k^p\ \rho^k(a)\ \exp(-bq \log k)\} \leq \exp(-\tilde{A} j^{p-1-bq}) \quad ,$$

where $\tilde{A}$ is a positive constant. If $p-1-bq > 0$, i.e. $b < (p-1)/q$, the latter terms are summable in $j$, which, by Borel-Cantelli, completes the proof.

## 3. Examples

Since, via Theorem 1, second order large deviation estimates are available now for rather general families of random variables, convergence rate improvements of certain specific Erdős-Rényi laws appearing in S. Csörgő [7] and Steinebach [21] are obvious. Some typical examples will be presented next.

Example 1 (Weighted sums). Let $S_n = \sum_{i=1}^{n} a_i X_i$ denote the weighted sum of a standardized i.i.d. sequence $\{X_i\}_{i=1,2,\dots}$ under nonnegative weights $\{a_i\}_{i=1,2,\dots}$ such that

(i) $M(s) = E \exp(s X_1) < \infty$ for $s \in (0, s_1)$,

(ii) there exist constants $\alpha$ and $\theta$, $0 < \alpha \leq 1$, $0 < \theta \leq 1$, such that, for large $k$ and all $i \in \mathbb{N}$ at least $\alpha k$ of the numbers $\{a_j : i+1 \leq j \leq i+k\}$ exceed or equal $\theta\, m_k(i)$, where $m_k(i) = \max\{a_j : i+1 \leq j \leq i+k\}$.

Put $A_n = \sum_{i=1}^{n} a_i$ and $U(n,k) = \max_{0\leq i\leq n-k} (S_{i+k}-S_i)/(A_{i+k}-A_i)$. Then the following improvement of Book's [1] "Erdős-Rényi law of large numbers for weighted sums" is covered by Theorem 2:

For $a \in \alpha\theta A := \{\alpha\ \theta M'(s)/M(s) : s \in (0,\alpha\,\theta^2 s_1)\}$ and certain functions $c_1(a)$, $c_0(a)$, $q(a) > 0$ we have with $k_j = [c_j(a) \log n]$ , $j = 0,1$ :

$$\limsup_{n\to\infty} (U(n,k_1)-k_1 a)/\log k_1 \leq 1/(2q(a)) \qquad \text{a.s.} \tag{3.1}$$

$$\liminf_{n\to\infty} (U(n,k_0)-k_0 a)/\log k_0 \geq -1/(2q(a)) \qquad \text{a.s.} \tag{3.2}$$

For a proof set

$$T_{i,k} = k\,\frac{S_{i+k}-S_i}{A_{i+k}-A_i}$$

$$\rho_j(a) = \exp(-1/c_j(a)) \qquad (j=0,1)$$

and observe that

$$P(T_{i,k} \geq ka) = P(k\,A_{ik}^{-1} \sum_{j=i+1}^{i+k} a_{ikj}\, X_j \geq k\,a)$$

with

$$a_{ikj} = a_j\Big(\sum_{l=i+1}^{i+k} a_l^2\Big)^{-1/2},\quad A_{ik} = \sum_{j=i+1}^{i+k} a_{ikj} \quad \text{and hence} \quad \sum_{j=i+1}^{i+k} a_{ikj}^2 = 1.$$

For fixed $i$, the sequence $\{T_{i,k}\}_{k=1,2,\ldots}$ satisfies the conditions of Theorem 1 and we get, if $k\,z_k^2 = o(1)$,

$$P(T_{i,k} \geq k(a+z_k)) \sim (2\pi k)^{-1/2}\,\rho_{ik}^{\,k}(a)\; B_{ik}(a+z_k)\,\exp(-k\,z_k\,h_{ik})$$

with

$$\rho_{ik}^{\,k}(a) = \prod_{j=i+1}^{i+k} M(k\,A_{ik}^{-1}\,a_{ikj}\,h_{ik})\;\exp(-k\,h_{ik}\,a)$$

where $h_{ik}$ is the unique solution of

$$(\log M_{ik})'(s) = ka \quad , \quad M_{ik}(s) = E\,\exp(s\,T_{ik})$$

and $\log B_{ik}(a+z_k) = O(1)$ uniformly in $\{z_k\}$.

The conditions (ii) on the weights imply uniformity (in i) of the large deviation bounds required in Theorem 2, with $p_1 = p_0 = -\frac{1}{2}$.

For the lim-inf-part we use a deterministic sequence $\{\nu_{i,n}\}_{i=1,2,\dots}$, i.e.

$$\nu_{i,n} = i[c \log n], \quad i=1,2,\dots \ .$$

Example 2. (Waiting-times). Let $W_n$ denote the waiting-time of the (n+1)-th customer in a queueing model G/G/1. If $X_n$ is defined as the difference between the n-th service-time and the interarrival-time between the n-th and (n+1)-th customer, then we have

$$W_n = \max(0, W_{n-1}+X_n), \quad n = 1,2,\dots , \qquad W_0 = 0.$$

We assume, that $\{X_n\}_{n=1,2,..}$ is i.i.d. with a nondegenerate distribution and $M(s) = E \exp(s X_1) < \infty$ for $s \in (0,s_1)$. If $EX_1 \geq 0$, Theorem 2 yields the following best convergence rate improvement of the first order Erdös-Rényi law for waiting-times presented in [19]:

$$\max_{0 \leq i \leq n-[c \log n]} \frac{W_{i+[c \log n]} - W_i}{a[c \log n]} = 1 + 0\left(\frac{\log \log n}{\log n}\right) \quad \text{a.s.} \tag{3.3}$$

for all $a \in \{M'(s)/M(s) : s \in (0,s_1)\}$ and $c$ such that $\exp(-1/c) = \rho(a)$, where $\rho(a)$ denotes the first order large deviation rate of the sequence $\{W_n\}_{n=1,2,..}$.

Observe that for the sequence $\{W_n\}$ the assumptions of Corollary 1 are fulfilled with $K(s) = \log M(s)$.
This follows from

$$M_n(s) = E \exp(s W_n) = M^n(s) \sum_{k=0}^{n} r_k(s) M^{-k}(s), \qquad s \in (0,s_1)$$

where

$$r_k(s) = \int_{\{W_{n-1}+X_n \leq 0\}} (1-\exp(s(W_{n-1}+X_n))dP, \quad k=1,2,\dots ,$$

$$r_0(s) \equiv 1.$$

Moreover the rate $\rho(a)$ of $\{W_n\}$ is the same as that for the sequence of partial sums $\{S_n\}$, $S_n = X_1+\dots+X_n$.

With $T_{i,k} = W_{i+k} - W_i$, it follows, if $kz_k^2 = o(1)$,
$P(T_{i,k} \geq k(a+z_k)) \leq P(W_k \geq k(a+z_k)) \leq A_1 k^{-1/2} \rho^k(a) \exp(-k z_k h_a)$ where $h_a$ is the unique solution of

$$M'(s)/M(s) = a,$$

i.e. the assumptions of Theorem 2a) and Remark 1b) are satisfied with $p_1=p_1' = -1/2$ und $\rho_1(a) = \rho(a)$. Hence, we have

$$\limsup_{n\to\infty} \max_{0\le i\le n-k} (T_{i,k}-ka)/\log k \le 1/(2h_a) \quad \text{a.s.} \tag{3.4}$$

and

$$\liminf_{n\to\infty} \max_{0\le i\le n-k} (T_{i,k}-ka)/\log k \le -1/(2h_a) \quad \text{a.s.} \tag{3.5}$$

where $k = [c \log n]$ and $\exp(-1/c) = \rho(a)$.

Since $W_{i+k} - W_i \ge S_{i+k} - S_i$, the improved Erdős-Rényi law for the partial sums $S_n$ in Deheuvels, Devroye and Lynch [10] yields

$$\limsup_{n\to\infty} \max_{0\le i\le n-k} (T_{i,k}-ka)/\log k \ge 1/(2h_a) \quad \text{a.s.} \tag{3.6}$$

and

$$\liminf_{n\to\infty} \max_{0\le i\le n-k} (T_{i,k}-ka)/\log k \ge -1/(2h_a) \quad \text{a.s.} \tag{3.7}$$

with $k = [c \log n]$ as above.

Remark 2. If, in Example 2, we assume $EX_1 < 0$ and $M(s_0) = 1$ for some $s_0 \in (0,s_1)$, we get the same statements for $a \in \{M'(s)/M(s) : s \in (s_0,s_1)\}$.
For $a \in \{M'(s)/M(s) : s \in (0,s_0)\}$ the assumptions of Theorem 1 are not fulfilled, but in this case uniform large deviation asymptotics can be proved by using "ladder"-variables.
Theorem 2 then yields the same convergence rate statement as in the latter case, but with different constants.
For the lim inf - part, however, the approximation by partial sums $\{S_n\}$ is not good enough. But, similar to [19] we can make use of a random sequence of indices, defined by

$\nu_{o,n} = 0$,

$\nu_{i,n}$ = smallest integer $> \nu_{i-1,n} + [c \log n]$ such that $W_{\nu_{i,n}} = 0$, $(i=1,2,\ldots)$.

Example 3. (Cumulative renewal processes)
Let $N(t)$ denote the number of renewals occuring up to time $t$ ($\ge o$) associated with an i.i.d. sequence $\{X_i\}_{i=1,2,\ldots}$ of nondegenerate "failure times" $X_i$ with $EX_i > 0$. Furthermore, let $\{Y_j\}_{j=1,2,\ldots}$ be a sequence of non-negative i.i.d. random variables, mutually independent of $\{X_i\}$. Then the "cumulative process" $\{Z_t\}_{t\ge o}$ is defined by

$$Z_t = \sum_{j=1}^{N(t)} Y_j \quad .$$

In [2o], the following Erdős-Rényi law has been proved for the process $\{Z_t\}_{t\geq 0}$ :

$$\lim_{T\to\infty} \sup_{o\leq t\leq T-c\log T} \frac{Z_{t+c\log T} - Z_t}{c\log T} = a \qquad \text{a.s.} \tag{3.8}$$

for a's in some interval $(EY_1/EX_1, a_1)$ and c such that $\exp(-1/c) = \rho(a)$, where $\rho(a)$ denotes the first order large deviation rate of $\{Z_t\}_{t\geq 0}$.

Using certain monotonicity properties of the process, it is obvious that the limit assertions need only be proved for T and t being integer-valued. So, Theorem 2 can be used to derive a convergence rate statement in the Erdős-Rényi type law (3.8). For the lim inf - part we take a sequence of random times defined by

$\tau_{0,T} = 0$

$\tau_{i,T}$ = first renewal time after $\tau_{i-1,T} + c\log T$ $(i=1,2,\dots)$,

which provides i.i.d. increments $N(\tau_{i,T} + c\log T) - N(\tau_{i,T})$, $i=1,2,\dots$

In the case of degenerate $Y_j$'s, the duality

$$P(N(t)\geq n) = P(\max(0,S_1,\dots,S_n)\leq t),\ n\in\mathbb{N},\ S_n = X_1+\dots+X_n,$$

yields, via Petrov's [15] theorem together with Corollary 1 in Deheuvels-Devroye[11], uniform large deviation bounds as needed in Theorem 2.

If $Y_1$ is nondegenerate, assume that the momentgenerating functions $M_{N(t)}$ of N(t) satisfy the condition (iii) of Theorem 1 (for t = 1,2,...). Then, for the sequence $\{Z_t\}_{t=1,2,\dots}$, the assumptions of Corollary 1 are fulfilled, and we get the desired large deviation bounds of Theorem 2, with $\rho_1(a) = \rho_o(a) = \rho(a)$ and $p_1 = p_0 = -1/2$. In particular, for compound Poisson processes, second order large deviation asymptotics can be derived from Corollary 1.

Remark 3. A number of further improvements of first order Erdős-Rényi laws can be derived from Theorem 2, e.g. for sums of independent, but not necessarily identically distributed random variables (see e.g. [2]), moving quantiles ([3],[12]) and linear stochastic processes ([23]), with exponential weights. In all these cases, Theorem 1 yields the required improvements on the large deviation estimates needed.

REFERENCES

[1] Book, S.A. (1973). The Erdős-Rényi new law of large numbers for weighted sums. Proc. Amer. Math. Soc. 38, 165-171.

[2] Book, S.A. (1976). Large deviation probabilities and the Erdős-Rényi law of large numbers. Canad. J. Statist. 4, 185-21o.

[3] Book, S.A., Truax, D.R. (1976). An Erdős-Rényi strong law for sample quantiles. J. Appl. Prob. 13, 578-583.

[4] Chaganty, N.R., Sethuraman, J. (1985). Large deviation local limit theorems for arbitrary sequences of random variables. Ann. Probab. 13, 97-114.

[5] Chernoff, H. (1952). A measure of asymptotic efficiency for tests of a hypothesis based on the sum of observations. Ann. Math. Statist. 23, 493-5o7.

[6] Csörgő, M., Steinebach, J. (1981). Improved Erdős-Rényi and strong approximation laws for increments of partial sums. Ann. Probab. 9, 988-996.

[7] Csörgő, S. (1979). Erdős-Rényi laws. Ann. Statist. 7, 772-787.

[8] De Acosta, A., Kuelbs, J. (1983). Limit theorems for moving averages of independent random vectors. Z. Wahrsch. Verw. Geb. 64, 67-123.

[9] Deheuvels, P. (1985). On the Erdős-Rényi theorem for random fields and sequences and its relationships with the theory of runs and spacings. Z. Wahrsch. Verw. Geb. 7o, 91-116.

[1o] Deheuvels, P., Devroye, L., Lynch, J. (1986). Exact convergence rates in the limit theorems of Erdős-Rényi and Shepp. Ann. Probab. 14, 2o9-223.

[11] Deheuvels, P., Devroye, L., (1987). Limit laws of Erdős-Rényi type. Ann.Probab. (to appear).

[12] Deheuvels, P., Steinebach, J. (1986). Exact convergence rate of an Erdős-Rényi strong law for moving quantiles. J. Appl. Prob. 23, 355-369.

[13] Dersch, E. (1986). Gleichmäßige Asymptotik für Wahrscheinlichkeiten großer Abweichungen. DMV-Tagung, Marburg, 14. - 19. Sept. 1986.

[14] Erdős, P., Rényi, A. (197o). On a new law of large numbers. J. Analyse Math. 23, 1o3-111.

[15] Petrov, V.V. (1965). On the probabilities of large deviations of sums of independent random variables. Theory Prob. Appl. 1o, 287-298.

[16] Retka, M. (1982). Gesetze vom Erdős-Rényi-Typ bei Erneuerungsprozessen. Diplomarbeit, Universität Marburg.

[17] Révész, P. (198o). How to characterize the asymptotic properties of a stochastic process by four classes of deterministic curves. Carleton Math. Ser. 164, 34 pp.

[18] Shepp, L.A. (1964). A limit theorem concerning moving averages. Ann. Math. Statist. 35, 424-428.

[19] Steinebach, J. (1976). Das Gesetz der großen Zahlen von Erdős-Rényi im Wartemodell G/G/1. Studia Sci. Math. Hungar. 11, 459-466 (198o)

[2o] Steinebach, J. (1978). A strong law of Erdős-Rényi type for cumulative processes in renewal theory. J. Appl. Prob. 15, 96-111.

[21] Steinebach, J. (1981). On general versions of Erdős-Rényi laws. Z. Wahrsch.Verw. Geb. 56, 549-554.

[22] Steinebach, J. (1982). Between invariance principles and Erdős-Rényi laws. Proc. Coll. "Limit Theorems in Probab. and Statist.", Veszprém, Hungary, June 21 - 26, 1982; 981-1005 (1984).

[23] Steinebach, J. (1983). On the partial sums of linear stochastic processes. Proc. "Third Prague Symp. on Asympt.Statist.", Kutná Hora, CSSR, Aug. 28 - Sept.2, 1983, 4o5-415 (1984).

[24] Steinebach, J. (1983). On the increments of partial sum processes with multi-dimensional indices. Z. Wahrsch. Verw. Geb. 63, 59-7o.

[25] Steinebach, J. (1986). Improved Erdös-Rényi and strong approximation laws for increments of renewal processes. Ann. Probab. 14, 547-559.

# Sequential Analysis: Exact Values for the Bernoulli Distribution

U. Dieter
Institut für Statistik
Technische Universität Graz
A-8010 Graz, Austria

SUMMARY: For the Bernoulli case of Wald's sequential analysis it is shown that the following quantities can be calculated with any desired precision: First, the Operating Characteristic (OC-function) $P(p_0 \text{ accepted} \mid p \text{ true})$, especially the errors of the first and second kind $\alpha$ and $\beta$. Furthermore, the exact distribution of the number of trials, especially the average sample number (ASN-function) and higher moments of the number of trials.

## 1. THE SEQUENTIAL PROBABILITY RATIO TEST

Assume that X has the density $f(x;p)$, then A. Wald has introduced the following test for deciding whether a hypothesis $H_0 : p = p_0$ or $H_1 : p = p_1$ is true (with given errors $\alpha$ and $\beta$):

0. Set $Z \leftarrow 0$, and let a and b be constants such that $b < 0 < a$.
1. Take a sample X with value x; set $Z \leftarrow Z + \log(f(x;p_1)/f(x;p_0))$.
2. If $Z \leq b$, accept $p_0$ ; if $Z \geq a$ , accept $p_1$ ; otherwise go to 1.

It is known that the algorithm terminates finitely with probability 1 provided that rather weak assumptions are fulfilled, i.e. $P(Z=0) < 1$. Furthermore, if the errors of the first and second kind $\alpha$ and $\beta$ are defined by

$$\alpha = P(H_0 \text{ is rejected} \mid H_0 \text{ is true}) , \quad \beta = P(H_1 \text{ is rejected} \mid H_1 \text{ is true}),$$

A. Wald showed that

$$\log(\beta/(1-\alpha)) \leq b < 0 < a \leq \log((1-\beta)/\alpha). \qquad (1)$$

Hence he recommended the bounds

$$b = \log(\beta/(1-\alpha)) \quad \text{and} \quad a = \log((1-\beta)/\alpha). \qquad (2)$$

For this choice of the parameters the new errors $\alpha'$ and $\beta'$ could not be determined by A. Wald. He could only show that $\alpha'+\beta' \leq \alpha+\beta$ holds.

Contributions to Stochastics.
Ed. by W. Sendler
© Physica-Verlag Heidelberg 1987

For calculating the *Operating Characteristic*

$$P(H_0 \text{ is accepted} \mid p \text{ is true}), \tag{3}$$

where p is any parameter value, he established a famous Lemma which leads to approximate determinations of the probabilities ( 3 ).

Finally, he asked for the *average sample number*, the *ASN-function*; he showed that the *stopping number* N fulfills the equations

$$E(N) = E(S_N)/E(Z) \text{ if } E(Z) \neq 0 \quad \text{and} \quad E(N) = E((S_N)^2)/E(Z^2) \text{ if } E(Z) = 0.$$

Here $S_N$ is the first $S_n = Z_1+Z_2+...+Z_n$ which lies outside the interval (b,a). These identities yield again approximate expressions for the average number of samples needed. However, since Wald equates $S_N$ either to b or a, overshooting is neglected and the ASN-function is usually to small. Since all expressions in Wald's *Sequential Analysis* are only approximately known, it is interesting to derive numerically exact values for these quantities. This will be done for one distribution.

## 2. THE BERNOULLI CASE OF SEQUENTIAL ANALYSIS

Here the variable Z becomes

$$Z = \log \frac{f(x;p_1)}{f(x;p_0)} = \begin{cases} \log(p_1/p_0) & \text{with probability } p \\ \log((1-p_1)/(1-p_0)) & \text{with probability } 1-p. \end{cases} \tag{4}$$

Henceforth we may assume that $p_1 > p_0$. Consequently, Z is positive with probability p, and negative with probability 1-p. Z is considered as a multiple of some unit U:

$$\log(p_1/p_0) = (v+\delta)U, \qquad -\log((1-p_1)/(1-p_0)) = (r-\delta)\,U\,. \tag{5}$$

Then

$$\theta = -\log(p_1/p_0)/\log((1-p_1)/(1-p_0)) = \frac{v+\delta}{r-\delta}, \tag{6}$$

and v and r are obtained as numerators and denominators of the convergents of the continued fraction for θ. For the error term δ one has

$$\delta = (r\theta - v)/(1+\theta)\,. \tag{7}$$

In the theory of Diophantine Approximations it is shown that the error term δ can be made arbitrarily small for any irrational number θ. We choose v and r so that $|\delta|$ becomes smaller than a given quantity, e.g. $|\delta| < 10^{-3}$. Calculations were also carried out for smaller values of δ. However, the numerical results were not affected considerably.

Since Z is a multiple of the unit U, $S = \Sigma\ Z_i$ becomes a multiple of U as well. Hence

$$S \rightarrow \begin{cases} S + v+\delta & \text{with probability } p \\ S - r+\delta & \text{with probability } 1-p. \end{cases} \tag{8}$$

Step (8) will be carried out as long as $b/U < S < a/U$ holds. In the remaining part of this article $\delta$ will be equated to 0; consequently $S$ become integers which lie between $[b/U] + 1 < 0$ and $-[-a/U] - 1 > 0$ .

We shift the interval $([b/U], -[-a/U])$ to $(0, -[b/U]-[-a/U])$. The right endpoint is denoted by $c$ and $M$ is the former midpoint 0.

$$c = -[b/U] - [-a/U] \quad , \quad M = -[b/U] .$$

Using (5) leads to

$$c = -[bv/\log(p_1/p_0)] - [-av/\log(p_1/p_0)] \quad , \quad M = -[bv/\log(p_1/p_0)]. \qquad (9)$$

At the beginning $S$ is one of the values $1,2,\ldots,c-1$; as soon as $S \leq 0$ the hypothesis $H_0$: $p = p_0$ is accepted, and if $S \geq c$, the alternative $H_1$: $p = p_1$ is taken. This terminates the test.

Now we introduce the probabilities that $H_0$ or $H_1$ is accepted in exactly $m$ steps. This is usually considered in *ruin games* or in the theory of *random walks*. For $m \geq 0$ we define

$$q_m(s) = P(S \text{ moves from } s \text{ into } (-\infty,0] \text{ in exactly } m \text{ steps; } S \text{ always } < c), \qquad (10)$$

$$p_m(s) = P(S \text{ moves from } s \text{ into } [c,\infty) \text{ in exactly } m \text{ steps; } S \text{ always } > 0). \qquad (11)$$

Since the shifts $s \to s+v$ and $s \to s-r$ happen with probability $p$ or $1-p$, the probabilities $q_m(s)$ and $p_m(s)$ are subject to the same simple recursions:

$$\begin{aligned} q_{m+1}(s) &= p\, q_m(s+v) + (1-p)q_m(s-r), \\ p_{m+1}(s) &= p\, p_m(s+v) + (1-p)p_m(s-r). \end{aligned} \qquad (12)$$

In accordance with (10) and (11) we must assume

$$q_m(s) = p_m(s) = 0 \text{ for } m \geq 1 \text{ and } s \leq 0,\ s \geq c. \qquad (13)$$

$p_m(s)$ and $q_m(s)$ are different by their initial values

$$\begin{aligned} q_0(s) &= 1 \text{ for } s \leq 0, \quad q_0(s) = 0 \text{ for } s > 0, \\ p_0(s) &= 0 \text{ for } s < c, \quad p_0(s) = 1 \text{ for } s \geq c. \end{aligned} \qquad (14)$$

Explicitely, (12) forms a set of recursive equations with side conditions (13) and (14) for determining $q_m(s)$ and $p_m(s)$ in (10) and (11). The calculation is carried out in the following way: For $m=1$ one calculates $q_m(s)$, $p_m(s)$ for $s = 1,\ldots, c-1$ and subsequently one does the same for $m = 2,3,\ldots$ . Simultaneously, one determines the following expressions:

$$Q_n(s) = \sum_{m=0}^{n} q_m(s) = P(S \text{ moves from } s \text{ into } (-\infty,0] \text{ in at most } n \text{ steps; } S < c) \qquad (15)$$

$$P_n(s) = \sum_{m=0}^{n} p_m(s) = P(S \text{ moves from } s \text{ into } [c,\infty) \text{ in at most } n \text{ steps; } S > 0) \qquad (16)$$

$$\sum_{m=1}^{n} m\, q_m(s) = E(\, N \leq n \mid S \to (-\infty,0],\ p\,) \qquad (17)$$

$$\sum_{m=1}^{n} m\, p_m(s) = E( N \leq n \mid S \to [c,\infty),\ p ), \tag{18}$$

and the second moments

$$\sum_{m=1}^{n} m^2\, q_m(s) \quad \text{and} \quad \sum_{m=1}^{n} m^2\, p_m(s)\ . \tag{19}$$

The calculations are carried out for all $m < m'$ until

$$\text{Max}\ \{q_{m'}(s)+p_{m'}(s) \mid 1 \leq s \leq c-1\} \leq \epsilon \tag{20}$$

holds with given $\epsilon$, e.g. $\epsilon = 10^{-9}$.

The limits for $n \to \infty$ yield the expressions:

$$Q(s|p) = \sum_{m=0}^{\infty} q_m(s) = P(S \to (-\infty,0] \mid p), \tag{21}$$

$$P(s|p) = \sum_{m=0}^{\infty} p_m(s) = P(S \to [c,\infty) \mid p) = 1 - Q(s|p), \tag{22}$$

$$D(s|p) = E(N \mid S \to (-\infty,0] \text{ or } S \to [c,\infty)\ ,\ p) = \sum_{m=1}^{\infty} m\, q_m(s) + \sum_{m=1}^{\infty} m\, p_m(s), \tag{23}$$

and the conditional expectations

$$D(S \to 0|p) = E(N|S \to (-\infty,0],p) = \sum_{m=1}^{\infty} m\, q_m(s)/Q(s|p) \tag{24}$$

$$D(S \to c|p) = E(N|S \to [c,\infty)\ ,p) = \sum_{m=1}^{\infty} m\, p_m(s)/P(s|p)\ . \tag{25}$$

The variances and the conditional variances of $N$ can be calculated analogously from (19).

Finally, for different values of $p$ one obtains the Operating Characteristic, the *OC-function*

$$L(p) = P(H_0 \text{ is accepted} \mid p \text{ is true}) = Q(M|p)\ . \tag{26}$$

Especially, the errors of the first and second kind are obtained as

$$\alpha = 1 - P(H_0 \text{ is accepted} \mid p_0 \text{ is true}) = 1 - Q(M|p_0) \tag{27}$$

$$\beta = \quad P(H_0 \text{ is accepted} \mid p_1 \text{ is true}) = Q(M|p_1)\ . \tag{28}$$

Anyone, who is only interested in the probabilities $Q(s|p)$ in (21) can obtain these expressions as solutions of systems of linear equations

$$\begin{aligned} &Q(s|p) = p\, Q(s+v|p) + (1-p)\, Q(s-r|p) \qquad \text{for } 1 \leq s \leq c-1 \\ &Q(s|p) = 1 \quad \text{for } s \leq 0, \qquad Q(s|p) = 0 \quad \text{for } s \geq c. \end{aligned} \tag{29}$$

(29) follows directly from the definition of $Q(s|p) = P(S\to(-\infty,0] \mid p)$ or by summing up the first equations in (12) for $m = 1,2,\ldots$

In the same way the average number of trials $D(s|p)$, — the expected value of the stopping number N, — can be calculated from

$$D(s|p) = 1 + p\,D(s+v|p) + (1-p)\,D(s-r|p) \quad \text{for } 1 \leq s \leq c-1$$
$$D(s|p) = 0 \qquad \text{for } s \leq 0 \text{ and } s \geq c \,. \qquad (30)$$

Its derivation is completely elementary: After one trial one comes from S to S+v with probability p and with probability 1-p to S-r. Afterwards one has to carry out further $D(s+v|p)$ or $D(s-r|p)$ trials.

The higher moments of the stopping number **N** are calculated in the same way: Let $N_s$ be the random variable counting the number of trials to be carried out if one starts at S. Obviously one has

$$N_s = 1 + \begin{cases} N_{s+v} & \text{with probability } p \\ N_{s-r} & \text{with probability } 1-p \,. \end{cases}$$

For integer k it follows that

$$(N_s-1)^k = \begin{cases} (N_{s+v})^k & \text{with probability } p \\ (N_{s-r})^k & \text{with probability } 1-p. \end{cases}$$

and for the expectations $D_k(s|p) = E(N_s)^k$ :

$$E(N_s-1)^k = E(N_s)^k - \binom{k}{1}E(N_s)^{k-1} + \binom{k}{2}E(N_s)^{k-2} - \ldots + (-1)^k = p\,E(N_{s+v})^k + (1-p)E(N_{s-r})^k$$

or

$$D_k(s|p) = \binom{k}{1}D_{k-1}(s|p) - \binom{k}{2}D_{k-2}(s|p) + . - (-1)^k + pD_k(s+v)|p) + (1-p)D_k(s-r)|p).$$

Finally, one can also calculate the conditional expectations

$$D(S \to 0|p) = E(N \mid S \to (-\infty,0],p) \text{ and } D(S \to c|p) = E(N \mid S \to [c,\infty),p)$$

as solutions of systems of inhomogeneous linear equations. Dividing the recursion (29) by $Q(s|p)$ yields

$$1 = \frac{p\,Q(s+v|p)}{Q(s|p)} + \frac{(1-p)Q(s-r|p)}{Q(s|p)} = P(S \to s+v \to 0) + P(S \to s-r \to 0).$$

This contains the conditional probabilities which have to be used instead of p and 1-p in (30). Consequently one has

$$D(S \to 0|p) = 1 + \frac{p\,Q(s+v|p)}{Q(s|p)}\,D(s+v|p) + \frac{(1-p)Q(s-r|p)}{Q(s|p)}\,D(s-r|p),$$

or, with the abbreviation $DL(s|p) = D(S \to 0|p)\,Q(s|p)$:

$$DL(s|p) = Q(s|p) + p\,DL(s+v|p) + (1-p)\,DL(s-r|p).$$

These formulae are well suited for computer calculation of the conditional expectations. Higher moments can be obtained in a similar way.

## 3. NUMERICAL RESULTS

The different formulae were used for computer calculations. The results are given in the final table of this note. First, the case $p_0 = .5$ and $p_1 = .7$ will be discussed in some details. The table below containes the convergents v/r from the continued fraction for

$$\theta = -\log(p_1/p_0)/\log((1-p_1)/(1-p_0)) = .65868316611,$$

the error terms $\delta$ and c-values for the pairs $(\alpha,\beta) = (.05,.05), (.01,.01), (.005,.005)$. The last line cannot be used since $c > 100{,}000$ in all three cases. This is too big for most large computers.

| v | r | $\lvert\delta\rvert$ | c(.05,.05) | c(.01,.01) | c(.005,.005) |
|---|---|---|---|---|---|
| 2 | 3 | .014439 | 36 | 54 | 64 |
| 27 | 41 | .003623 | 474 | 738 | 850 |
| 83 | 126 | .003570 | 1454 | 2268 | 2612 |
| 110 | 167 | .000053 | 1926 | 3006 | 3462 |
| 7453 | 11315 | .000020 | 130442 | 203568 | 234498 |

If one is interested to calculate $Q(s|p)$ for $\alpha = \beta = .05$ one has to choose one of the values of c, namely c = 36, 474, 1454 or 1926. For p = .1, .2, ..., .9 this will lead to nine curves shown on the next page. For c = 36 one calculates 34 points in each case. These points are connected by lines in the display. The essential phaenomenae are not displayed. But the situation changes for c = 474. If p <.6, one has discontinuities at s = c-13, c-27, c-41,.. If one enlarges c from 474 to 1454 or even 1926, the situation does not change anymore. However, one can see that the arcs between the main discontinuities contain smaller jumps too.

The discontinuities at c-13, c-27, c-41,... for c = 474 (or at c-43, c-83, c-126 ... for c = 1454) can be explained quite easily: With probability p the area $[c,\infty)$ is reached in one step if one starts between c-27 and c-1. This explaines the discontinuity at c-27. If $c-13 \leq s < c$ holds, and if one jumps first to the left, it is still possible to reach the area $[c,\infty)$ in two steps.

It is easy to see that

$$Q_n(s|p) = \sum_{m=0}^{n} q_m(s) \quad \text{is monotone non increasing in } s. \tag{31}$$

The second picture shows D(s,p) for p = .1, .2, ..., .9. Figures of this kind might be used for practical purposes. For testing .5 against .7 one has to start at the point **M**. The figure shows that one has to sample 36 times, in the average. After 36 trials one checks which point s has been reached. If s is on the left side of **M**, .5 is probably true and the figure shows the average number of trials which have to be executed in the future. Consequently, the number of trials might be predicted in *batches*.

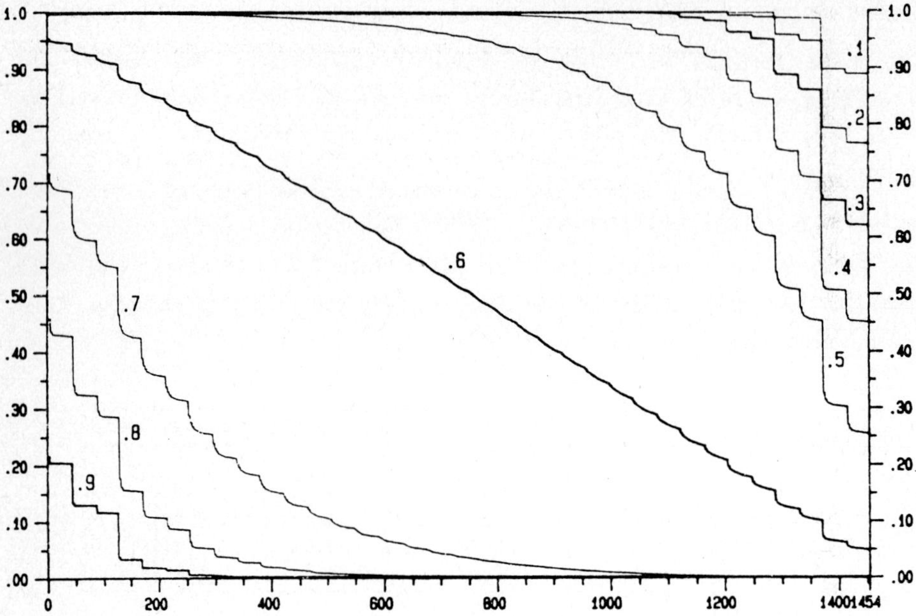


above: Q(s|p) for p = .1, .2, ..., .9

$p_0 = .5$ , $p_1 = .7$ , $\alpha = .05$ , $\beta = .05$
$v = 83$ , $r = 126$ , $\delta = -.003571$ , $c = 1454$

below: D(s|p) for p = .1, .2, ..., .9

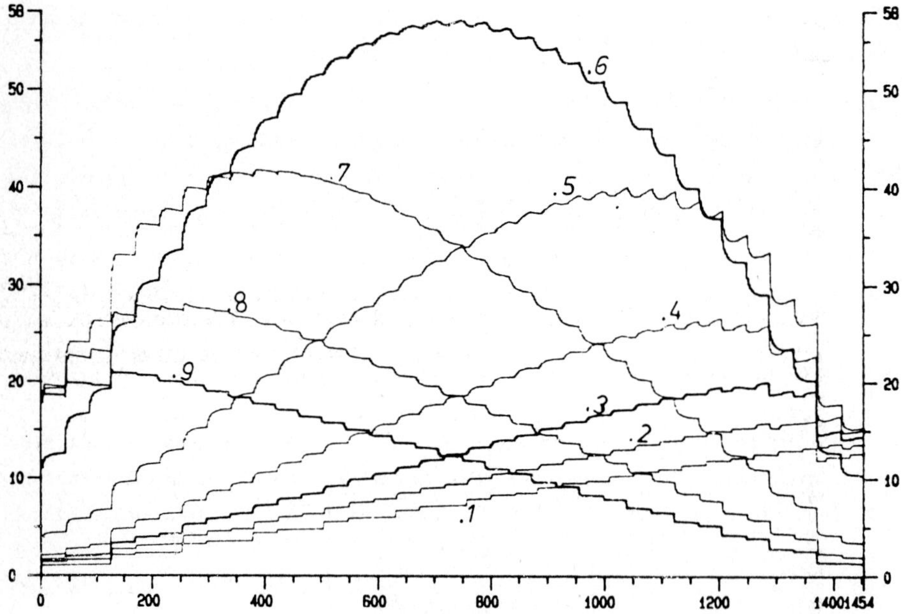

The exact calculation of Q(s|p) enables us to determine exact bounds a and b, so that the errors of the first and second kind are as close to the given values α and β as possible. Exact values of this kind were calculated by an iteration process. The Table below contains improved values for a and b in the first line, and in the second line A. Wald's original suggestions (2) are presented. Subsequently, the errors of the first and second kind are listed, and finally the expected values of the stopping times are printed.

| $p_0$ | $p_1$ | b | a | α | β | E(N; $p_0$) z → (-∞, b] | E(N; $p_1$) z → (-∞, b] | E(N; $p_0$) z → [a, ∞) | E(N; $p_1$) z → [a, ∞) |
|---|---|---|---|---|---|---|---|---|---|
| .1 | .2 | -2.882 | 2.706 | .04954 | .04984 | 73.09 | 73.40 | 59.29 | 59.53 |
| | | -2.944 | 2.944 | .03917 | .04827 | 75.36 | 74.78 | 64.68 | 64.98 |
| .1 | .3 | -2.836 | 2.549 | .04878 | .04817 | 23.46 | 23.65 | 17.69 | 17.42 |
| | | -2.944 | 2.944 | .03369 | .04499 | 24.71 | 24.69 | 19.68 | 19.93 |
| .1 | .4 | -2.840 | 2.473 | .04847 | .04203 | 12.73 | 13.19 | 8.57 | 8.79 |
| | | -2.944 | 2.944 | .02706 | .04049 | 13.33 | 13.55 | 11.26 | 10.68 |
| .1 | .5 | -2.660 | 2.450 | .04992 | .04770 | 7.46 | 7.31 | 5.12 | 5.26 |
| | | -2.944 | 2.944 | .02950 | .03445 | 8.66 | 9.12 | 5.99 | 6.45 |
| .1 | .6 | -2.440 | 1.969 | .04606 | .04033 | 5.29 | 5.60 | 4.40 | 3.80 |
| | | -2.944 | 2.944 | .02167 | .03842 | 5.64 | 5.49 | 4.77 | 4.79 |
| .1 | .7 | -2.204 | 1.953 | .03610 | .04647 | 3.63 | 4.02 | 3.44 | 2.99 |
| | | -2.944 | 2.944 | .01833 | .03440 | 4.07 | 3.93 | 3.76 | 3.66 |
| .1 | .8 | -2.436 | 2.083 | .03049 | .04365 | 2.56 | 2.31 | 2.92 | 2.51 |
| .1 | .9 | -2.944 | 2.944 | .01220 | .01220 | 2.44 | 2.43 | 2.43 | 2.44 |
| .2 | .3 | -2.885 | 2.784 | .04988 | .04992 | 103.00 | 103.00 | 94.23 | 93.90 |
| | | -2.944 | 2.944 | .04280 | .04696 | 107.31 | 107.78 | 99.66 | 99.77 |
| .2 | .4 | -2.812 | 2.669 | .04996 | .04908 | 29.39 | 28.45 | 25.40 | 25.34 |
| | | -2.944 | 2.944 | .03879 | .04371 | 31.27 | 31.48 | 27.92 | 28.01 |
| .2 | .5 | -2.822 | 2.680 | .04773 | .04878 | 14.16 | 14.25 | 11.10 | 12.17 |
| | | -2.944 | 2.944 | .03566 | .03676 | 15.85 | 16.55 | 14.01 | 13.70 |
| .2 | .6 | -2.775 | 2.605 | .04010 | .03879 | 8.92 | 9.40 | 8.18 | 7.73 |
| | | -2.944 | 2.944 | .03455 | .03550 | 9.28 | 9.31 | 8.03 | 8.16 |
| .2 | .7 | -2.973 | 2.508 | .04107 | .04221 | 5.45 | 5.11 | 5.38 | 5.11 |
| | | -2.944 | 2.944 | .02778 | .02561 | 6.52 | 7.17 | 6.17 | 5.87 |
| .2 | .8 | -2.944 | 2.944 | .01539 | .01538 | 4.85 | 4.84 | 4.84 | 4.85 |
| .3 | .4 | -2.873 | 2.830 | .05000 | .04994 | 123.05 | 122.87 | 116.89 | 117.19 |
| | | -2.944 | 2.944 | .04478 | .04673 | 127.34 | 127.16 | 122.27 | 122.75 |
| .3 | .5 | -2.808 | 2.744 | .04993 | .04983 | 32.40 | 31.97 | 30.51 | 30.47 |
| | | -2.944 | 2.944 | .04030 | .04401 | 34.49 | 34.16 | 33.63 | 33.25 |
| .3 | .6 | -2.691 | 2.672 | .04987 | .04869 | 14.69 | 14.72 | 13.67 | 13.90 |
| | | -2.944 | 2.944 | .03705 | .04033 | 16.03 | 16.05 | 15.89 | 15.69 |
| .3 | .7 | -2.944 | 2.944 | .03264 | .03263 | 9.35 | 9.34 | 9.34 | 9.35 |
| .4 | .5 | -2.852 | 2.852 | .04990 | .04984 | 131.91 | 131.51 | 129.68 | 129.97 |
| | | -2.944 | 2.944 | .04564 | .04627 | 136.66 | 136.23 | 134.65 | 135.20 |
| .4 | .6 | -2.944 | 2.944 | .03756 | .03755 | 36.99 | 36.95 | 36.94 | 36.99 |

The Table shows that the exact values for the *average sample numbers* are not too different for the improved values of $\alpha$ and $\beta$ from A. Wald's original suggestions. Furthermore, if one carries out the necessary calculations for A. Wald's approximations of the *average sample numbers*, the exact and the approximate expressions are pretty close to each other. In the case of the Bernoulli distribution A. Walds approximations can still be used. However, for other distributions the situation is not so favorable. This will be shown in forthcoming papers. It should be mentioned that more information for the Bernoulli distribution is given in a paper of the author and his student M. Unger, who carried out all the calculations which are presented here.

## REFERENCES


[ 1 ] Ahrens, J.H. and Dieter, U. (1983): Realistic and abstract roulette, a comparison. Proc. of the 4th Pannonian Symp. on Math. Stat., Bad Tatzmannsdorf, Austria. Ed. by W. Großmann, G. Pflug, I. Vincze, W. Wertz, North Holland Comp.

[ 2 ] Anscombe, F.J. (1949). Tables of sequential inspection schemes to control fraction defectives, J.Royal Stat.Soc. A **112**, 180-206.

[ 3 ] Anscombe, F.J. (1953). Sequential estimation (with discussion), J.Royal Stat.Soc. B **15**, 1-29.

[ 4 ] Aroian, L.A. (1968). Sequential analysis, direct method, Technometrics **10**, 125-132.

[ 5 ] Dieter, U. und Unger, M. (1987). Sequentielle Analysis: Genaue Werte für die Bernoulli-Verteilung. Eingereicht bei der österreichischen Zeitschrift für Statistik und Informatik.

[ 6 ] Feller, W. (1950). *An Introduction to Probability Theory and Its Applications*, Vol. I, Wiley, New York.

[ 7 ] Ghosh, B.K. (1970). *Sequential Tests of Statistical Hypotheses*, Addison-Wesley Publ. Comp.

[ 8 ] Govindarajulu, Z. (1975). *Sequential Statistical Procedures*, Academic Press, New York.

[ 9 ] Scheiber, V. (1971). Berechnung der OC- und ASN-Funktionen geschlossener Sequentialtests bei Binomialverteilung. Computing **8**, 107-112.

[ 10 ] Siegmund, D. (1985). *Sequential Analysis, Tests and Confidence Intervals*, Springer-Verlag, New York, Berlin, Heidelberg, Tokyo.

[ 11 ] Unger, M. (1985): Sequentialanalysis, Diplomarbeit, Institut für Statistik der TU Graz.

[ 12 ] Wald, A. (1947) *Sequential Analysis*, Wiley, New York.

# A Remark on Realizations of Stochastic Processes

Thomas Drisch
Fachbereich Mathematik
Universität Dortmund
D-4600 Dortmund, FRG

Wolfgang Sendler
Fachbereich IV - Mathematik
Universität Trier
D-5500 Trier, FRG

SUMMARY: Many with respect to their path properties unpleasant realizations of stochastic processes with continuous time parameter are universal, i.e. do not depend on the distributions of the processes. In particular, a process can be realized in each Baire class if the time parameter set contains a nonvoid interval.

Let $(P_I : I \subset T, I \text{ finite})$ be a projective spectrum of probability measures on $E^T$, where $T \subset \mathbb{R}$ and E is a Hausdorff space. With respect to the problem of finding a stochastic process $(X_t : t \in T)$ with state space E, defined on some probability space $(\Omega, \mathfrak{A}, P')$, having the $P_I$ as its finite dimensional distributions, the most important question is how "smooth" in view of its trajectories the process can be constructed, i.e. to realize the process on a set $H \subset E^T$ of functions $T \to E$ being as smooth as possible. Suppose the projective limit of the spectrum exists and denote it by P; then by [3], theorem 2, the process can be realized on H iff $P^*(H) = 1$ (* denotes the outer measure); in the terminology of Bauer (see [1]) this would mean that H has to be essential w.r.t. the projektive spectrum. This statement is the basis of nearly all examples of processes with unpleasant path properties, occuring in text books; clearly it refers to the special measure P under consideration. But actually very impressive examples of unpleasant realizations can be given without reference to P.

1 Definition. For an arbitrary measurable space $(\Omega, \mathfrak{A})$ a set B is called universally essential (= u.e.) on $(\Omega, \mathfrak{A})$ iff $\{A \in \mathfrak{A}: B \subset A\} = \{\Omega\}$.

Thus B is u.e. iff $\mu^*(B) = 1$ for each probability measure $\mu$ on $(\Omega, \mathfrak{A})$. Let us also note that each u.e. set except $\Omega$ is nonmeasurable. In the

Contributions to Stochastics.
Ed. by W. Sendler
© Physica-Verlag Heidelberg 1987

sequel we shall give some examples of (in the above sense) very unpleasant u.e. sets in $(E^T, \mathcal{B}^T)$, $\mathcal{B}$ denoting the Borel structure of E and $\mathcal{B}^T$ the Borel structure of $E^T$.

2 Proposition. Suppose $T = \mathbb{R}$ (for convenience) and let E contain at least two points. Then the functions $T \to E$ with no point of continuity are u.e. in $(E^T, \mathcal{B}^T)$.

Proof: For each $D \in \mathcal{B}^{\mathbb{R}}$ there is an at most countable set $S \subset \mathbb{R}$ and a $C \in \mathcal{B}^S$ such that $D = C \times \prod_{s \in \mathbb{R}-S} E_s$ (we write $E^A =: \prod_{s \in A} E_s$ for arbitrary $A \subset R$). Now let $f \in E^S$ be arbitrary and let us define $\tilde{f} \in E^{\mathbb{R}}$ as follows:

(i) $\tilde{f}(s) = f(s)$, for $s \in S$;

(ii) $\tilde{f}(t)$ arbitrary, if $t \in S^{cl} - S$ and $f(tr_S\, \mathfrak{U}(t))$ not convergent ($^{cl}$ means "closure" and $tr_S\, \mathfrak{U}(t)$ the trace in S of the neighbourhood filter $\mathfrak{U}(t)$ of t);

(iii) $\tilde{f}(t) \neq a$, if $t \in S^{cl} - S$ and $f(tr_S\, \mathfrak{U}(t)) \to a$;

(iv) $\tilde{f}(t) = g(t)$ for $t \in \mathbb{R} - S^{cl}$, where g is an arbitrary nowhere continuous function on $\mathbb{R}$.

Since f was arbitrary and $\tilde{f}$ is nowhere continuous we have proved $\P_S(D) = E^S$ ($\P_S$ is the projection on $E^S$), whence $D = E^{\mathbb{R}}$. □

Proposition 2 tells us that each stochastic process with values in E and index set $\mathbb{R}$ can be realized with nowhere continuous paths. A similar reasoning shows that the process can be realized in the set of "nowhere measurable" functions, i.e. functions whose restrictions to any open subset of $\mathbb{R}$ are not measurable.

3 Remark. Replace $\mathbb{R}$ by an arbitrary Hausdorff space T; the reasoning of proposition 2 leads to a realization of the process with paths discontinuous in all but the isolated points of T.

4 Remark. Replace E by a normed space V; the arguments of the proposition can be used to show that for arbitrary fixed $\gamma > 0$ the set N of all functions, having variation $\geq \gamma$ in all but the isolated points of T, is an u.e.set; the variation of f in the point x is defined as $\inf\{d(f(U)) : U \in \mathfrak{U}(x)\}$ where d(A) is the diameter

$\sup\{\|x-y\|\}: x,y \in A\}$ of $A \subset V$.

Let us now restrict to real valued processes and again, let us assume $T = \mathbb{R}$. It was realized by Mann ( [ 3 ], theorem 1; see also the remark in [ 2 ]) that $H_2$, the Baire class of index 2, is u.e. in $(\mathbb{R}^{\mathbb{R}}, \mathcal{B}^{\mathbb{R}})$. This statement is derived from Kolmogoroffs existency proof, but in fact it can be derived only from the assumption of the existence of the projective limit P (not utilizing the Kolmogoroff proof). More generally, if $\beta$ is an ordinal number less than the first nondenumerable ordinal number $\Omega_1$ and if $H_\beta$ denotes the Baire class of index $\beta$ (as defined in [ 4 ], p. 438 ff.) then we have the following result:

5 <u>Proposition.</u> $H_\beta$ is u.e. in $(\mathbb{R}^{\mathbb{R}}, \mathcal{B}^{\mathbb{R}})$ for $2 \leq \beta < \Omega_1$.

<u>Proof:</u> Let $G \subset \mathbb{R}$ be an arbitrary at most denumerable set and let f be a real function on G. As in the proof of proposition 2 it suffices to show that f can be extended to a function $\tilde{f} : \mathbb{R} \to \mathbb{R}$ with $\tilde{f} \in H_\beta$. We consider two cases:

(i) $\beta = 2$. Either there is a full interval $\mathfrak{J}$ with $\mathfrak{J} \cap G = \emptyset$, then choose the extension $\tilde{f}$ arbitrary but such that $f \cdot 1_{\mathfrak{J}} \in H_2$; or G is dense in $\mathbb{R}$. Then we defininе $G'_c := \{x \in G : f(x) \neq c\}$ for each $c \in \mathbb{R}$. Suppose first that $G'_{c_0}$ is dense in $\mathbb{R}$ for some $c_0$; we assume without loss of generality $G'^{+}_{c_0} := \{x \in G: f(x) > c_0\}$ to be dense; but then choose $c_1 < c_0$ and define $\tilde{f}(x) = c_1$ for $x \notin G$ and $\tilde{f}(x) = f(x)$ for $x \in G$. Since f is nowhere continuous and the set of discontinuity points of functions in $H_1$ is of first category we get $\tilde{f} \notin H_o \cup H_1$, and since $\tilde{f}$ is constant up to the countable set G we conclude $\tilde{f} \in H_\beta$ for some $\beta \leq 2$. On the other hand, assume that there exists no c such that $G'_c$ is dense; then define $\tilde{f}(x) = 0$ for $x \notin G$ and $\tilde{f}(x) = f(x)$ for $x \in G$. By assumption there exists an interval $\mathfrak{J}$ in the complememt of $G'_1$, whence $\tilde{f}$ has no continuity point in $\mathfrak{J}$ and the above arguments apply again.

(ii) $3 \leq \beta < \Omega_1$. By contradiction, assume the statement is wrong. Then for arbitrary fixed $\varphi \in H_\beta$ there is an ordinal number $\alpha = \alpha(\beta, \varphi) \neq \beta$ such that $\psi := \varphi 1_{\mathbb{R} - G} + f 1_G \in H_\alpha$. Since G is conuntable, $f 1_G \in H_\gamma$ for some $\gamma$ with $0 \leq \gamma \leq 2$. Thus by [ 4 ],

Satz 3, p. 441, we get $\alpha(\beta,\varphi) < \beta$ and
$\varphi 1_{\mathbb{R}-G} = \psi - f 1_G \in H_{\alpha'}$ for some $\alpha' = \alpha'(\varphi,\beta) \le \max(\alpha,2) < \beta$.
Again $\varphi 1_G \in H_{\gamma'}$, for some $\gamma' \le 2$; thus using the same argument once more we get $\varphi = \varphi 1_{\mathbb{R}-G} + \varphi 1_G \in H_{\max(\alpha',\gamma')}$ in contradiction to $\max(\alpha',\gamma') < \beta$. □

Thus every real valued stochastic process with $T = \mathbb{R}$ can be realized in each Baire class.

6 Remark. Though the smoothness properties of $H_\beta$ functions are unsatisfactory, proposition 5 goes (at least for $\beta = 2$) in a certain sense in the "opposite" direction as proposition 2, because the functions of $H_\beta$ are "almost" continuous, i.e. to each $f \in H_\beta$ there exists a $F \subset \mathbb{R}$ from second category such that the restriction of f to F is continuous.

## REFERENCES


[ 1 ] H. Bauer: Probability Theory and Elements of Measure Theory. Holt, Rinehart and Winston Inc., New York (1969).

[ 2 ] V. Mandrekar, H.B. Mann: On the Realization of Stochastic Processes by Probability Distributions in Function Spaces. Sankhya, Ser. A, 31, 477 - 480 (1969).

[ 3 ] H.B. Mann: On the Realization of Stochastic Processes by Probability Distributions in Function Spaces. Sankhya, Ser. A, 11, 3 - 8 (1951).

[ 4 ] I.P. Natanson: Theorie der Funktionen einer reellen Veränderlichen. Verlag Harri Deutsch, Zürich (1977).

# Nonparametric Methods: The History, the Reality, and the Future (with Special Reference to Statistical Selection Problems)

Edward J. Dudewicz
University Statistics Council
Department of Mathematics
Syracuse University
Syracuse, New York 13244, USA

SUMMARY: After giving definition(s) of "nonparametric" methods, we note that some problems are "naturally" nonparametric, and that for parametric problems there are strong reasons for desiring nonparametric solutions. "Nonparametric" and "robust" methods are contrasted, and relationships with "rank-based," "bootstrap," and "simulation and Monte Carlo" nonparametric procedures are noted. Work on nonparametric selection is reviewed: the needs for it, the early hopes for easy solution (and the Rizvi-Woodworth counterexample to Lehmann's work), Sobel's procedure (and the invalidity of more intuitive procedures), dual problems, tables, computer science applications, and robust two-stage solutions. Possible directions of new research are noted.

## 1. INTRODUCTION; DEFINITION(S) OF "NON-PARAMETRIC," WITH DISTRIBUTION-FREE, NATURAL NONPARAMETRICS, ORDER STATISTICS, RANKS, ROBUSTNESS, BOOTSTRAP, AND SIMULATION CONSIDERATIONS

In this paper our purpose is to provide up-to-date coverage with a broad overview and a specific focus, as well as possible new directions of research.

Key Words: Nonparametric, Distribution Free, Order Statistics, Ranks, Robustness, Bootstrap, Simulation, Selection, Decision Theory, Blocking, Vector Ranks, Heteroscedasticity, Quantiles

Contributions to Stochastics.
Ed. by W. Sendler
© Physica-Verlag Heidelberg 1987

As we know, ". . . statistics is an all-inclusive term, whose meaning now extends far beyond its original definition . . . the term statistics used to mean simply a mass of numbers. A statistician now calls such numbers data" (Gibbons (1985, p. 1)), and "statistics" is "the science of decision making" (Dudewicz (1976, pp. 341, 363)). As the start of our review of "nonparametric" statistics, it is appropriate to ask what this term originally meant, and what it now signifies.

The origin of the word "non-parametric" has been attributed to a paper by J. Wolfowitz in the 1942 volume of the now-defunct Annals of Mathematical Statistics, with an explanation that parametric procedures signify those where one makes ". . . the assumption that populations have distributions of known functional form . . . [with] a finite number of parameters," while ". . . methods . . . that did not require such specialized assumptions . . . became known as non-parametric" (Noether (1967, p. 41)). By 1967 the predominant spelling had changed to "nonparametric," which had meant

"short cuts for well-established parametric methods" . . . 1935-1945

"rough and ready (quick and dirty), inefficient methods wasteful of information . . . . . . . . . . . . . . . . . . 1950's

and at that time (1960's) was "outgrowing its early confines" and either needed a new name (Noether (1967)) or should simply not be differentiated from parametric statistics at all.

In 1972 we stated that ". . . we consider nonparametric statistics to be the science of providing statistical inference procedures which rely on weaker assumptions about the underlying distribution(s) of the population(s) than do the ordinary (e.g., normal distribution) procedures for the same problem (e.g., general location parameter instead of normal mean)" (Dudewicz and Geller (1972)). We also noted that much of the nonparametric literature of that time was "test-obsessed," with inadequate attention to other statistical inference problems. (This problem has become les severe in recent years (e.g., see Krishnaiah and Sen (1984)), though its residual effect still exists.)

The term distribution-free is sometimes used interchangeably with the term nonparametric, since under conditions (e.g., null hypothesis with a continuous population) the distribution of certain test statistics used in the area does not depend on population distributions. Due to these conditions and the relatively restricted setting, this "distribution-free" term has at best limited utility as a name for the area. Ury (1967) suggested the terms assumption-freer statistics and ISD (incompletely specified distributions) statistics, but admitted that "neither . . . is especially felicitous."

Currently, statisticians have an awareness that all procedures have assumptions, and that the real issues are satisfaction of assumptions, robustness, and realism of formulation. The best current definition of nonparametric statistics seems to be that of Gibbons (1985, p. 23): a statistical technique is called nonparametric if it satisfies at least one of the following five criteria:

1. The data are count data of number of observations in each category (or cross-category) [e.g., numbers of successes in treating 20 cancer patients with each of $k = 6$ therapies].
2. The data are nominal scale data [e.g., hair color].
3. The data are ordinal scale data [e.g., taste desirability of a product noted as 1,2,3,4, or 5].
4. The inference does not concern a parameter [e.g., an inference about whether a set of numbers are random on (0,1)].
5. The assumptions are general rather than specific [e.g., the assumption of a continuous population distribution].

We would call categories 1,2,3, and 4 above naturally nonparametric, while category 5 is arbitrarily nonparametric (or artificially nonparametric), since "generality" of assumptions is a matter on which statisticians will disagree (since this is a matter of judgment for which precise criteria are not established). Even for categories such as category 1 (count data), the relatively recent development of log-linear models (where asymptotics are used to justify a normality assumption) may blur the distraction as parametric procedures are now being developed for naturally nonparametric problems; for a recent account see Fingleton (1984).

If a sample $X_1, X_2, \ldots, X_n$ of size $n$ from a (specified or unspecified) population is arranged in nondecreasing order so that $X_1 \leq X_2 \leq \ldots \leq X_n$, then these are called the order statistics of the sample. As noted by Harter (1983a, p. 4), "Order statistics has often been considered (incorrectly) to be a branch of nonparametric statistics . . ." since order statistics play a large role in nonparametric statistics. In fact, order statistics is a much older field tracing its origins to Ptolemaeus (c.150), who was interested in such questions as estimation of $\sup(X_n - X_1)$ where $X_i$ is the length of year $i$, and has an extensive bibliography (Harter (1983a, 1983b)).

Nonparametric procedures are often desired for parametric problems due to a desire for robustness of the procedure to violation(s) of the assumptions under which it was developed. Some strong current themes in nonparametric research are use of rank-based procedures (see Hettmansperger (1984) or Meddis (1984)), and the so-called "bootstrap" nonparametric procedures which are related to the simulation and Monte Carlo procedures which have been well-known to operations researchers (but not to statisticians, who are rediscovering them) (see (Chernick and Murthy (1985)). (In simulation one often samples from a distribution function $\hat{F}$ which has been fitted to the data. Choice of $\hat{F} = F_n$, the empiric distribution function, yields what are now being called "bootstrap" methods.)

## 2. NONPARAMETRICS IN DECISION THEORY AND SELECTION

We noted in Section 1 that, until rather recently, nonparametrics was "test-obsessed": an extremely high proportion of the work in the area dealt with tests of hypotheses. Recently the area has included broader consideration of confidence procedures and multiple comparisons. However, there is still inadequate attention being given to nonparametric decision theory in general, and nonparametric selection in particular. Since nonparametric tests are now classical and well-covered elsewhere, while nonparametric selection is a difficult problem which has yet to receive definitive treatment, the latter will be our focus in the remainder of this paper.

The need for nonparametric treatment of the selection problem is the same as the need for nonparametric treatment of hypothesis testing, confidence interval, multiple comparisons, and other statistical problems: a desire for procedures which are valid under fewer assumptions, for use in situations where less is known about underlying populations. There were early hopes for easy solution of the nonparametric selection problem by an easy carryover of test-based considerations, as e.g. in Lehmann (1963), Puri and Puri (1969), and Bartlett and Govindarajulu (1968). However, these hopes vanished when Rizvi and Woodworth (1970) published their now-classic paper with counterexamples to this work, in the process embarrassing some of the most noted nonparametricians of the day, both authors and reviewers (e.g., Chernoff (1964)). (For related references, see Dudewicz and Koo (1982).)

## 3. THE FIRST NONPARAMETRIC SELECTION PROCEDURE

Suppose that $X_1,\dots,X_k$ are mutually independent observations from populations $\pi_1,\dots,\pi_k$ and that $X_i$ has a density $f_{X_i}(\cdot)$ $(1 \le i \le k)$ (hence $P[X_i = X_j] = 0$ for $i \ne j$). For $i = 1,\dots,k$ let

$$p_i = P[X_i > \max_{j \ne i} X_j], \qquad (1)$$

and let $p_{[1]} \le \dots \le p_{[k]}$ denote the ordered values of the $p_i$.

Bechhofer and Sobel (1958) proposed a nonparametric selection procedure for the problem of selecting a population which has probability $p_{[k]}$ of producing the largest observation; their procedure is nonparametric in that the density $f_{X_i}(\cdot)$ may be different for each $i$ $(1 \leq i \leq k)$ and unknown. This appears to be the first nonparametric selection procedure developed for general $k$ $(k \geq 2)$.

If the cumulative distribution functions (c.d.f.'s) of observations from each of $\pi_1, \ldots, \pi_k$ are known to have the same (unknown) form, differing only in location, the c.d.f. of $X_i$ may be taken as $F(x-\nu_i)$ $(i=1,\ldots,k)$. Here $\nu_1,\ldots,\nu_k$ are unknown. Denoting the unknown ranked $\nu_i$ by $\nu_{[1]} \leq \cdots \leq \nu_{[k]}$, a solution to the general problem of selecting a population which has probability $p_{[k]}$ of producing the largest observation, will yield a solution to problem (2):

> Problem: Given k populations $\pi_1,\ldots,\pi_k$ the observations from which are known to be identically distributed except for a location parameter (an observation $X_i$ from $\pi_i$ has c.d.f. $F(x-\nu_i)$; $F(\cdot)$ may be unknown), select a population associated with $\nu_{[k]}$ (the unknown largest location parameter). (2)

Dudewicz (1971) determined, for problem (2) and normal and uniform alternatives $F(\cdot)$, how much one pays (in terms of increased sample sizes) for the nonparametric method's certainty of guaranteeing a reasonable requirement on the probability of a correct selection. If the Nonparametric Procedure (NPP) of Bechhofer and Sobel is formalized as NPP: Take $N$ independent vectors $\underline{X}_j = (X_{1j},\ldots,X_{kj})$, $j=1,\ldots,N$, where $X_{ij}$ denotes the $j^{th}$ observation from population $\pi_i$. For each $\underline{X}_j$ form a new vector $\underline{Y}_j$ with elements

$$Y_{ij} = \begin{cases} 1, & X_{ij} = \max(X_{1j},\ldots,X_{kj}) \\ 0, & \text{otherwise.} \end{cases} \tag{3}$$

The $Y_{ij}$ $(i=1,\ldots,k)$ are now multinomial random variables with probabilities $p_i$ as in (1). The population associated with the largest $\Sigma_{j=1}^{N} Y_{ij}$ $(i=1,\ldots,k)$ is chosen as one associated with $p_{[k]}$, any ties for the largest sum being broken at random. We then select, as being a population associated with $\nu_{[k]}$, the population selected as being a population associated with $p_{[k]}$, and (following Bechhofer, Elmaghraby, and Morse (1959)) we set $\theta_{i+1,i} = p_{[i+1]}/p_{[i]}$ $(i=1,\ldots,k-1)$, let $\{\theta^*, P^*\}$ $(1 < \theta^* < \infty,\ 1/k < P^* < 1)$ be two specified constants, and limit consideration to procedures for problem (2) which guarantee the

Probability Requirement: We select a population associated with $\nu_{[k]}$, i.e., we make a correction selection (CS), with probability $P(CS) \geq P^*$ whenever $\theta_{k,k-1} \geq \theta^*$, (4)

then (Bechhofer, Elmaghraby, and Morse (1959, p. 104)) the configuration of the $p_{[i]}$'s which, for fixed $\{\theta^*, P^*\}$, minimizes P(CS) for fixed N subject to $\theta_{k,k-1} \geq \theta^*$ is

$$p_{[1]} = \cdots = p_{[k-1]}, \quad \frac{p_{[k]}}{p_{[k-1]}} = \theta^* ; \tag{5}$$

this is called the least favorable configuration (LFC) of the $p_i$'s. It will be clear below that the $p_{[i]}$'s are in the LFC (5) iff the $\nu_{[i]}$'s are in a configuration

$$\nu_{[1]} = \cdots = \nu_{[k-1]}, \quad \nu_{[k]} - \nu_{[k-1]} = \delta^*_F(\theta^*;k) \tag{6}$$

which is therefore an LFC for the $\nu_{[i]}$'s. Here $\delta^*_F(\theta^*;k) > 0$, say $\delta^*$, is a quantity which is, for each $F(\cdot)$, uniquely determined by k and $\theta^*$. Using this fact and sample sizes for the multinomial selection procedure it is possible to study the Relative Efficiency (RE) of any procedure $\mathcal{P}$ to the NPP (3), where

$$RE = \frac{(N \text{ using procedure } \mathcal{P})}{(N \text{ using the NPP})} . \tag{7}$$

It turns out that, for normal and uniform alternatives $F(\cdot)$, the NPP requires 30 percent to 300 percent larger samples, when compared to procedures specifically designed for normal and uniform samples (by Bechhofer's (1954) means procedure and Dudewicz's (1971) midrange procedure, respectively).

This study was extended to 2-point populations by Dudewicz and Fan (1973). Let $\pi_1, \pi_2, \ldots, \pi_k$ be k ($\geq 2$) populations such that if an observation X is drawn from $\pi_i$ then

$$P[X = c_i] = 1-p, \quad P[X = s + c_i] = p \qquad (i = 1, \ldots, k), \tag{8}$$

where s ($s > 0$) and p ($0 < p < 1$) are known and $c_1, \ldots, c_k$ are unknown. Assume that the association between $\pi_1, \ldots, \pi_k$ and $c_{[1]}, \ldots, c_{[k]}$ (where $c_{[1]} \leq \cdots \leq c_{[k]}$ denote the ordered $c_1, \ldots, c_k$) is completely unknown, and that the best population is that with the largest location parameter $c_{[k]}$.

They considered the problem of selecting the population associated with $c_{[k]} = \max(c_1, \ldots, c_k)$, with n independent observations per population in a single stage, where n is set so that the probability of a correct selection (CS) satisfies

$$P(CS) \geq P^* \text{ whenever } c_{[k]} - c_{[k-1]} \geq \delta^* \tag{9}$$

where $P^*$ ($1/k < P^* < 1$) and $\delta^*$ ($\delta^* > 0$) are specified in advance by the experimenter. Both a procedure based on sample means, and a new procedure designed specifically for two-point populations by Dudewicz and Fan (1973), were considered, as was the NPP of Bechhofer and Sobel (1958). With efficiency measure (7), it turned out that RE (which had ranged from 0.27 to 0.74 in the case of normal or uniform populations, i.e., 30 percent to 300 percent larger samples needed by the NPP) now ranged from 0.21 to 11.43 for the NPP vs. the new two-point procedure under two-point alternatives (the NPP could require as much as 400 percent larger samples *or* as little as 10 percent as much sampling--it could be *more* efficient than a parametric procedure designed for the parametric alternatives under study). This led Dudewicz and Fan (1973) to propose a new nonparametric procedure, discussed in the next section.

## 4. THE VR NONPARAMETRIC SELECTION PROCEDURES

As we have seen, the nonparametric selection procedure (NPP) of Bechhofer and Sobel (1958) took $n$ vectors of observations $(X_{ij},\ldots, X_{kj})$ $(j=1,2,\ldots,n)$. In each vector the largest observation was assigned a 1 (the others 0's), and the population which amassed the most 1's was selected. Tables which could be used to implement $\mathcal{P}_{BS}$ were given by Bechhofer, Elmaghraby and Morse (1959). Dudewicz (1971) showed that NPP has reasonable efficiency relative to certain alternatives, and Dudewicz and Fan (1973) found results indicating that $\mathcal{P}_{BS}$ may be surprisingly good in certain instances. Now, replacing $(X_{1j},\ldots,X_{kj})$ by $(0,0,\ldots,0,1,0,\ldots,0)$ (the 1 replacing the largest of $X_{1j},\ldots,X_{kj})$ intuitively seems a great loss of information. If instead one replaced the largest observation by $k$, the next largest by $k-1$, $\ldots$, the smallest by 1, and selected that population achieving the largest rank sum, how would one do? It seems intuitively that this procedure VR should be better than NPP (at least under mild restrictions).

Procedure VR had also been given, in the special context of normal populations, by Matsui (1972). Matsui (1974) postulated that a slippage configuration would minimize the P(CS) of procedure VR, and all his results were based on that postulate. Now, given $k$ $(k\geq 2)$ populations $\pi_1,\ldots,\pi_k$ such that observations from $\pi_i$ have distribution function $F(x-\theta_i)$ with probability density function $f(x-\theta_i)$, let $\theta_{[1]}\leq\cdots\leq\theta_{[k]}$ denote the ordered unknown values of $\theta_1,\ldots,\theta_k$, and let $\pi_{(i)}$ denote the population associated with $\theta_{[i]}$. For the goal of selecting $\pi_{(k)}$ a number of authors have considered procedures based on the joint ranks of the observations; see Matsui (1974) for some references. A procedure based on ranks within blocks was proposed and studied independently, and for different reasons (to take account of block effects and to simplify computations, respectively) by: Dudewicz and Fan (1973) in a general case, and by Matsui (1972) under normality. Consider the single-stage rule of Dudewicz-Fan-Matsui. Take $n$ independent vectors $X_j = (X_{j1},\ldots,X_{jk})$, $j = 1,\ldots,n$ ($X_{ji}$ denotes the jth observation from $\pi_i$); compute $T_i = \Sigma_{j=1}^{n} R(X_{ji})$ $(i = 1,\ldots,k)$, where $R(X_{ji})$ is the number of $X_{ji'} \leq X_{ji}$, $1 \leq i' \leq k$, and base the terminal decision on $T_1,\ldots,T_k$, selecting the population which yields $\max(T_1,\ldots,T_k)$ as being $\pi_{(k)}$. (Any ties for the maximum are broken by randomization.) Denote the event of selecting $\pi_{(k)}$ by CS and its probability by P(CS). Dudewicz and Lee (1979) showed that the P(CS) is not always minimized by the slippage configuration $\theta_{[1]} = \ldots = \theta_{[k-1]} = \theta_{[k]} - \delta^*$ (when one sets $n$ to guarantee $P(CS) \geq P^*$ whenever $\theta_{[k]} - \theta_{[k-1]} \geq \delta^*$), and that in fact sometimes the slippage configuration maximizes the P(CS).

Dudewicz and Lee (1980) applied a generalization of procedure VR to awardee selection, in the specific setting of the Frank Wilcoxon and Jack Youden Prizes (awarded annually since 1969 for the best practical application and expository papers in Technometrics, by the American Society for Quality Control). In that context, each referee (judge) is a block. To set the stage, suppose that we have $k$ contenders, and that each expert ranks at least $t$ $(1 \leq t \leq k)$ contenders from best to worst. Denote the $i^{th}$ contender by $C_i$. Let $\phi_{i\ell}$ denote the probability that contender $C_i$ will be rated as the $(k-\ell+1)^{st}$ best (among the $k$ contenders $C_1,\ldots,C_k$) by any expert. Then define

$$\mu_i(t) = \sum_{\ell=k-t+1}^{k} (\ell+t-k)\phi_{i\ell} \qquad (10)$$

and call the contender associated with the largest of $\mu_1(t),\ldots,\mu_k(t)$ the best contender (by ranking) t-out-of-k; for ease of reference, let us denote that contender as C-best(t).

A reasonable selection procedure VR(t) for finding the C-best(t) contender is: Assign scores $t, t-1,\ldots,2,1$ and $k-t$ 0's to the $k$ contenders, using the rank score matrix of the $n$ judges. (The higher the score, the better a contender is rated.) Denote by $R_{ji}(t)$ the score assigned to $C_i$ by the $j^{th}$ judge. Let $V_i(t) = \Sigma_{j=1}^{n} R_{ji}(t)$, and select the contender with the largest of $V_1(t),\ldots,V_k(t)$, i.e. $\max(V_1(t), V_2(t),\ldots,V_k(t))$, as the best. If two or more tie either split the first place or select one based on other considerations (or break the tie by randomization); the first two options are recommended for practice, while the last option is required for the theoretical mathematical development summarized below.

To explain the rationale of the adopted selection procedure, some additional notations are needed. Namely, let $\mu_{[1]}(t) \le \mu_{[2]}(t) \le \ldots \le \mu_{[k]}(t)$ denote ordered unknown values of $\mu_1(t), \mu_2(t), \ldots, \mu_k(t)$ and let $\phi_{(i)\ell}$ be the probability associated with $\mu_{[i]}(t)$. The contestant associated with $\mu_{[i]}(t)$ is of course unknown for each $i$, and is denoted by $C_{(i)}(t)$. The strength of a selection procedure $R$ is measured by the <u>probability of correct selection $P(CS|R)$</u>. If in our setting $V_{(i)}(t)$ denotes the statistic associated with $C_{(i)}(t)$, then here a correct selection (CS) means the event $V_{(k)}(t) = \max(V_1(t), \ldots, V_k(t))$; and, if $P(CS|VR(t)) \le P(CS|VR(t'))$ when we have $n$ judges and $k$ contenders, then we conclude that procedure $VR(t')$ is more efficient.

For notational convenience only, and without loss of generality, assume that the $k^{th}$ contestant is the best. Thus the statistic and parameters associated with $C_k$ are simply $V_k(t)$ and $\phi_{k\ell}$. Now suppose that

$$\phi_{kk} \ge \phi_{k\ell} \text{ and } \phi_{kk} \ge \phi_{\ell' k}, \quad 1 \le \ell, \ \ell' \le k-1, \tag{11}$$

that is, suppose that the probability that the best contestant is rated best by a judge is greater than the probability that he is rated as $(k-\ell+1)^{st}$ best $(1 \le \ell \le k-1)$, <u>and</u> is greater than the probability that another contestant is rated best.

Let $n, k, t$, and $\lambda^*$, $1 < \lambda^* < \infty$, be fixed and suppose (11) satisfies

$$\phi_{kk} \ge \lambda^* \phi_{k\ell} \text{ and } \phi_{kk} \ge \lambda^* \phi_{\ell' k}, \quad 1 \le \ell, \ \ell' \le k-1 . \tag{12}$$

Then for $t \ge 2$ Lee and Dudewicz (1986a) show that

$$\inf_{\phi_{ij}} P[CS|\phi_{ij}, VR(t)] \le P[CS|VR(t), \phi_{kk} = \lambda^*\phi_{k\ell} = \lambda^*\phi_{\ell' k}, \phi_{\ell\ell'} = \phi_{\ell'\ell}, \quad 1 \le \ell, \ \ell' \le k-1] \tag{13}$$

and for $t=1$ it is known (Kesten and Morse (1959)) that

$$\inf_{\phi_{ij}} P[CS|\phi_{ij}, VR(1)] = P[CS|VR(1), \phi_{kk} = \lambda^*\phi_{\ell k}, 1 \le \ell \le k-1] . \tag{14}$$

(Note that without randomization the equality (14) is true only asymptotically, where it follows since (as $n \to \infty$) the probability that randomization is required goes to zero.)

We could now compare the selection procedures $VR(t)$ for various values of $t$ by computing (13) and (14) for each $t$ with fixed $n, k$, and $\lambda^*$ :

P[CS] comparisons with $n$ judges, $k$ papers, and distance measure $\lambda^*$ : complete ranking VR(k) vs. bullet-vote VR(1).

| n | k | $\lambda^*$ | P[CS \| VR(k)] | P[CS \| VR(1)] |
|---|---|---|---|---|
| 30 | 5 | 2.0 | 0.731 | 0.787 |
| 30 | 6 | 2.0 | 0.559 | 0.728 |
| 30 | 7 | 2.0 | 0.470 | 0.673 |

However, instead of computing $P[CS|VR(t)]$ to compare $VR(t)$'s for given $n,k$, and $\lambda^*$, equate (13) and (14) to a given $P^*(1/k < P^* < 1)$ and obtain an asymptotic expression for the smallest $n$ needed to satisfy the equation. Denote that $n$ by $n_{k,t}(\lambda^*,P^*)$. Then the ratio $n_{k,t}(\lambda^*,P^*)/n_{k,t'}(\lambda^*,P^*)$, $1 \leq t \neq t' \leq k$, is called the relative efficiency of $VR(t')$ with respect to $VR(t)$, denoted by $Eff[VR(t'),VR(t)]$. If $Eff[VR(t'),VR(t)] \geq 1$, then procedure $VR(t')$ is at least as efficient as $VR(t)$. In particular focus on $Eff[VR(1),VR(t)]$. Since $Eff[VR(1), VR(t)]$ requires a computation for each combination of $(k,\lambda^*,P^*)$, for simplicity instead compute and obtain its limit for "close" alternatives,

$$\lim_{\lambda^* \to 1} Eff[VR(1),VR(t)] = \frac{t+1}{3}\left[\frac{(k-1)(4tk + 2k - 3t^2 - 3t)}{t(2k-t-1)^2}\right]. \tag{15}$$

(See Lee and Dudewicz (1986a).) For the 1976 Wilcoxon contest example with $k=6$, (15) is evaluated as:

Relative Efficiency: number of judges needed by $VR(t)$ ranking the $t$ best, divided by number of judges needed by bullet-vote $VR(1)$, with $k=6$, any $P^*$, limit on $\lambda^*$.

| $t$ | 2 | 3 | 4 | 5 | 6 |
|---|---|---|---|---|---|
| $\lim_{\lambda^* \to 1} Eff[VR(1),VR(t)]$ | 1.30 | 1.67 | 2.04 | 2.33 | 2.33 |

from which we see that $VR(2)$ is less efficient than $VR(1)$: it requires at least 30 percent more judges to obtain the same $P(CS)$ guarantee. Procedures $VR(3),\ldots,VR(6)$ are even worse, requiring from at least 67 percent to at least 133 percent more judges than does $VR(1)$. Thus, Dudewicz and Lee (1980) and Lee and Dudewicz (1986a) have demonstrated that the selection of the C-best(1) contender requires a smaller number of judges than the best contender of any other definition considered. Even though the best contender may differ from definition to definition, it is normally expected that the best contender will be independent of definition if the majority of referees are well-qualified, fair and unbiased. Such is, in general, true in the n-block-k-population model where the rank score in a block is generated by order statistics of stochastically increasing random variables. These considerations are the basis of the adopted procedure for determining the winners of the annual Technometrics prizes.

## 5. ROBUSTNESS IN SELECTION: THE k-POPULATION-n-BLOCK MODEL

Suppose there are k populations and that n independent blocks of k observations, one from each population, are taken. (For example, each block may consist of k continuous independent random variables with unknown functional form for their distribution function.) Several inference methods are then available, including robust methods designed to insure against departures from the assumed underlying c.d.f.'s.

Let $X_{ji}$ $(i=1,\ldots,k;\ j=1,\ldots,n)$ be independent random variables such that $P(X_{ji} \leq x) = F(x-\eta_j-\theta_i)$ where $\eta_j$ is the $j^{th}$ block's nuisance parameter and selection on $\theta_i$'s is desired. We may compute

$$\bar{X}_i = \sum_{j=1}^{n} X_{ji}/n \qquad (i=1,\ldots,k), \tag{16}$$

or

$$V_i = \sum_{j=1}^{n} R_{ji} \qquad (i=1,\ldots,k) \tag{17}$$

where

$$R_{ji} = \{\# \text{ of } X_{ji'} \leq X_{ji}\ (i'=1,\ldots,k)\},$$

or

$$Y_i = \sum_{j=1}^{n} Y_{ji} \qquad (i=1,\ldots,k) \tag{18}$$

where

$$Y_{ji} = \begin{cases} 1 & \text{if } X_{ji} = \max(X_{j1},\ldots,X_{jk}) \\ 0 & \text{otherwise,} \end{cases}$$

or

$$\bar{H}_i = \sum_{\ell=1}^{k} H_{i\ell}/k \qquad (i=1,\ldots,k) \tag{19}$$

where

$$H_{i\ell} = \underset{1\leq j\leq j'\leq n}{\text{med}} \{(X_{ji}-X_{j\ell}) + (X_{j'i}-X_{j'\ell})\}/2 \qquad (i,\ell=1,\ldots,k).$$

Selection of the population yielding $\max(\bar{X}_1,\ldots,\bar{X}_k)$ is Bechhofer's (1954) Means Procedure MP; selection of the population yielding $\max(V_1,\ldots,V_k)$ is the Vector Rank Procedure VR of Dudewicz and Fan (1973); selection of the population yielding $\max(Y_1,\ldots,Y_k)$ is the Bechhofer and Sobel (1958) procedure NPP; and selection of the population yielding $\max(\bar{H}_1,\ldots,\bar{H}_k)$ is the Hodges-Lehmann estimator procedure HLP.

Using the asymptotic $(\delta^*\to 0)$ version of relative efficiency (RE) in (7), Lee and Dudewicz (1985) compare these four procedures utilizing results of Lee and Dudewicz (1986b) as well as new results under mixture models. They conclude that NPP is a reasonable choice over MP (in the k-population-n-block location model) iff $k = 2,3,$ or $4$. Asymptotically VR and HLP are virtually indistinguishable as their ARE is $k/(k+1)$ (with HLP in the numerator of (7) and VR in the denominator). VR is more efficient than NPP, for all $k$, for logistic and normal populations. However, VR can be inefficient in mixture settings where the distribution of $\pi_{(i)}$ is

$$(1-\Theta_{[i]})F(x) + \Theta_{[i]}G(x) . \tag{20}$$

For real data sets, Lee (1980) has shown instances where NPP outperforms VR (possibly because the data was not normal, nor even from a location family). In general choice of a selection procedure can be properly made only after careful review of the data and its satisfaction of statistical model assumptions.

## 6. ROBUSTNESS OF BECHHOFER'S MEANS PROCEDURE AND ITS GENERALIZATIONS TO HETEROSCEDASTICITY

The RE of Bechhofer's Means Procedure (MP) was addressed above in Section 3 with regard to a nonparametric competitor of Bechhofer and Sobel (NPP), and its ARE against robust procedures was addressed in Section 5 above. Since Hall (1959) has shown that MP is most economical under normality, its robustness is of great importance. In this regard, Dudewicz and Mishra (1984) studied robustness of the P(CS) of MP in the LFC under short-tailed (uniform) and long-tailed (Student's t with three degrees of freedom) alternatives. Over a broad range of $\delta^*$, for $k = 2,3,6$, and for $n = 1(1)10$, they found that MP never delivered less than .02 below the desired $P^* = P(CS|LFC)$ over the range studied (which included $P^*$'s from .189 through .999). Thus, Bechhofer's MP is robust to both short-tailed and long-tailed (symmetric) alternatives over a broad range of possible specifications of number of populations, preference zone, and desired probability of correct selection.

Since Bechhofer's 1954 MP assumed a common known variance, while in practice variances are hardly ever known or equal, Dudewicz and Dalal (1975) developed a two-stage procedure to solve this problem, which had eluded solution for decades (and for which a single-stage solution does not exist). That procedure is, when $k = 1$, in a setting where $X = (X_1, X_2, X_3, \ldots)$ is a sequence of independent $N(\mu, \sigma^2)$ random variables (normal with mean $\mu$ and variance $\sigma^2 > 0$). Let $n_0$ ($\geq 2$) and $h > 0$ be fixed. Using $X_1, \ldots, X_{n_0}$, calculate

$$\overline{X}(n_0) = \sum_{i=1}^{n_0} X_i/n_0 , \quad s^2 = \sum_{i=1}^{n_0} (X_i - \overline{X}(n_0))^2/(n_0 - 1) , \tag{21}$$

$$n_1 = \max\{n_0 + 1, [s^2h^2]\} \tag{22}$$

where [y] denotes the smallest integer $\geq y$. Let $a_1, \ldots, a_{n_1}$ be random variables (through $n_1$ and $s^2$ only) such that

$$a_1 + \ldots + a_{n_1} = 1, \quad a_1 = \ldots = a_{n_0}, \quad s^2 \sum_{i=1}^{n_1} a_i^2 = (1/h)^2 . \tag{23}$$

Define

$$\tilde{\tilde{X}} = \sum_{i=1}^{n_1} a_i X_i . \tag{24}$$

Then

$$\frac{\tilde{\tilde{X}}-\mu}{1/h} = \frac{\tilde{\tilde{X}} - \mu}{s\sqrt{\sum_{i=1}^{n_1} a_i^2}} \tag{25}$$

has Student's-t distribution with $n_0-1$ degrees of freedom, i.e. is $t_{n_0-1}$.

In the more general setting of non-normality, let $X(n^*) = (X_1(n^*), X_2(n^*), \ldots)$ be a sequence of independent random variables such that $X_i(n^*) \xrightarrow{\mathcal{D}} X_i$ as $n^* \to \infty$ (i.e., $X_i(n^*)$ converges in distribution to a $N(\mu, \sigma^2)$ random variable as $n^* \to \infty$). Then Dudewicz and van der Meulen (1983, p. 134) show that, as $n^* \to \infty$,

$$\frac{\tilde{\tilde{X}}(n^*) - \mu}{1/h} \tag{26}$$

converges to a $t_{n_0-1}$ random variable, hence providing robustness (consistency) properties of the two-stage solution of Dudewicz and Dalal (1975) and other related solutions of other statistical inference problems developed since that work (e.g., in heteroscedastic confidence intervals, multiple comparisons, multiple comparisons with a control, analysis of variance, and multivariate analysis).

## 7. NONPARAMETRIC QUANTILE SELECTION

Proportions are an area of natural nonparametricity in the scheme of Gibbons outlined in Section 1 above, and an area of great interest in applications, e.g. recently in computer science (e.g., Amer and Dudewicz (1980) consider the computer performance evaluation area). For continuous populations, an area of natural parametricity, one may desire to concentrate on quantiles. The "natural" selection procedure for the $\alpha$-quantile involves requiring $P(CS) \geq P^*$ whenever the $\alpha$-quantiles are separated by at least $d^*$. However, it can be shown that a procedure (based, e.g., on sample quantiles) cannot guarantee this requirement since c.d.f.'s can be arbitrarily close arbitrarily near the $\alpha$-quantile (and yet separated by at least $d^*$ at that quantile.) To solve this problem, Sobel (1967) gave his celebrated nonparametric selection procedure. Recently Dhariyal and Dudewicz (1985) extended these procedures to nonparametric procedures for simultaneous selection and estimation of the largest $\alpha$-quantile, a development which we will now summarize.

Let $x_\alpha(F_i)$ denote the $\alpha$-quantile of the df $F_i(x) \equiv F_i$, $i = 1,\ldots,k$. Assume that each $x_\alpha(F_i)$ is unique and write $F_{[i]} \leq F_{[j]}$ if $x_\alpha(F_{[i]}) \leq x_\alpha(F_{[j]})$ where $x_\alpha(F_{[1]}) \leq \ldots \leq x_\alpha(F_{[k]})$ are the $k$ $\alpha$-quantiles in numerical order. Let $\pi_{(i)}$ denote the population associated with $x_\alpha(F_{[i]})$, $i = 1,\ldots,k$. Assume that no prior information is available concerning the true pairing of the $\pi_i$ and the $\pi_{(i)}$, $i = 1,\ldots,k$. Let "CS" denote the event of correctly selecting a population $\pi_{(k)}$, and "CI" the event $x_\alpha(F_{[k]}) \in I$, where $I$ is the proposed confidence interval. Following Sobel (1967), consider four formulations.

Formulation 1A. Let $\epsilon^*_\gamma > 0$, $\gamma = 1,2$, be given numbers such that $\epsilon^*_1 \leq \alpha \leq 1 - \epsilon^*_2$. Let $\underline{x}_\beta(F_i) = \inf\{x : F_i(x) = \beta\}$, $\overline{x}_\beta(F_i) = \sup\{x : F_i(x) = \beta\}$, and $\underline{F} = \underline{F}(x) = \min(F_{[1]}(x),\ldots,F_{[k-1]}(x))$ for all real $x$. Define the closed interval $J$ by $J = [\overline{x}_{\alpha-\epsilon^*_1}(F_{[k]}), \underline{x}_{\alpha+\epsilon^*_2}(F_{[k]})]$. Let $d = \inf_{x \in J}(\underline{F}(x) - F_{[k]}(x))$. With these notations the goal is to construct a procedure $R$ that selects a population $\pi_{(k)}$ and simultaneously estimates $x_\alpha(F_{[k]})$ by a confidence interval $I$ such that $P(CS \cap CI|R) \geq P^*$ for all $F_1,\ldots,F_k$ with $d \geq d^*$, where $P^*(\frac{1}{k} < P^* < 1)$ and $d^*$ $(> 0)$ are preassigned given numbers.

Formulation 1B. This is the same as Formulation 1A except with an additional assumption that, for all $x$, $F_{[i]}(x) \geq F_{[k]}(x)$, $i = 1,\ldots, k-1$.

Formulation 2A. Let $\epsilon^*_\gamma > 0$, $\gamma = 1,2$, be two given numbers such that $\epsilon^*_1 \leq \alpha \leq 1-\epsilon^*_2$. Let $d' = \underline{x}_{\alpha-\epsilon^*_1}(F_{[k]}) - \overline{x}_{\alpha+\epsilon^*_2}(\underline{F})$. The goal is to find a procedure $R$ such that $P(CS \cap CI|R) \geq P^*$ when $d' \geq 0$, $P^*$ being a preassigned given number, $1/k < P^* < 1$.

Formulation 2B. This is the same as Formulation 2A except with an additional assumption that for all $x$ $F_{[i]}(x) \geq F_{[k]}(x)$, $i = 1,\ldots,k-1$.

The Dhariyal-Dudewicz Procedure R is: Take independent random samples of size $n$ from populations $\pi_1,\ldots,\pi_k$, say $X_{i1},\ldots,X_{in}$ from $\pi_i$ $(1 \leq i \leq k)$. Let $X_{i(j)}$ denote the $j^{th}$ ordered $X_{ij}$ in the sample from

$\pi_i$. Assume that n is large enough that $1 \leq (n+1)\alpha \leq n$, and that n takes on only values in a set of integers N such that $(n+1)\alpha = r$, an integer. (For detailed explanation, see Sobel (1967), p. 1806.) Consider the $r^{th}$ order statistic in each sample, $X_{1(r)}, \ldots, X_{k(r)}$, and let $X_{[1](r)} < \ldots < X_{[k](r)}$ denote these $r^{th}$ order statistics in their numerical order. Note that, since the df's are continuous, $P(X_{[i](r)} = X_{[j](r)}) = 0$ for $i \neq j$. Let the other order statistics associated with $X_{[k](r)}$ be denoted by $X_{\{k\}(j)}$, $j = 1, \ldots, n$, $j \neq r$. For arbitrary $c_1, c_2$, $0 \leq c_1 \leq r$, $0 \leq c_2 \leq n-r$, let $I = [X_{\{k\}(r-c_1)}, X_{\{k\}(r+c_2)}]$. Then assert that $x_\alpha(F_{[k]}) \in I$ and select the population which yielded $X_{[k](r)}$ as having the largest $\alpha$-quantile.

For each formulation, Dhariyal and Dudewicz (1985) calculate a lower bound on the $P(CS \cap CI|R)$. For example, in Formulation 1A their final expression is

$$P(CS \cap CI|R) = \int_{\alpha-\epsilon_1^*}^{\alpha} G_{n-r-c_2+1,c_2}\left(\frac{1-\alpha}{1-u}\right) G_{r,n-r+1}^{k-1}(u+d^*)\, dG_{r,n-r+1}(u)$$
$$+ \int_{\alpha}^{\alpha+\epsilon_2^*} G_{r-c_1,c_1}\left(\frac{\alpha}{u}\right) G_{r,n-r+1}^{k-1}(u+d^*)\, dG_{r,n-r+1}(u) \qquad (27)$$
$$+ \left\{\int_{\alpha+\epsilon_2^*}^{1} G_{r-c_1,c_1}\left(\frac{\alpha}{u}\right) dG_{r,n-r+1}(u)\right\} G_{r,n-r+1}^{k-1}(\alpha+\epsilon_2^* + d^*) .$$

By equating $\inf P(CS \cap CI|R)$ to $P^*$, the probability requirement is met. Although theoretically this solves the problem, in practice, from the expressions for $\inf P(CS \cap CI|R)$ for given constants $\epsilon_1^*$, $\epsilon_2^*$, $d^*$, and $P^*$ it will be difficult to find the smallest n and "optimal" $c_1$, $c_2$ such that the probability requirement is attained. Therefore, they suggest an approximate modified method as follows. Since

$$P(CS \cap CI|R) = P(X_{\{k\}(r-c_1)} \leq x_\alpha(F_{[k]}) \leq X_{\{k\}(r+c_2)}, X_{(i)(r)} < X_{(k)(r)}, \; i = 1, \ldots, k-1)$$
$$= P(X_{(k)(r-c_1)} \leq x_\alpha(F_{[k]}) \leq X_{(k)(r+c_2)} | X_{(i)(r)} < X_{(k)(r)}, \; i = 1, \ldots, k) \qquad (28)$$
$$\times P(X_{(i)(r)} < X_{(k)(r)}, \; i = 1, \ldots, k-1),$$

and from Sobel (1967), under Formulation 1A,

$$B(n) \equiv \inf P(X_{(i)(r)} < X_{(k)(r)}, \; i = 1, \ldots, k-1)$$
$$\geq \underline{B}(n)$$
$$= \int_{\alpha-\epsilon_1^*}^{\alpha+\epsilon_2^*} G_{r,n-r+1}^{k-1}(u+d^*)\, dG_{r,n-r+1}(u)$$
$$+ G_{r,n-r+1}^{k-1}(\alpha+\epsilon_2^*+d^*)\,[1-G_{r,n-r+1}(\alpha+\epsilon_2^*)] \qquad (29)$$

while

$$\begin{aligned} P(X_{(k)(r-c_1)} \le x_\alpha(F_{[k]}) \le X_{(k)(r+c_2)} | X_{(i)(r)} < X_{(k)(r)},\ i=1,\dots,k-1) \\ \approx P(U_{(r-c_1)} \le \alpha \le U_{(r+c_2)}) \\ = G_{r-c_1,n-r+c_1+1}(\alpha) - G_{r+c_2,n-r-c_2+1}(\alpha) \\ = A(c_1,c_2,n), \text{ say,} \end{aligned} \tag{30}$$

where $U_{(\ell)}$ denotes the $\ell^{th}$ order statistic in a random sample of size $n$ from the uniform distribution on the interval $(0,1)$, therefore, subject only to the goodness of the approximation $A(c_1,c_2,n)$,

$$P(CS \cap CI|R) \ge A(c_1,c_2,n) \cdot \underline{B}(n) . \tag{31}$$

Now for a given $P^*$ one takes two numbers $P_1^*$ and $P_2^*$ such that $P^* < P_1^*$, $P_2^* < 1$ and $P_1^* P_2^* = P^*$. (The choice of $P_1^*$, $P_2^*$ depends on the experimenter's preference as to reliability on "CS" and "CI" given "CS." Therefore, if one puts $\underline{B}(n) = P_2^*$ and $A(c_1,c_2,n) = P_1^*$, then one may prefer to take $P_2^* > P_1^*$ since a better assurance about "CS" may be preferable as "CI" depends on "CS.") Now one can find (see Sobel (1967)) the smallest $n$ satisfying $\underline{B}(n) = P_2^*$. With this $n$ one can find (see Gibbons (1971), p. 40) $c_1$, $c_2$ such that $A(c_1,c_2,n) \ge P_1^*$ and $c_1 + c_2$ is minimized.

As an example, let us use the data given in Bishop and Dudewicz (1978), modifying their example in an appropriate manner, to illustrate the proper use of the modified method (under Formulation 1A). An experiment is to be conducted to study the effect of solvent on the ability of the fungicide methyl-2-benzimidozole-carbamate to destroy the fungus Penicillium expansum. The fungicide is to be diluted in exactly the same manner in three different types of solvent and sprayed on the fungus, and the percentage of fungus destroyed will be measured. The experimenter desires to select the solvent for which the median percentage of fungus destroyed is the greatest and simultaneously wants to estimate this largest median. He is not willing to assume a functional form for the associated distribution functions. Here, we have $k=3$, $\alpha=0.5$, and suppose the experimenter has specified $d^* = \epsilon_1^* = \epsilon_2^* = 0.20$, $P^* = 0.800$, $P_2^* = 0.850$ and hence $P_1^* = 0.941$.

The first step is to determine the sample size $n$ that should be chosen. For this choice we may use the tables given in Sobel (1967). If we do so, with the specifications and $P_2^* = 0.850$, we find $n = 23$. Now suppose the experiment is conducted and the following observations are obtained (we are using the first 23 observations from the first three populations of Bishop and Dudewicz (1978)):

| <u>Population (Solvent) 1</u> | <u>Population (Solvent) 2</u> | <u>Population (Solvent) 3</u> |
|---|---|---|
| 96.44, 96.87, 97.24, | 96.63, 93.99, 94.61, | 93.58, 93.02, 93.86. |
| 95.41, 95.29, 95.61, | 91.69, 93.00, 94.17, | 92.90, 91.43, 92.68, |
| 95.28, 94.63, 95.58, | 92.62, 93.41, 94.67, | 91.57, 92.87, 92.65, |
| 98.20, 98.29, 98.30, | 95.28, 95.13, 95.68, | 95.31, 95.33, 95.17, |
| 98.65, 98.43, 98.41, | 97.52, 97.52, 97.37, | 98.59, 98.00, 98.79, |
| 98.59, 98.20, 98.37, | 96.97, 97.21, 97.44, | 96.36, 96.69, 96.89, |
| 98.57, 98.42, 98.29, | 96.86, 97.26, 98.27, | 96.13, 97.65, 97.81 |
| 98.51, 98.89 | 97.57, 97.81 | 97.71, 97.48. |

Now from the three samples we find that $r = (n+1)/2 = 12^{th}$ smallest (the median) observations are

$$X_{1(12)} = 98.29, \quad X_{2(12)} = 95.68, \quad X_{3(12)} = 95.33.$$

<u>Since $X_{1(12)}$ is the largest we select solvent 1 as the best</u>. (If one uses all the observations and the generalized means $\tilde{X}_1 = 97.553$, $\tilde{X}_2 = 95.250$, $\tilde{X}_3 = 94.754$ of Bishop and Dudewicz (1978) one arrives at the same conclusion.) Now in order to estimate the largest population median we must find $c_1$ and $c_2$. Here, since $\alpha = 0.5$, we have $c_1 = c_2$ minimizes $c_1 + c_2$. Also

$$\begin{aligned} A(c,c,n) &= G_{r-c,n-r+c+1}\left(\tfrac{1}{2}\right) - G_{r+c,n-r-c+1}\left(\tfrac{1}{2}\right) \\ &= B(r+c-1;\, n, \tfrac{1}{2}) - B(r-c-1;\, n, \tfrac{1}{2}) , \end{aligned} \qquad (32)$$

where $B(x;n,p)$ denotes the binomial df. Therefore using tables of the binomial df, we choose the largest integer $\delta$ such that

$$1 - B(\delta;\, n, \tfrac{1}{2}) \geq (1+P_1^*)/2 , \qquad (33)$$

and have $c = r-\delta-1$ .

For ease of experimenters, Dhariyal and Dudewicz (1985, Section 6) provide tables of $(n,c)$ pairs for different values of $P^*$, $P_1^*$ and $P_2^*$ for $k = 2(1)10$. Using their Table 4 we find for $k = 3$, $P^* = 0.800$, $P_1^* = 0.941$, $P_2^* = 0.850$, $c = 5$. Hence we find from amongst the observations on solvent 1, the $r-c = 7^{th}$ and $r+c = 17^{th}$ smallest observations are $X_{1(7)} = 96.44$ and $X_{1(17)} = 98.42$. <u>Hence the interval estimate is (96.44, 98.42)</u>.

## 8. <u>POSSIBLE DIRECTIONS OF NEW RESEARCH</u>

While there are many possible avenues of new research, let us state several of the ones that arise from the work just detailed in Section 7 above.

<u>Open Problem I</u>: The operational meaning of $\epsilon_1^*$ and $\epsilon_2^*$ and $d^*$ here, and in Sobel (1967), is not clear. What considerations should

the experimenter take into account to arrive at numerical values of these quantities?

Open Problem II. For some standard distributions, give tables and graphs to show exactly what $\epsilon_1^*$, $\epsilon_2^*$ signify in terms of the mean of the distribution (and how far apart the means must be to guarantee a particular d*).

Open Problem III: Since $X_{\{k\}(r-c_1)}$ and $X_{\{k\}(r+c_2)}$ are associated with $X_{[k](r)}$ and since we do not necessarily have $X_{[k](r)}$ associated with $F_{[k]}(x)$, we cannot find the probability

$$P(X_{\{k\}(r-c_1)} \leq x_\alpha(F_{[k]}) \leq X_{\{k\}(r+c_2)})$$

without conditioning on $X_{[k](r)} = X_{(k)(r)}$. Hence the "naive" conservative method of considering

$$P(CS \cap CI|R) \geq P(X_{\{k\}(r-c_1)} \leq x_\alpha(F_{[k]}) \leq X_{\{k\}(r+c_2)})$$
$$+ P(X_{(i)(r)} < X_{(k)(r)},\quad i=1,\ldots,k-1) - 1$$

fails at (28). Might conditioning on the $r-c_1{}^{st}$ and $r+c_2{}^{nd}$ order statistics yield a simpler solution?

Open Problem IV: These procedures require n large enough for $1 \leq \alpha(n+1) \leq n$ and $(n+1)\alpha = r$ an integer. So, if you are interested in the $95^{th}$ percentile, you need 19 observations. What do you do if you are interested in the $95^{th}$ percentile but have only 10 observations? Or, the $90^{th}$ percentile but have 2 observations? Some inference should be possible (without going to a parametric, e.g. normal, setting).

## REFERENCES

Amer, P. D. and Dudewicz, E. J. (1980). Two new goals for selection based on proportions. Communications in Statistics, 9A, 1461-1472.

Bartlett, N. S. and Govindarajulu, Z. (1968). Some distribution-free statistics and their application to the selection problem. Annals of the Institute of Statistical Mathematics, 20, 79-97.

Bechhofer, R. E. (1954). A single-sample multiple decision procedure for ranking means of normal populations with known variances. Annals of Mathematical Statistics, 25, 16-39.

Bechhofer, R. E., Elmaghraby, S. and Morse, N. (1959). A single-sample multiple-decision procedure for selecting the multinomial event which has the highest probability. Annals of Mathematical Statistics, 30, 102-119.

Bechhofer, R. E. and Sobel, M. (1958). Non-parametric multiple-decision procedures for selecting that one of k populations which has the highest probability of yielding the largest observation (preliminary report). Abstract, Annals of Mathematical Statistics, 29, 325.

Bishop, T. A. and Dudewicz, E. J. (1978). Exact analysis of variance with unequal variances: test procedures and tables. Technometrics, 20, 419-430.

Chernick, M. R. and Murthy, V. K. (1985). Properties of bootstrap samples. American Journal of Mathematical and Management Sciences, 5, 161-170.

Chernoff, H. (1964). Review of "A class of selection procedures based on ranks" by E. L. Lehmann. Mathematical Reviews, 27, Review #6350.

Dhariyal, I. D. and Dudewicz, E. J. (1985). Nonparametric simultaneous selestimation (selection and estimation) of the largest $\alpha$-quantile (with discussion). The Frontiers of Modern Statistical Inference Procedures, American Sciences Press, Inc., Columbus, Ohio, 185-214.

Dudewicz, E. J. (1971). A nonparametric selection procedure's efficiency: largest location parameter case. Journal of the American Statistical Association, 66, 152-161.

Dudewicz, E. J. (1976). Introduction to Statistics and Probability. Holt, Rinehart and Winston, New York.

Dudewicz, E. J. and Dalal, S. R. (1975). Allocation of observations in ranking and selection with unequal variances. Sankhyā, 37B, 28-78.

Dudewicz, E. J. and Fan, C.-ℓ. (1973). Further light on nonparametric selection efficiency. Naval Research Logistics Quarterly, 20, 737-744.

Dudewicz, E. J. and Geller, N. L. (1972). Book Review of "Nonparametric Statistical Inference" by Jean Dickinson Gibbons. Journal of the American Statistical Association, 67, 962-963.

Dudewicz, E. J. and Koo, J. O. (1982). The Complete Categorized Guide to Statistical Selection and Ranking Procedures. American Sciences Press, Inc., Columbus, Ohio.

Dudewicz, E. J. and Lee, Y. J. (1979). On the least favorable configuration of a selection procedure based on rank sums: counterexample to a postulate of Matsui. Journal of the Japan Statistical Society, 9, 65-69.

Dudewicz, E. J. and Lee, Y. J. (1980). Awardee selection procedures, with special reference to the Frank Wilcoxon and Jack Youden prizes. Technometrics, 22, 121-124.

Dudewicz, E. J. and Mishra, S. N. (1984). The robustness of Bechhofer's normal means selection procedure. Design of Experiments, Ranking and Selection: Essays in Honor of Robert E. Bechhofer (edited by T. J. Santner and A. C. Tamhane), Marcel Dekker, Inc., New York, 35-45.

Dudewicz, E. J. and van der Meulen, E. C. (1983). Entropy-based statistical inference, II: selection-of-the-best/complete ranking for continuous distributions on (0,1), with applications to random number generators. Statistics & Decisions, 1, 131-145.

Fingleton, B. (1984). Models of Category Counts. Cambridge University Press, Cambridge, England.

Gibbons, J. D. (1971). Nonparametric Statistical Inference. McGraw-Hill, New York.

Gibbons, J. D. (1985). Nonparametric Methods for Quantitative Analysis (Second Edition). American Sciences Press, Inc., Columbus, Ohio.

Hall, W. J. (1959). The most-economical character of some Bechhofer and Sobel decision rules. Annals of Mathematical Statistics, 30, 964-969.

Harter, H. L. (1983a). The Chronological Annotated Bibliography of Order Statistics, Volume I: Pre-1950. American Sciences Press, Inc., Columbus, Ohio.

Harter, H. L. (1983b). The Chronological Annotated Bibliography of Order Statistics, Volume II: 1950-1959. American Sciences Press, Inc., Columbus, Ohio.

Hettmansperger, T. P. (1984). Statistical Inference Based on Ranks. John Wiley & Sons, Inc., New York.

Kesten, H. and Morse, N. (1959). A property of the multinomial distribution. Annals of Mathematical Statistics, 30, 120-127.

Krishnaiah, P. R. and Sen, P. K. (Editors) (1984). Handbook of Statistics, Vol. 4: Nonparametric Methods. North-Holland, Amsterdam.

Lee, Y. J. (1980). Nonparametric selections in blocked data: application to motor vehicle fatality rate data. Technometrics, 22, 535-542.

Lee, Y. J. and Dudewicz, E. J. (1985). Robust/nonparametric selection methods in blocked data: relative efficiency study. Tamkang Journal of Mathematics, 16, 29-38.

Lee, Y. J. and Dudewicz, E. J. (1986a). On selecting the best contender. Journal of Computational and Applied Mathematics, 16, in press.

Lee, Y. J. and Dudewicz, E. J. (1986b). Robust selection procedures based on vector ranks. Metrika, International Journal for Theoretical and Applied Statistics, to appear.

Lehmann, E. L. (1963). A class of selection procedures based on ranks. Mathematische Annalen, 150 (III), 268-275.

Matsui, T. (1972). On selecting the best one of k normal populations based on ranks. Journal of the Japan Statistical Society, 2, 71-81.

Matsui, T. (1974). Asymptotic behavior of a selection procedure based on rank sums, Journal of the Japan Statistical Society, 4, 57-64.

Meddis, R. (1984). Statistics Using Ranks: A Unified Approach. Blackwell, Oxford, England.

Noether, G. E. (1967). Needed--a new name! The American Statistician, 21 (2), 41.

Puri, M. L. and Puri, P. S. (1969). Multiple decision procedures based on ranks for certain problems in analysis of variance. Annals of Mathematical Statistics, 40, 619-632.

Rizvi, M. H. and Woodworth, G. G. (1970). On selection procedures based on ranks: counterexamples concerning least favorable configurations. Annals of Mathematical Statistics, 41, 1942-1951.

Sobel, M. (1967). Nonparametric procedures for selecting the t populations with the largest $\alpha$-quantiles. Annals of Mathematical Statistics, 38, 1804-1816.

Ury, H. (1967). Letter to the Editor. The American Statistician, 21 (4), 53.

# On the Power of Entropy-Based Tests Against Bump-Type Alternatives

Edward J. Dudewicz
University Statistics Council
Department of Mathematics
Syracuse University
Syracuse, New York 13244, USA

Edward C. van der Meulen
Department of Mathematics
Katholieke Universiteit Leuven
Leuven, Belgium

SUMMARY : Recently bump-type alternatives to uniformity have generated interest among practitioners, especially in physics and reliability, and hence statistical tests for uniformity versus bump-type alternatives are currently of strong interest. In this paper we report the power of a new entropy-based test of uniformity under a bump alternative. The results are contrasted to previous statements of Cressie based on a statistic involving logarithms of high-order spacings.

## 1. INTRODUCTION

Given a random sample $X_1,\ldots,X_n$ $(n \geq 3)$ from a population with absolutely continuous density function $f(x)$ concentrated on the interval $[0,1]$, distribution function $F(x)$, and variance $\sigma^2(f)$, consider the problem of testing the hypothesis $H_0$ that the $X_i$ are uniformly distributed, denoted by $U(0,1)$. Denote the order statistics by $Y_1 \leq Y_2 \leq \ldots \leq Y_n$ and define $Y_i = Y_1$ if $i < 1$, $Y_i = Y_n$ if $i > n$. Using the Vasicek [4] entropy estimate

$$H(m,n) \equiv \frac{1}{n} \sum_{i=1}^{n} \{\log \frac{n}{2m} (Y_{i+m} - Y_{i-m})\}, \tag{1}$$

where m is a positive integer smaller than n/2, Dudewicz and van der

---

Research supported by the NATO Research Grants Programme (NATO Research Grant N° 1674) and the Ministerie van Wetenschapsbeleid, Brussels, Belgium (Project GOA 83/88-53).
*AMS 1980 subject classifications.* Primary 62G10; secondary 62B10, 62G30.
*Key words and phrases.* Entropy-based statistical tests, tests for uniformity, power comparisons, bump-type alternatives, Monte Carlo.

Contributions to Stochastics.
Ed. by W. Sendler
© Physica-Verlag Heidelberg 1987

Meulen [3] defined and studied the entropy-based test ENT(H(m,n);$\alpha$) which rejects $H_0$ if and only if

$$H(m,n) \leqslant H_\alpha^*(m,n), \tag{2}$$

where $H_\alpha^*(m,n)$ is set so that the test has the desired level $\alpha$ for given m and n. For $\alpha$ = .005, .01, .025, .05, .10, .20, .30, .40 they tabled $-H_\alpha^*(m,n)$ for n = 10(10)50, 100 for all m = 1(1)10(5)20(10)40 for which m < m/2. They compared the power of the ENT(H(m,n);$\alpha$) test, under seven alternatives, with the power of the Kolmogorov-Smirnov D, Crámer-von Mises $W^2$, Kuiper V, Watson $U^2$, Anderson-Darling $A^2$, log-statistic Q, and $\chi^2$ tests of uniformity, and found ENT(H(m,n);$\alpha$) possesses good power properties for many alternatives. Asymptotic distributions were also derived, and a theory of entropy-based tests outlined.

## 2. POWER EVALUATION, BUMP ALTERNATIVE

Using Monte Carlo methods (10,000 trials), we evaluated the power of ENT(H(m,n);$\alpha$), over the m and n specified above and for $\alpha$ = .10, under alternative

$$D : F(z) = \begin{cases} \frac{z}{2} & \text{if} \quad 0 \leqslant z \leqslant \frac{1}{4}, \\ \frac{1}{8} + \frac{3}{2}(z - \frac{1}{4}) & \text{if} \quad \frac{1}{4} \leqslant z \leqslant \frac{3}{4}, \\ \frac{7}{8} + \frac{1}{2}(z - \frac{3}{4}) & \text{if} \quad \frac{3}{4} \leqslant z \leqslant 1, \end{cases} \tag{3}$$

called a "bump" type of alternative by Cressie [2]. (For a discussion of "bump hunting" as a "major current activity of experimental physicists" and others, see Cressie [2, p. 216].) In Table 1 we give, for each n, the best m (of those studied) and the associated power under alternative D at $\alpha$ = .10. Comparison with unpublished results of Vasicek shows our power under alternative D to substantially exceed both that of another test proposed by Vasicek and the power of the Kolmogorov-Smirnov test.

TABLE 1

Power of ENT(H(m,n);α), α = .10

| n | Best m | Power |
|---|---|---|
| 10 | 4 | .4315 |
| 20 | 9 | .5679 |
| 30 | 10 | .6525 |
| 40 | 15 | .7748 |
| 50 | 20 | .8749 |
| 100 | 40 | .9947 |

## 3. COMPARISONS

Cressie [2] investigated the power of another test of uniformity (based on a statistic which he denoted $L_N^{(m)}$, but which we will denote by L(m,n) below; see equation (5)) based on logarithms of $m^{th}$ order gaps, especially for the class of bounded positive step functions on (0,1) (to which alternative D belongs). It can be shown (see Section 4 of Dudewicz and van der Meulen [3]) that

$$\begin{aligned} \Delta'(m,n) &\equiv nH(m,n) - L(2m,n) \\ &= \sum_{i=1}^{m} \log(Y_{i+m} - Y_i) + \sum_{i=n-m+1}^{n} \log(Y_n - Y_{i-m}) \\ &\quad - \log(1 - Y_{n-2m+1}) - \log(Y_{2m}) + n \log(n/(2m)). \end{aligned} \tag{4}$$

For fixed m, as $n \to \infty$, nH(m,n) and L(2m,n) have asymptotic normal distributions with the same variance and linearly related means (under all bounded positive step function alternatives on (0,1)), hence both tests have (under this class of alternatives) the same asymptotic properties.

We have found that for n = 10,20 the "best" m is also the largest m (= n/2 - 1). For n = 30 our investigations extend up to m = 10, and of those m the best one is m = 10. For n = 40 the best m also coincides with the highest m investigated (m = 15), and similarly for n = 50 (best m = highest m = 20) and for n = 100 (best m = highest m = 40). Thus there is reason to conjecture that if all m ($1 \leq m \leq n/2 - 1$) are investigated, we will always find that the best m is n/2 - 1 for alternative D.

On the other hand, Cressie [2] concluded that for his test statistic L(m,n), based on logarithms of $m^{th}$ order spacings and defined by

$$L(m,n) = \sum_{i=0}^{n-m+1} \log(Y^*_{i+m} - Y^*_i), \tag{5}$$

where $Y^*_0 = 0$, $Y^*_j = Y_j$ $(1 \leq j \leq n)$, and $Y^*_{n+1} = 1$, the choice of m = 3 is probably optimal for testing Birnbaum's [1] bump-type alternative "... for finite sample sizes down to N = 25." This choice of m = 3 corresponds with our m being 1.5, but our simulation studies show a very different result : for n = 20 and n = 30 (both close to the n = 25 studied by Cressie [2]) we find that higher values of m (in fact the highest possible ones, m = 9 for n = 20 and $m \geq 10$ for n = 30) will yield considerably higher power against our choice of bump alternative D.

Several explanations for these different findings are possible. (i) The two test statistics ENT(H(m,n);α) and L(m,n) are different. However, from the relationship between the two test statistics obtained at and after equation (4), it does not appear that they are so substantially different that their difference could account for this change in the optimal value of m. (ii) The alternatives studied are different. Indeed our alternative has a greater bump width (and smaller bump height) than the one considered by Cressie [2] (Birnbaum's alternative), which could have an effect on the optimal m. While at one point Cressie [2] writes "... we might expect that as the bump width increased and N remained fixed, gaps of higher order than m = 3 should be used ...", he next concludes that "... our results show that the best choice is m = 3 ...". Regarding bump height, he concludes that it "... should have no real effect upon ..." the optimal m. Thus our power results contradict his conclusions. (iii) The simulations were carried out differently. This seems to be the real reason for the contradictory results. Cressie [2] determined his cutoff points as follows : first choose α and n; then determine the rejection region from the normal distribution, viz. $R_\alpha = \{x: x \leq Z_\alpha\}$; next for each m determine by Monte Carlo (2500 uniform samples) the fraction of samples falling in this critical region; subsequently pick those m's for which this fraction is close to the desired α; finally continue to find, among those m's, the one with highest simulated power. Thus, Cressie's [2] choice of m depends largely (almost solely, in fact) on the normal approximation and behavior of the test statistic under the null hypothesis only. Our Monte Carlo estimates, on the other hand, first determined the exact critical region for each m (and fixed α) for various choices of n (using 10,000 random samples), yielding precise percentage points $H^*_\alpha(m,n)$. For each critical region the power was determined by observing the proportion of 10,000 samples (from the alternative) falling in that critical region. We could therefore compare the power (for each fixed $\alpha$ and n) for various values of m in a meaningful way, thus reaching our (valid) conclusions. From (i), (ii), (iii) (especially (iii)) we conclude that the optimal choice of m is not 3 (as claimed by Cressie), but much higher (probably $m = n/2 - 1$) when $10 \leqslant n \leqslant 100$ and testing against bump alternative D; this contradicts Cressie's [2] general claims.

## REFERENCES


[1] Birnbaum, Z.W. (1975), "Testing for intervals of increased mortality," in *Reliability and Fault Tree Analysis*, Philadelphia, SIAM, 413-426.

[2] Cressie, N. (1978), "Power results for tests based on high-order gaps," *Biometrika*, 65, 214-218.

[3] Dudewicz, E.J. and van der Meulen, E.C. (1981), "Entropy-based tests of uniformity," *J. Amer. Statist. Assoc.*, 76, 967-974.

[4] Vasicek, O. (1976), "A test for normality based on sample entropy," *J. Roy. Statist. Soc. Ser. B*, 38, 54-59.

# Robust Statistical Methods Applied in the Analysis of Geochemical Variables

R. Dutter
Institut für Statistik und
Wahrscheinlichkeitstheorie
Technische Universität Wien
A-1040 Wien, Austria

SUMMARY: A preliminary case study of the analysis of geochemical variables is reported. Some ideas of resistant analysis and robust statistics are incorporated in certain statistical methods employed in geoscience: Descriptive statistics, principal components analysis, grouping of data, two-dimensional presentation, multiple regression analysis, canonical correlation, outlier detection. The main objective in this study is the outlier-resistent analysis of the data and the report of the outliers which are interesting for exploration.

## 1. Introduction

This paper closely follows the ideas of a case study which was undertaken with geochemical variables measured on different places in the Austrian alps. Approximately 32.000 samples of about 36 variables were to be expected, however, most of the discussion is motivated by a study of 649 measurements in a smaller region.

The variables measured were AG, AL, AS, BA, BE, CA, CE, CO, CR, CU, FE, GA, K, LA, LI, MG, MN, MO, NA, NB, NI, P, PB, RB, SC, SN, SR, TH, TI, U, V, W, Y, ZN, ZR. The measurements were taken in river beds of at least 4 geological units in Styria in Austria.

In Sections 2 to 4 we state some basic tools of robust statistics. These are incorporated in methods which were then applied successfully in the analysis of geochemical variables. Sections 5 to 10 describe possible robustifications of classical statistical methods. The results of the pilot project are contained in several technical reports, the whole data set is still under study.

Contributions to Stochastics.
Ed. by W. Sendler
© Physica-Verlag Heidelberg 1987

## 2. Robust Location and Scale

We are guided by the lines of Huber and we consider the M-estimator (see Huber, 1964[15], 1981[17]). Suppose that we have n (univariate) observations ($x_i$, i=1,...,n). Then the location and the scale are estimated by (Huber Proposal II) the solution of the system of equations

$$\sum_{i=1}^{n} \psi\left(\frac{x_i - \mu}{\sigma}\right) = 0$$

$$\sum_{i=1}^{n} \left[\psi\left(\frac{x_i - \mu}{\sigma}\right)\right]^2 = \beta$$

where $\beta$ is the expected value in the normal case

$$\beta = E_{\Phi} \, [\psi(X)]^2$$

and $\psi$ is a robustifying function, e.g.

$$\psi(t) = \begin{cases} -c & \text{if} \quad t < -c \\ t & \text{if} \quad |t| \leq c \\ c & \text{if} \quad c < t \end{cases}$$

where c is a certain constant greater zero. A suggested value is c = 1.345 (corresponding to 95% efficiency in the normal case). It might also be found from the minimax solution in the gross-error model if a certain portion $\varepsilon$ of non-normal observations is assumed (see also the next section).

## 3. Robust Covariance Matrices

Here we refer to Maronna (1976[26]) and Huber (1977[16], 1981[17]). See also Dutter (1983a[4]) and Marazzi (1980b[23]). Suppose that we have n (complete, multivariate) observations ($\underset{\sim}{x}_i$, i=1,...,n) of dimension p, i.e. a data matrix

$$(x_{ij}), \quad j=1,\ldots,p, \quad i=1,\ldots,n.$$

The aim is the computation of the location and the scatter (the empirical covariance matrix) of the elements of $\underset{\sim}{x}_i$.

We first try to find a transformation $(A,\underset{\sim}{b})$ where A is a (p x p)-matrix and $\underset{\sim}{b}$ a (p x 1)-vector. The data $\underset{\sim}{x}_i$ should be transformed to

$$\underset{\sim}{y}_i = A(\underset{\sim}{x}_i - \underset{\sim}{b}), \quad i=1,\ldots,n,$$

such that

$$\underset{i}{\text{ave}} \{w(||\underset{\sim}{y}_i||)\underset{\sim}{y}_i\} = \underset{\sim}{0},$$

$$\underset{i}{\text{ave}} \{u(||\underset{\sim}{y}_i||) \frac{\underset{\sim}{y}_i \underset{\sim}{y}_i^T}{||\underset{\sim}{y}_i||^2} - v(||\underset{\sim}{y}_i||) I \} = \sigma,$$

where u, v and w are certain weight functions. A robust location estimate is then defined by

$$\underset{\sim}{b}$$

and a (pseudo-) covariance matrix by

$$C = (A^T A)^{-1} \tau^2$$

where $\tau$ is a correction factor.

Suggested forms of the weight functions u, v and w are

$$u(t) = \begin{cases} a^2 & \text{if } t < a \\ t^2 & \text{if } a \leq t \leq b \\ b^2 & \text{if } b < t, \end{cases}$$

$$w(t) = \begin{cases} 1 & \text{if } t < c \\ c/t & \text{if } c \leq t, \end{cases}$$

and

$$v(t) = \frac{1}{p} E_\Phi [u(||X||)] = \text{const.}$$

The constants a, b and c could be found from the model assumptions that we have a portion of non-normal data (using the gross-error model) and we look for the minimax solution (see Huber, 1977[16]). The only open parameter then is $\varepsilon$ which should have some meaning to the practitioner (e.g. $\varepsilon = .05$).

## 4. Robust and Bounded Influence Regression

We follow the lines of Hampel (1986[10]) and Hampel et al. (1986[11]). The computer program mainly used in this work is described in Dutter (1987[6]). See also Marazzi (1980a[22]), Marazzi and Randriamiharisoa (1986[24]).

Suppose that we have n observations of a "dependent" (univariate) variable, $y_i$, i=1,...,n, and of p "independent" variables, $(x_{i1},\ldots, x_{ip})^T = \underset{\sim}{x}_i$, i=1,...,n. A linear relationship between $y_i$ and $\underset{\sim}{x}_i$ is assumed, i.e.

$$y_i \sim \underset{\sim}{x}_i^T \underset{\sim}{b} = \sum_{j=1}^{p} x_{ij} b_j$$

where the coefficients $\underset{\sim}{b}$ are unknown. They are found by the solution of the system of equations

$$\sum_{i=1}^{n} \psi\left(\frac{y_i - \underset{\sim}{x}_i^T \underset{\sim}{b}}{\sigma p_i}\right) v_i x_{ij} = 0, \quad j=1,\ldots,p,$$

where $p_i$ and $v_i$ are weights responsible for bounding the influence of points in the factor space and $\sigma$ is a scale, also to be estimated from the data (see also Section 2). If $p_i \equiv v_i \equiv 1$ and $\psi(t)=t$ we obtain the usual least squares estimate. For different options on $p_i$, $v_i$ and $\psi$ the reader is referred to Dutter (1987[6]).

## 5. Primary Statistics (Univariate Analysis)

At the beginning the data for each variable at a time is analyzed. The variables of our case study are regionalized ones, i.e. they are space-dependent. This suggests the presentation of the data in coded form in some geographical map.

For the coding we need some outlier resistent description of the data (see Tukey, 1977[30], Hoaglin et al., 1983[12], 1985[13], Velleman and Hoaglin, 1981,[31]). Let $x_{(i)}$ denote some ordered data values $x_i$, $i=1,\ldots,n$. $i$ in $x_{(i)}$ stands for "upward rank" and, analogously, $n+1-i$ for "downward rank". With "depth" we mean the smaller of the upward and the downward rank. Then we define the "median" $\tilde{x}$ by the value with depth $(n+1)/2$, more precisely by $\tilde{x} = x_{((n+1)/2)}$ if n is odd and by $\tilde{x} = \frac{1}{2}(x_{(k)}+x_{(k+1)})$ if n is even and $n=2k$. Then "forth's" are defined by the values with

$$\text{depth} = \frac{[\text{depth of median}] + 1}{2}$$

where [.] means the integer smaller or equal the argument. We call them upper and lower forth's $F_U$ and $F_L$, respectively. The "spread" or "forth-spread" is defined by the difference

$$d_F = F_U - F_L$$

which, at the normal distribution, corresponds to

$$d_F = 1.349\,\sigma$$

where $\sigma$ is the standard deviation. Furthermore, "outlier cutoffs" are defined by $F_L - 1.5d_F$ and $F_U + 1.5d_F$, respectively.

Our tool of resistantly describing the data is the boxplot which is a graphical medium: Draw a box with ends at lower and upper forths and a cross bar at the median. Then, draw a line from each end of the box to the most remote point that is not an outlier. Outliers are points outside the outlier cutoffs and these usually are drawn separately.

## 6. Principal Component Analysis

The main aim of principal component analysis is the reduction of the dimensionality of the data (compare Jolliffe, 1986[19]). The basis of this as well as of the other multivariate techniques (as e.g. factor analysis, see Jöreskog et al., 1976[20]) is the covariance (or correlation) matrix. For our case study it is natural to use a matrix which has been computed in a robust manner. On the basis of this covariance matrix (and a location vector) it is also relatively easy to look for outliers.

Denote by C a (somehow) computed (p x p) covariance matrix and by $\underset{\sim}{b}$ a location of the data $\underset{\sim}{x}_i$, i=1,...,n. Then decompose C into

$$C = U \Lambda U^T$$

where U is orthogonal ($U^T U = I$) and $\Lambda$ is a diagonal matrix with values of the diagonal $\lambda_1,\ldots,\lambda_p$ such that $\lambda_1 \geq \lambda_2 \geq \ldots \geq \lambda_p$. Multiplying the equation by U from the right, that is,

$$CU = U\Lambda,$$

shows that the $\lambda_i$'s are the eigenvalues of C and U contains the eigenvectors $\underset{\sim}{u}_i$, i=1,...,p (columnwise). Then, the $\underset{\sim}{u}_i$'s are also called the <u>principal axes</u> which, in practice, should be interpretable, e.g. the first axis is responsible for the environment, the second one for some mineralization, etc.

The projection of the data point $\underset{\sim}{x}_i$ on the j-th principal axis is called <u>principal</u> component and computed by

$$y_{ij} = \underset{\sim}{u}_j^T (\underset{\sim}{x}_i - \underset{\sim}{b}) \qquad (j=1,\ldots,p, \quad i=1,\ldots,n).$$

The data point in the new coordinate system of principal axis is given by

$$\underset{\sim}{y}_i = U^T(\underset{\sim}{x}_i - \underset{\sim}{b}).$$

If $\lambda_{q+1}$ $(1<q\leq p)$ is relatively small in respect to $\lambda_1,\ldots,\lambda_q$ then the components $y_{i,q+1},\ldots,y_{i,p}$, i=1,...,n, for reasons of reduction of dimensionality might be neglected because the first q components explain reasonably well the variability of the data.

Using the robustly calculated covariance C and the decomposition in principal components it is easy to construct some outlier decomposition procedure. The (squared) Mahalanobis distance of the i-th data point to the center is

$$r_i^2 = (\underset{\sim}{x}_i - \underset{\sim}{b})^T C^{-1} (\underset{\sim}{x}_i - \underset{\sim}{b})$$

which is

$$r_i^2 = (\underset{\sim}{x}_i - \underset{\sim}{b})^T U \Lambda^{-1} U^T (\underset{\sim}{x}_i - \underset{\sim}{b})$$

$$= \sum_{j=1}^{p} y_{ij}^2 / \lambda_j .$$

If we only use the first q components for further analysis the distance squared is

$$r_i^2 = \sum_{j=1}^{q} y_{ij}^2 / \lambda_j .$$

The statistical distribution of $r_i^2$ on the basis of normally distributed data and large sample size n, is approximately $\chi_p^2$ in the first case and $\chi_q^2$ in the second one.

The values of $r_i$ can of course be analyzed by any exploratory technique like boxplots, and geographical maps can be drawn from these values.

## 7. Grouping of Data

The data may be grouped according to some criteria by any clustering algorithm (see e.g. Friedman and Rafsky, 1981[7], Huber, 1985[18]). In the case study the main aim is the grouping in geographically connected geological units.

The evaluation of the goodness of the clustering may be based on different measures (statistics). The question whether two groups are significantly different (in respect to location) is classically answered by Hotelling's $T^2$-Test (see e.g. Afifi and Azen, 1979[1], p. 287, Afifi and Clark, 1984[2]). Suppose that $C_1$ is the estimated covariance of the first group (size $n_1$) and $C_2$ of the second one (size $n_2$). Then the pooled covariance is

$$C = \frac{1}{n_1+n_2-2} [(n_1-1)C_1 + (n_2-1)C_2]$$

and the $T^2$-statistic

$$T^2 = \frac{n_1 n_2}{n_1+n_2} (\bar{\underset{\sim}{x}}_1-\bar{\underset{\sim}{x}}_2)^T C^{-1} (\bar{\underset{\sim}{x}}_1-\bar{\underset{\sim}{x}}_2).$$

If the usual assumptions hold, i.e. the data of the first and second groups are normally distributed with the same covariance and means $\underset{\sim}{\mu}_1$ and $\underset{\sim}{\mu}_2$, and the means are identical ($\underset{\sim}{\mu}_1 = \underset{\sim}{\mu}_2$) then

$$F = \frac{n_1+n_2-p-1}{(n_2+n_2-2)p} \; T^2$$

is distributed $F_{p,n_1+n_2-p-1}$. In case of different covariance matrices (the Behrens-Fisher problem) similar expressions can be deduced (see Srivastava and Khatri, 1979[29], p. 129, and Srivastava and Carter, 1983[28]).

In order to robustify this procedure it is naturally suggested to replace the empirical means $\bar{x}_1$ and $\bar{x}_2$ by some robust location estimate and the empirical covariance matrices by robustly estimated ones.

If more than one groups are to be tested on equality of their means, simultaneous tests may be adequate (see also Srivastava and Carter, 1983[28]). The statistic Wilks' λ is used to test on the equality of means of, say, m groups. The definition is

$$\lambda = \frac{|W|}{|T|}$$

where T denotes the empirical covariance matrix of all the data and W the covariance "within the groups", that is

$$W = \frac{1}{n-m} \sum_{j=1}^{m} \sum_{\ell=1}^{n_j} (x_{i_\ell} - \bar{x}_j)(x_{i_\ell} - \bar{x}_j)^T$$

where $\bar{x}_j$ is the mean of the j-th group which has $n_j$ samples. n is the total number of samples.

A decision on the number of groups has been suggested after some experimentation by Mariott (1971[25]). He proposes to plot Wilks' λ times $m^2$ versus this number m of groups when successively augmenting the number of groups. This graph (see Fig. 1) should increase if the partition into groups is somewhat artificial and decrease rapidly if the partition is according important clusters (see Sinding-Larson, 1975[27]). Of course, the procedure is not unique.

An obvious robustification of the procedure would be the use of robust means and covariance matrices of the groups and of the total.

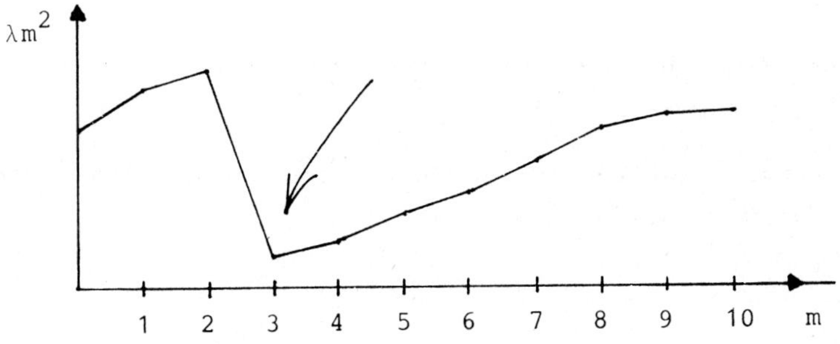


Fig. 1: Plot of Wilks' λ depending on the number of groups.

## 8. Two-dimensional Presentation

In this section we refer to the two-dimensional presentation of more-dimensional data by Gabriel (1981[8]) called bi-plot.

A (n x p)-matrix X of rank r may be factorized into

$$X = G H^T,$$

where G is n x r and orthogonal. A biplot display is constructed by using a reduced rank approximation $X_{[2]}$ of order 2, which is factorized by

$$X_{[2]} = G H^T$$

where G has 2 columns, and by plotting the rows of G and the columns of $H^T$. The rows $\underset{\sim}{g}_i$ of G show the structure of the data values and the rows $\underset{\sim}{h}_j$ of H the variance-covariance structure of the variables.

The robustification would be as follows: Suppose that we have obtained a robust covariance matrix C of X (resp. of $X_{[2]}$) which might be written as weighted inner product of the centered data matrix

$$\tilde{X} = X - \underset{\sim}{1}\, \underset{\sim}{b}^T$$

by itself,

$$C = \tilde{X}^T W \tilde{X}$$

where W denotes the diagonal matrix containing the weights. C is decomposed (as in principal components analysis) into

$$C = U \Lambda U^T.$$

Then we define

$$G = \tilde{X} U \Lambda^{-1/2}$$

and

$$H = U \Lambda^{1/2}.$$

The matrix G weighted by W is orthogonal, namely

$$\begin{aligned} G^T W G &= \Lambda^{-1/2} U^T \tilde{X}^T W \tilde{X} U \Lambda^{-1/2} \\ &= \Lambda^{-1/2} U^T U \Lambda U^T U \Lambda^{-1/2} = I, \end{aligned}$$

and H grasps the covariance structure of X because of

$$C = \tilde{X}^T W \tilde{X} = H G^T W G H^T = H I H^T = H H^T.$$

It follows that we should plot the rows of H for the variables and the rows of G, or better of $W^{1/2}$ G, for the data.

## 9. Multiple Regression Analysis

Evidently, the appearence of certain elements depend on the appearence of others. Therefore it is natural to try to find models showing these dependences. The first step is to take a variable (an element) which is relatively well understood and which shows up in a measurable size.

In this context e.g., the variable ZN has been chosen and modelled in dependence of all other variables. Certain problems arise because regression analysis with 35 variables is not easy. E.g. the program did not work with 200 observations and 35 variables. The robust covariance algorithm did not converge. So the first analysis was non-robust, i.e. we could use the Huber-type estimator but no bounding of influence, i.e.

$$\Sigma\ \psi\left(\frac{y_i - \underset{\sim}{x}_i^T \underset{\sim}{b}}{\sigma}\right) \underset{\sim}{x}_i = \underset{\sim}{0}.$$

In reality, we use the classically robust approach, i.e. clean the data univariately and apply carefully least squares. Check the distributions of the errors, try a different model, etc. In this sense we end up with a model which fits reasonably well a subset of data. With this model the residuals $r_i$ of all data (of a group, say) were computed together with their approximated standard deviations $s(r_i)$.

These standardized residuals

$$r_i \ / \ s(r_i)$$

can again be subjected to the usual EDA-procedure, namely, produce a boxplot of the residuals, code the values somehow and produce a geographic map using these symbols.

## 10. Canonical Correlation

The computational procedure in canonical correlation analysis is the finding of a linear combination of one specific set of variables and of a linear combination of another specific set of variables such that the correlation is maximized.

Suppose that we have p variables $X_1,\ldots,X_p$ and q variables $Y_1,\ldots,Y_q$. Then consider the linear combination

$$Z_X = u_1\ X_1 + \ldots + u_p\ X_p$$

and

$$Z_Y = v_1 Y_1 + \ldots + v_q Y_q .$$

Then the coefficients $u_1,\ldots,u_p$ and $v_1,\ldots,v_q$ are classically to be found such that the correlation coefficient

$$r_{XY} = \frac{\Sigma(Z_X-\bar{Z}_X)\ (Z_Y-\bar{Z}_Y)}{\sqrt{\Sigma(Z_X-\bar{Z}_X)^2\ \Sigma(Z_Y-\bar{Z}_Y)^2}}$$

is maximized. The centered values $Z_X-\bar{Z}_X$ and $Z_Y-\bar{Z}_Y$ are called canonical variables, for simplicity $CV_1$ and $CV_2$. The values of $CV_1$ and $CV_2$ might be put into a scatter plot to show some peculiarities of the data.

The word "classical" was used because this procedure should be efficient only in the case of normally distributed data. Otherwise we would use some robust or "pseudo-robust" procedure. Pseudo-robust means "cleaning" the data first and applying the usual procedure then. If the coefficients in the linear combinations have been found, all the data (including strange observations) may be plotted and the behavior may be analyzed.

A "modern" robustification could look like it follows. Observe that the correlation between the canonical variates may be written in terms of correlation matrices of their respective sets of variables. Here, the correlation matrices should be replaced by robust ones, and the usual canonical correlation analysis should be continued.

The question now arises which observations are interesting outliers. Suppose that the first canonical variable represents the "mineralisation" of the samples and the second one some other characteristics (e.g. the environment). The values of these variables of course can now be calculated of all data values and be plotted.

Naively one would think that an outlier is marked if the point is "too far" from the center. In this application, however, points which are close to the $45^o$-line are not very interesting. The same happens with points with small values of $CV_1$ relatively to $CV_2$.

Moreover, it is interesting to look at the influence of an additional point at the correlation. It is remarked that a point along the $45^o$-line increases the correlation, a point along the perpenticular line (through the center) would decrease it. Therefore some points between would not change anything. The influence function computed by Devlin et al. (1975[3]) (using the theoretical considerations by Hampel, 1974[9]) just shows this important behavior. Values of the influence function at the observations can be used for a further sophisticated analysis, coding and plotting the points in a geographic map. For a worked-out example, see Karnel (1986[21]).

References

[1] Afifi A.A. and Azen S.P. (1979): Statistical Analysis. A Computer Oriented Approach. Acad. Press, New York.

[2] Afifi A.A. and V. Clark (1984): Computer-aided Multivariate Analysis. Wadsworth, Inc., London.

[3] Devlin S.J., R. Gnanadesikan and J.R. Kettenring (1975): Robust Estimation and Outlier Detection with Correlation Coefficients. Biometrika 62, 3, 531-545.

[4] Dutter R. (1983a): COVINTER: A Computer Program for Computing Robust Covariances and for Plotting Tolerance Ellipses. Res. Rep. No. 10, Inst. f. Statist., Techn. Univ. Graz.

[5] Dutter R. (1983b): Robust and Bounded Influence Regression· Realization and Application. In: Proc. 4th Pann. Symp. Math. Statist., Bad Tatzmannsdorf, Austria. Grossmann et al. (eds.), Reidel Publ. Comp., Dortrecht-Holland.

[6] Dutter R. (1987): Computer Program BLINWDR Bounded Influence Regression. Manual. Res. Rep. TS87-1, Inst. Statist. and Prob., Techn. Univ. Vienna.

[7] Friedman J.H. and L.C. Rafsky (1981): Graphics for the Multivariate Two-sample Problem. J. Amer. Statist. Soc. 76, 374, 277-294.

[8] Gabriel K.R. (1981): Biplot Display of Multivariate Matrices for Inspection of Data and Diagnoses. In: Interpreting Multivariate Data. V. Barnett (ed.). J. Wiley & Sons, New York.

[9] Hampel F.R. (1974): The Influence Curve and Its Role in Robust Estimation. J. Amer. Statist. Assoc., 69, 383-393.

[10] Hampel F.R. (1980): Optimally Bounding the Gross-error-sensitivity and the Influence of Position in Factor Space. Proc. Statist. Comp. Sect., Amer. Statist. Assoc., 59-64.

[11] Hampel F.R., E.M. Ronchetti, P.J. Rousseeuw and W.A. Stahel (1986): Robust Statistics. The Approach Based on Influence Functions. J. Wiley & Sons, New York.

[12] Hoaglin D.C., F. Mosteller and J.W. Tukey (1983): Understanding Robust and Exploratory Data Analysis. J. Wiley & Sons, New York.

[13] Hoaglin D.C., F. Mosteller and J.W. Tukey (1985): Exploring Data Tables, Trends and Shapes. J. Wiley & Sons, New York.

[14] Howarth R.J. and R. Sinding-Larson (1983): Multivariate Analysis. In: Statistics and Data Analysis in Geochemical Prospecting. (Handbook of Exploration Geochemistry, Vol. II.).

[15] Huber P.J. (1964): Robust Estimation of a Location Parameter. Ann. Math. Statist. 35, 73-101.

[16] Huber P.J. (1977): Robust Covariances. In: Statistical Decision Theory and Related Topics, II. Gupta and Moore (eds.). Acad. Press, New York.

[17] Huber P.J. (1981): Robust Statistics. J. Wiley & Sons, New York.

[18] Huber P.J. (1985): Projection Pursuit. Invited Paper. Ann. Statist. 13, 2, 435-475.

[19] Jolliffe I.T. (1986): Principal Component Analysis. Springer-Verlag, Berlin.

[20] Jöreskog K.G., J. E. Klovan and R.A. Reyment (1976): Geological Factor Analysis. Elsevier, Amsterdam.

[21] Karnel G. (1986): Robust Canonical Correlation. Res. Rep. TS-1986-

3, Inst. Statist. and Prob., Techn. Univ. Vienna.

[22] Marazzi A. (1980a): Robust Linear Regression Programs in ROBETH. ROBETH Document No. 2. Unpublished manuscript.

[23] Marazzi A. (1980b): Robust Affine Invariant Covariance in ROBETH. ROBETH Document No. 3. Unpublished manuscript.

[24] Marazzi A. and A. Randriamiharisoa (1986): ROBETH-ROBSYS: A Software for Robust Statistical Computing. Div. Statist. et Inform., Inst. Univ. Med. Soc. et Prev., Lausanne.

[25] Mariott F.H.C. (1971): Practical Problems in a Method of Cluster Analysis. Biometrics 27, pp. 501-514.

[26] Maronna R. (1976): Robust M-estimators of Multivariate Location and Scatter. Ann. Statist. 4, 51-67.

[27] Sinding-Larson R. (1975): A Computer Method for Dividing a Regional Survey Area into Homogeneous Subareas Prior to Statistical Interpretation. In: Geochemical Exploration 1974. Elliott and Fletcher (eds.). Elsevier, Amsterdam, pp. 191-217.

[28] Srivastava M.S. and E.M. Carter (1983): An Introduction to Applied Multivariate Statistics. North Holland, Amsterdam.

[29] Srivastava M.S. and C.G. Khatri 1979): An Introduction to Multivariate Statistics. North Holland, Amsterdam.

[30] Tukey J.W. (1977): Exploratory Data Analysis, Addison-Wesley, Reading, Mass.

[31] Velleman P.F., D.C. Hoaglin (1981): Applications, Basics, and Computing of Exploratory Data Analysis. Duxbury Press, Boston, Mass.

Acknowledgements

This paper would not have seen the light of the day without the stimulating work with Prof. Wolfbauer and his team, especially Dr. Kürzl, from Leoben. Extremely valuable also were the discussions with R. Sinding-Larson and H. Omre from Norway.

# Invariant Sufficiency, Equivariance and Characterizations of the Gamma Distribution

Walther Eberl, jr.
Fachbereich Mathematik und Informatik
Fernuniversität Hagen
D-5800 Hagen, FRG

SUMMARY: In the first section it is shown for any sub-$\sigma$-algebra $\mathcal{G}$ of the $\sigma$-algebra of all scale invariant Borel subsets of $\mathbb{R}^n$ that an equivariant statistic $S$ is $\mathcal{G}$-partially sufficient iff the generated $\sigma$-algebra $S^{-1}(\mathcal{B})$ and $\mathcal{G}$ are independent and that $S$ being invariantly sufficient and equivariant, the Pitman estimator is given by $S\,E_1(S)/E_1(S^2)$. For independent $X_1,\ldots,X_n$ the existence of an invariantly sufficient statistic $\Sigma c_j X_j^k$ is characterized by $X_1^k,\ldots,X_n^k$ having gamma distributions. In the second section there are established some characterizations of the gamma distribution by properties (admissibility, optimality) of the minimum variance unbiased linear estimator where $X_1,\ldots,X_n$ are required to be independent. Finally, the independence of $X_1,\ldots,X_n$ is replaced by a certain linear framework and a method is presented for carrying over the characterizations previously stated for the independent case to this linear set-up.

## 0. NOTATIONS

As usual, we write $\mathbb{N}$ for the set of all positive integers and $\mathbb{N}^0 = \mathbb{N} \cup \{0\}$. For $n \in \mathbb{N}$ using the standard notation $\mathbb{R}^n$ for the space of all real n-tuples, we put $\mathbb{R}_+^n = \{(x_1,\ldots,x_n)\} \in \mathbb{R}^n \mid x_j > 0 \ (1 \le j \le n)\}$; especially, we write $\mathbb{R} = \mathbb{R}^1$, $\mathbb{R}_+ = \mathbb{R}_+^1$. Moreover, $\mathcal{B}^n$ denotes the $\sigma$-algebra of the Borel sets in $\mathbb{R}^n$, $\mathcal{B}_+^n = \{B \cap \mathbb{R}_+^n \mid B \in \mathcal{B}^n\}$, $\mathcal{B} = \mathcal{B}^1$, $\mathcal{B}_+ = \mathcal{B}_+^1$. For any $n \in \mathbb{N}$ and all $\gamma \in \mathbb{R}_+$ let $S_{n,\gamma} = \gamma \cdot \mathrm{Id}_{\mathbb{R}^n}$ be the corresponding scale transformation where $\mathrm{Id}_{\mathbb{R}^n}$ denotes the identity mapping on $\mathbb{R}^n$. For any probability measure $P_1$ on $\mathcal{B}^n$ let $\mathcal{W} = \{P_\gamma = (P_1)_{S_{n,\gamma}} \mid \gamma \in \mathbb{R}_+\}$ ($(P_1)_{S_{n,\gamma}}$: image measure of $P_1$ under $S_{n,\gamma}$) be the corresponding n-dimensional scaling class. For $\gamma \in \mathbb{R}_+$ we denote by $E_\gamma$ and $V_\gamma$ the expectation and the variance w.r.t. $P_\gamma$, respectively. For any inte-

Contributions to Stochastics.
Ed. by W. Sendler
© Physica-Verlag Heidelberg 1987

grable random variable $X: (\mathbb{R}^n, \mathcal{B}^n) \to (\mathbb{R}, \mathcal{B})$ and any sub-$\sigma$-algebra $\mathcal{G}$ of $\mathcal{B}^n$ we use the symbol $E_\gamma^{\mathcal{G}}(X)$ for the conditional expectation of X under $\mathcal{G}$ w.r.t. $P_\gamma$. Correspondingly, for any statistic $S: (\mathbb{R}^n, \mathcal{B}^n) \to (\mathbb{D}, \mathcal{D})$ we write $E_\gamma^S(X)$ for the conditional expectation of X under $S^{-1}(\mathcal{D})$ w.r.t. $P_\gamma$.

Finally, we introduce for any $n \in \mathbb{N}$ the $\sigma$-algebra $\mathcal{B}_S^n$ of the Borel subsets of $\mathbb{R}^n$ being invariant w.r.t. all scale transformations:

$$\mathcal{B}_S^n = \{B \in \mathcal{B}^n \mid S_{n,\gamma}^{-1}(B) = B (\gamma \in \mathbb{R}_+)\}.$$

## 1. EQUIVARIANT AND PARTIALLY SUFFICIENT STATISTICS

For scaling classses the concept of the (scale) equivariant statistic is fundamental. This was first introduced in [30].

We call a statistic $S: (\mathbb{R}^n, \mathcal{B}^n) \to (\mathbb{R}, \mathcal{B})$ (scale) equivariant w.r.t. a n-dimensional scaling class $\mathcal{W} = \{P_\gamma | \gamma \in \mathbb{R}_+\}$ if it holds

$$S > 0 \quad P_1\text{-a.e.} \tag{1.1}$$

and

$$S \circ S_{n,\gamma} = \gamma \cdot S \qquad (\gamma \in \mathbb{R}_+) . \tag{1.2}$$

The set of all equivariant statistics will be denoted by $\mathcal{R}_n = \mathcal{R}_n(\mathcal{W})$. If $P_1(\mathbb{R}_+^n) = 1$, then we shall restrict tacitly the domain of definition of (equivariant) statistics to $\mathbb{R}_+^n$.

DEFINITION 1.1: Let $\mathcal{W} = \{P_\gamma | \gamma \in \mathbb{R}_+\}$ be a n-dimensional scaling class and $\mathcal{G}$ a sub-$\sigma$-algebra of $\mathcal{B}_S^n$. A statistic $S: (\mathbb{R}^n, \mathcal{B}^n) \to (\mathbb{D}, \mathcal{D})$ is called $\mathcal{G}$-partially sufficient (w.r.t. $\mathcal{W}$), if for any $G \in \mathcal{G}$ there exists a measurable function $h_G: (\mathbb{R}, \mathcal{B}) \to (\mathbb{R}, \mathcal{B})$ such that

$$h_G \circ S = E_\gamma^S (1_B) \qquad (\gamma \in \mathbb{R}_+) .$$

In case of $\mathcal{G} = \mathcal{B}_S^n$ the statistic S is called invariantly sufficient.

Our first theorem represents a simple, useful criterion for the $\mathcal{G}$-partial sufficiency of an equivariant statistic. Besides its self-reliant significance it will be useful for characterizations of the gamma distribution by sufficiency. Actually, only partial sufficiency for indicator functions in the sense of Definition 1.1 is needed.

THEOREM 1.2: Let $\mathcal{W} = \{P_\gamma | \gamma \in \mathbb{R}_+\}$ be a n-dimensional scaling class and $\mathcal{G}$ a sub-$\sigma$-algebra of $\mathcal{B}^n_S$. Then, for any $S \in \mathcal{R}_n$ the following two statements are equivalent:

(I) S is $\mathcal{G}$-partially sufficient;

(II) $S^{-1}(\mathcal{B})$ and $\mathcal{G}$ are independent (under $P_1$).

The proof of Theorem 1.2 may be accomplished quite analogously like that for translation classes in [8], [9], [12] and is omitted.

REMARKS 1.3: (a) Our concept of invariant sufficiency differs from that introduced in [16, p. 579] inasmuch as we do not require the invariance of an invariantly sufficient statistic.
(b) Assuming $P_1(\mathbb{R}^n_+) = 1$, we consider two linear statistics $L_1 = \sum_{j=1}^{n} X_j$ and $L_2 = \sum_{j=1}^{n} b_j X_j$, where $b_j \geq 0$ $(j = 1,\ldots,n)$. Then, for $\mathcal{G}=(L_2/L_1)^{-1}(\mathcal{B})$ the $\mathcal{G}$-partial sufficiency of $L_1$ is related to the partial sufficiency of $L_1$ considered in [20] which is more restrictive. By Theorem 1.2 we know that the $\mathcal{G}$-partial sufficiency of $L_1$ yields the independence of $L_1$ and $L_2/L_1$. Note that for the verification of this independence in the proof of Theorem 3 in [20] actually only the weaker requirement of our definition of partial sufficiency was needed, too.
(c) Replacing $\mathcal{G}$-partial sufficiency by sufficiency, the implication (I) $\Rightarrow$ (II) of Theorem 1.2 is covered in a more general framework by [7, Proposition 7.19], see also [4], [6], [16].
However, while evidently any sufficient statistic is $\mathcal{G}$-partially sufficient for all sub-$\sigma$-algebras $\mathcal{G}$ of $\mathcal{B}^n_S$, the converse does not hold true provided that $P_1(\mathbb{R}^n_+) = 1$. In fact, for any statistic $S: (\mathbb{R}^n,\mathcal{B}^n) \to (D,\mathcal{D})$ with $S^{-1}(\mathcal{D}) = \mathcal{B}^n_S$ it holds for every $G \in \mathcal{B}^n_S$ that

$$E^S_\gamma(1_G) = 1_G \qquad (\gamma \in \mathbb{R}_+) ,$$

i.e., S is invariantly sufficient. Now, suppose that S is sufficient w.r.t. $\mathcal{W}$. Then, for every $B \in \mathcal{B}^n$ there exists a measurable function $h_B: (\mathbb{D},\mathcal{D}) \to (\mathbb{R},\mathcal{B})$ such that

$$P_\gamma(A\cap B) = \int_G h_B \circ S \, dP_\gamma = \int_G h_B \circ S \, dP_1 \qquad (G \in \mathcal{B}^n_S;\ \gamma \in \mathbb{R}_+) .$$

Particularly, choosing $G = \mathbb{R}^n$ and $B = \mathop{X}_1^n (0,1]$, we obtain

$$P_1(\mathop{X}_1^n (0,\gamma]) = P_1(\mathop{X}_1^n (0,1]) \qquad (\gamma \in \mathbb{R}_+) .$$

Due to $P_1(\mathbb{R}^n_+) = 1$ this relation leads to a contradiction. Thus, for $P_1(\mathbb{R}^n_+) = 1$ a statistic S with $S^{-1}(\mathcal{D}) = \mathcal{B}^n_S$ is invariantly sufficient, but not sufficient. (Typical such statistics are the maximal invariant ones.)

(d) Replacing $G$-partial sufficiency by sufficiency, the implication (I) $\Rightarrow$ (II) of Theorem 1.2 is related to Theorem 2 in [2], where the equivariance of S is substituted by its bounded completeness. Note that in general an equivariant (invariantly) sufficient statistic need not be boundedly complete. E.g., consider the one-dimensional scaling class $\mathcal{W} = \{P_\gamma \mid \gamma > 0\}$ with $P_1$ given by the Lebesgue density

$$f_1(x) = \frac{1}{x} 1_{(1,e)}(x) \qquad (x \in \mathbb{R}) .$$

Clearly, $S = Id_{\mathbb{R}}$ is equivariant and sufficient. Letting $g: (\mathbb{R},\mathcal{B}) \to (\mathbb{R},\mathcal{B})$ be any measurable, periodic function with period 1 such that

$$\int_0^1 g(x) d\lambda(x) = 0$$

($\lambda$: Lebesgue measure on the real axis), for the function $h: \mathbb{R} \to \mathbb{R}$ given by

$$h(x) = \begin{cases} g(\ln x) & \text{for } x > 0 \\ 0 & \text{for } x \leq 0 \end{cases}$$

it holds that

$$E_\gamma(h) = \int_\gamma^{\gamma e} x^{-1} h(x)\, d\lambda(x) = \int_{\ln \gamma}^{1+\ln\gamma} h(e^y) d\lambda(y) =$$

$$= \int_{\ln\gamma}^{1+\ln\gamma} g(y) d\lambda(y) = 0 \qquad (\gamma > 0).$$

Thus, S is not boundedly complete. (This example corresponds to Example 3.7 in [27] for translation classes.)

Before formulating the next theorem we provide two useful lemmas on the best equivariant estimator under squared error loss. An estimator $\tilde{S} \in \mathcal{R}_n \cap \mathcal{L}^2(P_1)$ is called the Pitman estimator for $\gamma$ (w.r.t. $\mathcal{W}$), if it holds true that

$$E_\gamma[(\tilde{S}-\gamma)^2] = \min_{S \in \mathcal{R}_n} E_\gamma[(S-\gamma)^2] \qquad (\gamma > 0) .$$

The first lemma says that for $\mathcal{R}_n \cap \mathcal{L}^2(P_1) \neq \emptyset$ there exists a Pitman estimator and that it is essentially unique.

LEMMA 1.4: Let $\mathcal{W} = \{P_\gamma | \gamma > 0\}$ be any scaling class and let $S \in \mathcal{R}_n \cap \mathcal{L}^2(P_1)$. Then a Pitman estimator for $\gamma$ is given by

$$\tilde{S} = S \cdot E_1^{\mathcal{G}}(X)/E_1^{\mathcal{G}}(S^2) ,$$

where $\mathcal{G} = \mathcal{B}_S^n$. Moreover, the Pitman estimator is unique up to $P_1$-equivalence.

For the proof see Lemmas 1 and 2 in [22] or Theorem 3.3 in [19], respectively. Note that the assumptions required there in addition to ours are not used in the corresponding proofs.

LEMMA 1.5: Let $\mathcal{W} = \{P_\gamma | \gamma > 0\}$ be any scaling class. Then, $\tilde{S} \in \mathcal{R}_n \cap \mathcal{L}^2(P_1)$ is the Pitman estimator for $\gamma$ iff it holds true that

$$E_1(\tilde{S}S) = E_1(S) \qquad (S \in \mathcal{R}_n \cap \mathcal{L}^2(P_1)). \tag{1.3}$$

PROOF: First, let $\tilde{S}$ be the Pitman estimator. We write $\mathcal{G} = \mathcal{B}_S^n$ and $Z = E_1^{\mathcal{G}}(S)/E_1^{\mathcal{G}}(S^2)$. Furthermore, fix some $S \in \mathcal{R}_n \cap \mathcal{L}^2(P_1)$.
Using Lemma 1.4 and standard arguments for conditional expectations we obtain

$$\text{(i)} \quad \begin{aligned} E_1[(S-\tilde{S})(\tilde{S}-1)] &= E_1[S(1-Z)(\tilde{S}-1)] = \\ &= E_1\{(1-Z)E_1^{\mathcal{G}}[S(\tilde{S}-1)]\} = \\ &= E_1[(1-Z)E_1^{\mathcal{G}}(S^2Z-S)] = \\ &= E_1\{(1-Z)[ZE_1^{\mathcal{G}}(S^2)-E_1^{\mathcal{G}}(S)]\} = 0 \end{aligned}$$

where the last equality follows from the definition of Z. Applying Lemma 1.4 we obtain $E_1(\tilde{S}) = E_1(\tilde{S}^2)$. Therefore (i) establishes the validity of $E_1(\tilde{S}S) = E_1(S)$.

Now, let (1.3) be fulfilled. For $S \in \mathcal{R}_n \cap \mathcal{L}^2(P_1)$ it holds true that

$$\text{(ii)} \quad E_1[(S-1)^2] = E_1[(S-\tilde{S})^2] + 2E_1[(S-\tilde{S})(\tilde{S}-1)] + E_1[(\tilde{S}-1)^2] .$$

Since the application of (1.3) to S and $\tilde{S}$ yields

$$E_1[(S-\tilde{S})(\tilde{S}-1)] = 0 \qquad (S \in \mathcal{R}_n \cap \mathcal{L}^2(P_1)) ,$$

(ii) shows that $\tilde{S}$ is the Pitman estimator for $\gamma$. □

Lemma 1.5 corresponds with Lemma 1 in [5]. But note that the proof indicated there does not work in our situation, since for $\tilde{S}, S \in \mathcal{R}_n$ and

$\lambda < 0$ it does not hold that $\tilde{S} + \lambda S \in \mathcal{R}_n$, in general (due to our requirement (1.1) for equivariant statistics).

As a consequence of Theorem 1.2 we get

<u>THEOREM 1.6:</u> Let $\mathcal{W} = \{P_\gamma | \gamma > 0\}$ be any scaling class.

(a) If $S \in \mathcal{R}_n \cap L^2(P_1)$ is invariantly sufficient, then the statistic

$$\tilde{S} = c \cdot S \quad \text{with} \quad c = E_1(S)/E_1(S^2)$$

is the Pitman estimator.

(b) There exists essentially (i.e. up to $P_1$-equivalence and up to a constant factor) at most one invariantly sufficient statistic in $\mathcal{R}_n \cap L^2(P_1)$.

<u>PROOF:</u> Theorem 1.2 ensures the independence of $S^{-1}(\mathcal{B})$ and $\mathcal{G}$ which together with Lemma 1.4 yields claim (a). Statement (b) being an immediate consequence of part (a) and the uniqueness of the Pitman estimator, the theorem is proved. □

In the sequel $X_j: \mathbb{R}^n \to \mathbb{R}$ denotes the j-th projection ($j = 1,\ldots,n$). Our interest will focus on statistics $S = \sum_{j=1}^{n} c_j X_j^k$ with $k > 0$ and $c_j \geq 0$ ($j = 1,\ldots,n$). Evidently, for any such statistic the statistic $S^{1/k}$ is equivariant, if $P_1(\mathbb{R}_+^n) = 1$ and if at least one coefficient $c_{j_0}$ is positive. Later on we will restrict our considerations to linear statistics, i.e. to the case $k = 1$.

For $X_1,\ldots,X_n$ being i.i.d. in [23], [1], and [9], respectively, it was shown that $\bar{X} = \frac{1}{n} \sum_{j=1}^{n} X_j$ is sufficient iff $X_1,\ldots,X_n$ have a (possibly degenerate) gamma distribution. Theorems 1.7 and 1.9 extend this result. Note that Theorems 1.7 and 1.9 are related to Theorem 2 in [25], too. However, in the proof of this theorem a result in [15] is quoted whose proof is not stringent. Moreover, the degenerate case is there excluded in advance. Since from now on our considerations are confined to scaling classes with $P_1(\mathbb{R}_+^n) = 1$, the set $\mathbb{R}^n - \mathbb{R}_+^n$ will be ignored tacitly.

In the following, $G(\alpha,\xi)$ denotes the gamma distribution with the parameters $\alpha,\xi > 0$ ($\alpha$ being a scale parameter), i.e., $G(\alpha,\xi)$ is the probability measure (on $\mathcal{B}_+$) with the (Lebesgue) density given by

$$f(y) = f_{\alpha,\xi}(y) = \frac{\alpha^\xi}{\Gamma(\xi)} y^{\xi-1} e^{-\alpha y} \qquad (y \in \mathbb{R}_+) .$$

THEOREM 1.7: Let $\mathcal{W} = \{P_\gamma | \gamma > 0\}$ be a n-dimensional scaling class with $n \geq 2$ and $P_1(\mathbb{R}^n_+) = 1$ such that $X_1, \ldots, X_n$ are independent (under $P_1$). Furthermore, let the statistic $S = \sum_{j=1}^n c_j X_j^k$ with $k \in \mathbb{R}_+$ and $c_j \geq 0$ $(j = 1,\ldots,n)$ be invariantly sufficient. If there are at least two (different) $j_1, j_2 \in \{1,\ldots,n\}$ with $c_{j_1}, c_{j_2} > 0$, then either one of the following two statements holds true:

(a) For all $j \in \{1,\ldots,n\}$ with $c_j > 0$ the distribution of $X_j$ (under $P_1$) is degenerate;

(b) for all $j \in \{1,\ldots,n\}$ with $c_j > 0$ the distribution of $X_j^k$ (under $P_1$) is a gamma distribution $G(\alpha/c_j, \xi_j)$ with $\alpha, \xi_j > 0$.

PROOF: Assume w.l.o.g. that $c_1 > 0$ and that the distribution of $X_1$ is degenerate if there is any degenerate $X_j$ with $c_j > 0$. Writing $W = c_1 X_1^k$ and $V = S-W = \sum_{j=2}^n c_j X_j^k$, due to the assumptions W and V are independent. Further, since S and $S^{1/k}$ induce the same $\sigma$-algebra (over $\mathbb{R}^n_+$), the equivariant statistic $S^{1/k}$ is invariantly sufficient. Thus, due to Theorem 1.2 $S^{1/k}$ and W/V are independent which implies that S=W+V and W/V are independent, too.
If first the distribution of $X_1$ and therewith that of W is nondegenerate, then according to the above requirements the same holds true for V. Therefore, in this case W and V fulfill the conditions of a theorem in [28] (cf. also [13]), which in turn implies that W and V have gamma distributions with the same scale parameter. Now, if the distribution of $X_1$ and therewith that of W is degenerate, then the above stated independence of W+V and W/V yields the same for the distribution of V which implies that the distributions of all $X_j$ with $c_j > 0$ are degenerate. □

REMARKS 1.8: (a) Due to Theorem 1.7 it is clear that the distribution of the invariantly sufficient statistic S (under $P_1$) is a point mass or a gamma distribution, too.
(b) If in Theorem 1.7 the invariant sufficiency is replaced by sufficiency and if the degenerate case is excluded in advance, then the distributions of all $X_j^k$ $(j = 1,\ldots,n)$ are gamma distributions. In fact, in this case all coefficients $c_j$ $(j = 1,\ldots,n)$ are positive. For, if e.g. $c_1 = 0$, then the distributions $(P_\gamma)_S$ of $S = \sum_{j=2}^n c_j X_j^k$ are gamma di-

stributions obeying the overlapping property (i.e. for any $\gamma_1,\gamma_2 > 0$ and any $B \in \mathcal{B}$ with $(P_{\gamma_1})_S(B) = 1$ it follows that $(P_{\gamma_2})_S(B) > 0$). Thus, by a theorem in [3] the distributions $(P_\gamma)_{X_1}$ turn out being independent of $\gamma$.

THEOREM 1.9: Let $\mathcal{W} = \{P_\gamma | \gamma > 0\}$ be a n-dimensional scaling class with $n \geq 2$ and $P_1(\mathbb{R}_+^n) = 1$ such that $X_1,\ldots,X_n$ are independent (under $P_1$). Then the following statements are equivalent:

(I) For some $k>0$ there exists an invariantly sufficient statistic

$$S = \sum_{j=1}^{n} c_j X_j^k \quad \text{with} \quad c_j > 0 \ (j = 1,\ldots,n) \ ;$$

(II) for some $k > 0$ either the distributions of $X_j^k$ $(j = 1,\ldots,n)$ all are point masses or they all are gamma distributions.

If (I) and (therefore) (II) hold true, then (I) and (II) are satisfied for the same k, the statistic $S = \sum_{j=1}^{n} c_j X_j^k$ with

$$c_j = \begin{cases} E_1(X_j^k)/V_1(X_j^k) & \text{for } V_1(X_j^k) > 0 \quad (j = 1,\ldots,n) \\ 1 & \text{for } V_1(X_j^k) = 0 \quad (j = 1,\ldots,n) \end{cases}$$

is complete and sufficient and $S^{1/k}$ is the only (up to a positive factor and up to $P_1$-equivalence) invariantly sufficient equivariant statistic in $L^2(P_1)$.

PROOF: The implication (I) $\Rightarrow$ (II) is an immediate consequence of Theorem 1.7.
Considering for the converse first the degenerate case and assuming $P_1\{X_j^k = a_j\} = 1$ with $a_j \in \mathbb{R}_+$ $(1 \leq j \leq n)$ for some $k > 0$, we will establish the sufficiency and completeness of the statistic

$$S = \sum_{j=1}^{n} X_j^k .$$

To this end fix any $B \in \mathcal{B}_+^n$ and consider $g = 1_{(h\circ S)^{-1}(B)}$ with $h\colon \mathbb{R}_+ \to \mathbb{R}_+^n$ given by

$$h(y) = (a_1 y^{1/k},\ldots,a_n y^{1/k})\cdot(\sum_{j=1}^{n} a_j^k)^{-1/k} \qquad (y \in \mathbb{R}_+) .$$

Then, writing $\gamma a = (\gamma a_1,\ldots,\gamma a_n)$, for $D \in \mathcal{B}_+$ and $\gamma > 0$ it follows that

(i) $$\int_{S^{-1}(D)} 1_B \, dP_\gamma = 1_B(\gamma a)\cdot 1_{S^{-1}(D)}(\gamma a)$$

(ii) $$\int_{S^{-1}(D)} g \, dP_\gamma = 1_{(h\circ S)^{-1}(B)}(\gamma a)\cdot 1_{S^{-1}(D)}(\gamma a) .$$

It is readily verified that $\gamma a \in B$ holds iff $\gamma a \in (h \circ S)^{-1}(B)$. Thus, for all $D \in \mathcal{B}_+$ and all $\gamma > 0$ the right hand sides of (i) and (ii) coincide. Therefore, the same holds true for the left hand sides which yields the sufficiency of S.
For any function $f: \mathbb{R}_+ \to \mathbb{R}$ with $E_\gamma(f \circ S) = 0$ $(\gamma > 0)$ it follows that

$$E_\gamma(f \circ S) = f(\gamma \cdot \sum_{j=1}^{n} a_j) = 0 \qquad (\gamma > 0) .$$

Thus, S is complete.
If now $X_j^k$ is distributed under $P_1$ according to $G(\alpha_j, \xi_j)$ $(j = 1,\dots,n)$ for some $k > 0$, then it is well known that $\{Q_\gamma | \gamma > 0\}$ with $Q_\gamma = (P_\gamma)_{(X_1^k,\dots,X_n^k)}$ being the distribution of $(X_1^k,\dots,X_n^k)$ under $P_\gamma$ constitutes a 1-parameter exponential family in the statistic $\sum_{j=1}^{n} \alpha_j X_j$. From this it is deduced by elementary transformations of the densities under consideration that $\mathcal{W} = \{P_\gamma | \gamma > 0\}$ itself is a 1-parameter exponential family in $S = \sum_{j=1}^{n} \alpha_j X_j^k$. Therefore, in view of $\alpha_j = E_1(X_j^k)/V_1(X_j^k)$ $(j = 1,\dots,n)$ the proof for the nondegenerate case is accomplished. Finally, the claim concerning the uniqueness follows from Theorem 1.6 (b). □

<u>REMARK 1.10:</u> Theorem 1.9 shows that under the corresponding assumptions the only scaling classes which are at the same time 1-parameter exponential families in some statistic $\sum_{j=1}^{n} c_j X_j^k$ are those for which the distributions of $X_j^k$ $(j = 1,\dots,n)$ are gamma distributions.

## 2. CHARACTERIZATION BY OPTIMALITY PROPERTIES

This section is devoted to characterizations of the gamma distribution in scaling classes by optimality properties of certain estimators under the assumption that $X_1,\dots,X_n$ are independent (under $P_1$).

Basing on the set-up introduced in section I, for $j = 1,\dots,n$ we denote the distribution of $X_j$ under $P_1$ by $Q_j$, i.e. $Q_j = (P_1)_{X_j}$, and we assume throughout this section that $X_1,\dots,X_n$ are independent under $P_1$, i.e. $P_1 = \bigotimes_1^n Q_j$. Moreover, we write

$$\mu_{jk} = E_1(X_j^k) = \int y^k \, dQ_j(y) \qquad (j = 1,\dots,n;\ k \in \mathbb{N})$$

and assume that

$$0 < \sigma_j^2 = \mu_{j2} - \mu_{j1}^2 < \infty \qquad (j = 1,\dots,n) .$$

Then from Theorem 1.9 it is clear that there exists some linear statistic U being complete and sufficient w.r.t. $\mathcal{W}$ iff $Q_1,\ldots,Q_n$ are gamma distributions and that this statistic is given by

$$U = \sum_{j=1}^{n} c_j X_j \quad \text{with} \quad c_j = \mu_{j1}/\sigma_j^2 \quad (j = 1,\ldots,n) . \tag{2.1}$$

In this case by the Lehmann-Scheffé Theorem any statistic $g\circ U \in L^2(\mathcal{W}) := \bigcap_{\gamma>0} L^2(P_\gamma)$ is the minimum variance unbiased (MVU) estimator for $E_\gamma(g\circ U)$ w.r.t. $\mathcal{W}$. In the following theorem we state some converse which says that $Q_1,\ldots,Q_n$ turn out being gamma distributions, if some $R\circ U$ with R fulfilling a certain differentiability assumption is the MVU estimator for $E_\gamma(R\circ U)$.
The announced theorem generalizes Theorem 3 in [22], Theorem 1.3 in [18], Theorem 3.2 in [19], Theorem 1 in [24] (see also Theorem 7.13.1 in[21]), Theorem 1 in [5], and Theorem 1 in [11].

Inasmuch as $\tilde{U} = U/\tilde{c}$ with U given by (2.1) and with $\tilde{c} = \sum_{j=1}^{n} \mu_{j1}^2/\sigma_j^2$ is the MVU linear estimator for $\gamma$ w.r.t. $\mathcal{W}$ irrespective of the underlying distributions (cf. the proof of Theorem 7.12.2 in [21]), Theorem 2.1 may be viewed as a characterization of the MVU linear estimator, too.

<u>THEOREM 2.1:</u> Let $\mathcal{W} = \{P_\gamma \mid \gamma \in \mathbb{R}_+\}$ be a n-dimensional scaling class with $n \geq 2$, $P_1(\mathbb{R}_+^n) = 1$, $P_1 = \bigotimes_1^n Q_j$ and $0 < \sigma_j^2 < \infty$ $(1 \leq j \leq n)$. Let the statistic U be given by (2.1). Further, let $R: \mathbb{R} \to \mathbb{R}$ be a function, which is $m \geq 1$ times continously differentiable at the origin such that $R^{(j)}(o) = 0$ $(j = 1,\ldots,m-1)$, $R^{(m)}(o) \neq 0$, and such that there exist some $\lambda_o > 0$ and some $Y \in L^2(\mathcal{W}) = \bigcap_{\gamma>0} L^2(P_\gamma)$ with

$$|R(\lambda U)-R(0)|/\lambda^m \leq Y \quad \mathcal{W}\text{-a.e.} \qquad (\lambda \in (0,\lambda_o)). \tag{2.2}$$

Then the following two statements are equivalent:

(I) $Q_1,\ldots,Q_n$ are gamma distributions;

(II) $R\circ U$ is the MVU estimator for $E_\gamma(R\circ U)$ w.r.t. $\mathcal{W}$.

<u>REMARKS 2.2:</u> (a) If $R(t) = \sum_{j=o}^{\infty} a_j t^j$ is a power series not vanishing identically and converging on the whole real axis, then all the assumptions on R in Theorem 2.1 are satisfied, if it holds true that

(i) $\qquad \sum_{j=o}^{\infty} |a_j| U^j \in L^2(\mathcal{W})$ .

(Choose $m = \inf\{j \in \mathbb{N} \mid a_j \neq 0\}$, $\lambda_o = 1$ and $Y = \sum_{j=o}^{\infty} |a_j| U^j$.)

(b) If in the previous remark it holds that $\nu := \sup\{j \in \mathbb{N} | a_j \neq 0\} < \infty$, i.e., if R is a polynomial of degree $\nu$, then condition (i) is equivalent to

$$\mu_{j,2\nu} < \infty \qquad (j = 1,\dots,n).$$

We point out that herewith the question formulated at the end of [11] is answered in the affirmative.

For $\nu = \infty$ the validity of (i) is guaranteed by the conidition

$$\sum_{j=o}^{\infty} |a_j| E_1(U^{2j}) t^j < \infty \qquad (t \in \mathbb{R}_+) .$$

This follows from the relation

$$E_\gamma[(\sum_{j=o}^{\infty} |a_j| U^j)^2] = \sum_{j=o}^{\infty} \gamma^j E_1(U^j) \sum_{k=o}^{j} |a_k| \cdot |a_{j-k}| \le$$

$$\le [\sum_{j=o}^{\infty} |a_j| \gamma^j]^2 + [\sum_{j=o}^{\infty} |a_j| E_1(U^{2j}) \gamma^j]^2 \qquad (\gamma \in \mathbb{R}_+) ,$$

where the inequality is readily verified with the help of Hölder's inequality.

The proof for the implication (II) $\Rightarrow$ (I) will base heavily on the following lemma which represents a characterization of gamma distributions by a functional equation for the 'quotient functions' of its Mellin-Stieltjes transforms (MST-s). For any probability measure Q on $\mathcal{B}_+$ let

$$\varphi_Q(\xi) = \int x^{\xi-1} \, dQ(x) \tag{2.3}$$

denote its MST and

$$\varphi_{Q,j}(\xi) = \varphi_Q(\xi+j)/\varphi_Q(\xi) \tag{2.4}$$

with $j \in \mathbb{N}^o$ its quotient functions. Clearly, if $\int x^m dQ(x) < \infty$ for some $m \in \mathbb{N}$, then $\varphi_Q$ is well defined and finite on the strip $\{\zeta | 1 \le Re\,\xi \le 1+m\}$ and $\varphi_{Q,j}$ is well defined and finite at least on the real interval $[1,1+m-j]$.

<u>LEMMA 2.2:</u> Let two nondegenerate probability measures $Q_j$ on $\mathcal{B}_+$ with $\int x^m dQ_j(x) < \infty$ for some $m \in \mathbb{N}$ with $m \ge 2$ $(j=1,2)$ be given. If $\varphi$ and $\psi$ are the MST-s of $Q_1$ and $Q_2$, respectively, and $\varphi_j, \psi_j$ the corresponding quotient functions, then the following two statements are equivalent:

(I) $Q_1, Q_2$ are gamma distributions;

(II) there exist some regular matrix $C = (c_{jk})_{j,k=0,\dots,m-1}$ with (real) nonnegative entries where $c_{m-1,m-1} = 0$ as well as nondegenerate intervals $I_1, I_2 \subset (1,2)$ such that

$$\sum_{j=0}^{m-1} \sum_{k=0}^{m-1} c_{jk}\, \varphi_j(\xi_1)\, \psi_k(\xi_2) = h(\xi_1+\xi_2) \quad ((\xi_1,\xi_2) \in I_1 \times I_2) \tag{2.5}$$

defines a function h of $\xi_1 + \xi_2$, only.

For a proof of this lemma see Theorem 2 in [10].

PROOF OF THEOREM 2.1: We assume w.l.o.g. $R(0) = 0$; otherwise consider $\tilde{R} = R - R(0)$ instead of R.

From the differentiability assumptions on R we have by Taylor's formula

$$R(t) = \frac{t^m}{m!} R^{(m)}(0) + r_m(t) \qquad (t \in \mathbb{R})$$

with

$$\lim_{t \to 0} r_m(t)/t^m = 0.$$

Therefore, it holds

$$\lim_{\lambda \to 0} R(\lambda t)/\lambda^m = \frac{t^m}{m!} R^{(m)}(0) \qquad (t \in \mathbb{R}) .$$

Using (2.2), this relation implies with the aid of dominated convergence the validity of

(i) $\qquad U^m \in L^2(\mathcal{W})$

and of

(ii) $$\lim_{\lambda \to 0} \frac{1}{\lambda^m} E_\gamma[R(\lambda U) \cdot T] = \frac{1}{m!} R^{(m)}(0) E_\gamma(U^m T) \quad (\gamma \in \mathbb{R}_+;\ T \in L^2(\mathcal{W})).$$

To prove first the validity of the implication (I) ⇒ (II), let $Q_1, \dots, Q_n$ be gamma distributions. Then U is complete and sufficient. Additionally, by (i) it holds $U \in L^2(\mathcal{W})$. Therefore, it follows by the Lehmann-Scheffé Theorem that $R \circ U$ is the MVU estimator for $E_\gamma(R \circ U)$ w.r.t. $\mathcal{W}$, i.e., (II) is valid.

Conversely, let $R \circ U$ be the MVU estimator for $E_\gamma(R \circ U)$ w.r.t. $\mathcal{W}$. Obviously, by symmetry it suffices to show that $Q_1$ and $Q_2$ are gamma distributions.

For any $\xi_1, \xi_2, \eta_1, \eta_2 \in (0,1)$ with $\xi_1 + \xi_2 = \eta_1 + \eta_2$ we consider the estimator

$$T_{\xi_1,\xi_2,\eta_1,\eta_2} = X_1^{\xi_1}X_2^{\xi_2}/E_1(X_1^{\xi_1}X_2^{\xi_2}) - X_1^{\eta_1}X_2^{\eta_2}/E_1(X_1^{\eta_1}X_2^{\eta_2}) .$$

Having in mind that $X_1$ and $X_2$ have finite second moments under $P_1$ and that $X_1$ and $X_2$ are independent under $P_1$, it gets clear that $T_{\xi_1,\xi_2,\eta_1,\eta_2}$ are unbiased estimators for zero belonging to $L^2(\mathcal{W})$. Thus, by the celebrated MVU criterion of Lehmann-Scheffé ([27]) and Rao ([31]) it holds true that

$$\text{(iii)} \qquad E_\gamma[(R\circ U)\cdot T_{\xi_1,\xi_2,\eta_1,\eta_2}] = 0 \qquad (\gamma \in \mathbb{R}_+)$$

for all $\xi_1,\xi_2,\eta_1,\eta_2 \in (0,1)$ with $\xi_1+\xi_2 = \eta_1+\eta_2$. From (iii) we conclude that

$$\text{(iv)} \qquad \begin{aligned} &E_\gamma[R(\lambda U)\cdot T_{\xi_1,\xi_2,\eta_1,\eta_2}] = \\ &=\lambda^{-(\xi_1+\xi_2)}\cdot E_{\lambda\gamma}[(R\circ U)\cdot T_{\xi_1,\xi_2,\eta_1,\eta_2}] = 0 \quad (\lambda,\gamma \in \mathbb{R}_+)\end{aligned}$$

holds for all $\xi_1,\xi_2,\eta_1,\eta_2 \in (0,1)$ with $\xi_1+\xi_2 = \eta_1+\eta_2$, too. Combining (iii) and (iv) we obtain in spite of (ii) and $R^{(m)}(0) \neq 0$ the validity of

$$E_1[U^m\cdot T_{\xi_1,\xi_2,\eta_1,\eta_2}] = 0$$

for all $\xi_1,\xi_2,\eta_1,\eta_2 \in (0,1)$ with $\xi_1+\xi_2 = \eta_1+\eta_2$. Thus,

$$h(\xi_1+\xi_2) = E_1[U^m X_1^{\xi_1-1} X_2^{\xi_2-1}]/E_1[X_1^{\xi_1-1} X_2^{\xi_2-1}]$$

for all $\xi_1,\xi_2 \in (1,2)$ defines a function h which depends on $\xi_1+\xi_2$, only.

Denoting the MST-s of $Q_1$ and $Q_2$ by $\varphi$ and $\psi$, respectively, and its quotient functions by $\varphi_j$ and $\psi_j$ (see (2.3) and (2.4)), we get from (iv) after elementary transformations

$$\text{(v)} \qquad h(\xi_1+\xi_2) = \sum_{j=0}^{m}\sum_{k=0}^{m-j} \alpha_{jk}\varphi_j(\xi_1)\psi_k(\xi_2) \qquad (\xi_1,\xi_2 \in (1,2))$$

with

$$\text{(vi)} \qquad \alpha_{jk} = c_1^j c_2^k m!E_1[(U-c_1X_1-c_2X_2)^{m-j-k}]/[j!k!(m-j-k)!] \quad (j=0,\ldots,m;\ k=0,\ldots,m-j).$$

Now, we introduce the matrix $C = (c_{jk})_{j,k=0,\dots,m}$ with the entries

$$c_{jk} = \alpha_{jk} \quad (j = 0,\dots,m;\ k = 0,\dots,m-j), \quad c_{jk}=0 \text{ otherwise.}$$

From (vi) it is clear that $c_{j,m-j} > 0$ $(j = 0,\dots,m)$. Thus, C is a regular matrix with nonnegative entries such that $c_{mm} = 0$. Moreover, (v) may be rewritten as

$$h(\xi_1+\xi_2) = \sum_{j=0}^{m} \sum_{k=0}^{m} c_{jk}\varphi_j(\xi_1)\psi_k(\xi_2) \qquad (\xi_1,\xi_2 \in (1,2)) .$$

Therefore, Lemma 2.2 implies that $Q_1$ and $Q_2$ are gamma distributions. □

REMARK: From the proof of Theorem 2.1 it gets clear that the implication (II) ⇒ (I) retains its validity, if U is replaced by any linear statistic $V = \sum_{j=1}^{n} d_j X_j$ with $d_j > 0$ $(j = 1,\dots,n)$ such that (2.2) is fulfilled for V instead of U.

Next we give characterizations of the gamma distribution in scaling classes by the linearity of the Pitman estimator for $\gamma$ and of the MVU equivariant estimator as well as by the admissibility of linear statistics. We point out that $\hat{U} = U/(1+\tilde{c})$ (with U given by (2.1) and with $\tilde{c} = \sum_{j=1}^{n} \mu_{j1}^2/\sigma_j^2)$ is the best estimator for $\gamma$ among all estimators of the form dU with $d \in \mathbb{R}$ irrespective of the distributions $Q_1,\dots,Q_n$. Clearly, $\hat{U}$ is not unbiased.

First we provide a lemma which ensures the equivalence of statements (III) and (V) in Theorem 2.4 below.

LEMMA 2.3: If $\mathcal{W}$ is a n-dimensional scaling class and $\tilde{S}$ is the Pitman estimator for $\gamma$, then there is a constant d such that $d\tilde{S}$ is the MVU equivariant estimator for $\gamma$.

This lemma corresponds to Lemma 2 in [5] (without our condition (1.1) for equivariant statistics). The elementary, short proof can be accomplished in the same way by using Lemma 1.4 instead of Lemma 1 of [5].

THEOREM 2.4: Let $_n\mathcal{W}$ be a n-dimensional scaling class with $n \geq 3$, $P_1(\mathbb{R}_+^n) = 1$, $P_1 = \bigotimes_1^n Q_j$ and $0 < \sigma_j^2 < \infty$ $(j = 1,\dots,n)$. Then the following statements are pairwise equivalent:

(I) $Q_1,\ldots,Q_n$ are gamma distributions;

(II) $\tilde{U} = U/\tilde{c}$ is admissible among all unbiased estimators for $\gamma$;

(III) $\tilde{U} = U/\tilde{c}$ is the MVU equivariant estimator for $\gamma$;

(IV) $\hat{U} = U(1+\tilde{c})$ is absolutely admissible (i.e. admissible among all estimators for $\gamma$);

(V) $\hat{U} = U/(1+\tilde{c})$ is the Pitman estimator for $\gamma$.

Clearly, if the MVU equivariant estimator is linear, then it must be given by $\tilde{U}$. Correspondingly, if the Pitman estimator or the absolutely admissible estimator is linear, then it is necessarily given by $\hat{U}$. Thus, statements (II) to (V) can be reformulated e.g. like 'the Pitman estimator for $\gamma$ is linear' instead of (V).

PROOF: If (I) is valid, then $\tilde{U}$ is the MVU estimator for $\gamma$ which evidently implies (II). By the equivariance of $\tilde{U}$, from (II) follows (III). The proof for the implication (I) ⇒ (IV) may be accomplished by means of the method developed in [17], see the corresponding part of the proof of Theorem 7.12.1 in [21]. Again, by the equivariance of $\tilde{U}$ the validity of (IV) yields that of (V). It remains to establish the implications (III) ⇒ (I) and (V) ⇒ (I). Since it gets clear with the aid of Lemma 2.3 that (III) and (V) are equivalent, it suffices to settle (V) ⇒ (I).

To this end, let $\hat{U}$ be the Pitman estimator. We show w.l.o.g. that $Q_1$ and $Q_2$ are gamma distributions. We consider for any $\zeta_1,\zeta_2 \in [0,1/2]$ the estimator

$$T_{\zeta_1,\zeta_2} = X_1^{\zeta_1} X_2^{\zeta_2} X_3^{1-\zeta_1-\zeta_2}$$

which evidently is equivariant and square integrable w.r.t. $P_1$, i.e. $T_{\zeta_1,\zeta_2} \in \mathcal{R}_n \cap L^2(P_1)$. Thus, denoting the MST-s of $Q_1,Q_2,Q_3$ (the distributions of $X_1,X_2,X_3$ under $P_1$) by K,L,M and having in mind that $\tilde{c} = \sum_{j=1}^{n} \mu_{j1}^2/\sigma_j^2$, we obtain as an application of Lemma 1.5

$$\text{(i)}\qquad c_1 \frac{K(\zeta_1+1)}{K(\zeta_1)} + c_2 \frac{L(\zeta_2+1)}{L(\zeta_2)} = 1 + \sum_{j=1}^{3} \mu_{j1}^2/\sigma_j^2 - c_3 \frac{M(5-\zeta_1-\zeta_2)}{M(4-\zeta_1-\zeta_2)}$$

$$(\zeta_1,\zeta_2 \in [1,3/2])\ .$$

Since relation (i) is of the form (2.5) for $m = 2$, $C = (c_{jk})_{j,k=0,1}$ with $c_{01} = c_2 > 0$, $c_{10} = c_1 > 0$, $c_{00} = c_{11} = 0$ and $I_1 = I_2 = [1,3/2]$, it follows by Lemma 2.2 that $Q_1$ and $Q_2$ are gamma distributions. We mention that this could also be concluded by means of Lemma 1 in [26] (and the corollary following it). (Note that by relation (i) the functions $K_1(\zeta) = K(\zeta+1)/K(\zeta)$ and $L_1(\zeta) = L(\zeta+1)/L(\zeta)$ as solutions of Pexider's functional equation are linear.) This completes the proof of the theorem. □

For the i.i.d. case (i.e. $Q_1 = \dots = Q_N$) the equivalences (I) ⇔ (V) and (I) ⇔ (III) in Theorem 2.4 correspond to Theorems 2 and 3 in [5]. Moreover, the equivalences (I) ⇔ (IV) and (I) ⇔ (II) extend Theorems 7.12.1 and 7.12.2 in [21], where the optimality properties are required for two sample sizes $n_2 > n_1 \geq 3$. Whence Theorem 2.4 verifies the conjecture of Kagan-Linnik-Rao that this condition may be avoided (see the remark which follows Theorem 7.12.2). We mention that Theorem 2.4 retains its validity if in the definition of an equivariant statistic the condition (1.1) of positiveness is left. Moreover, we point out that some of the equivalences in Theorem 2.4 hold true for $n = 2$, too.

## 3. EXTENSION OF CHARACTERIZATIONS TO LINEAR PROCESSES

This section is devoted to the extension of characterizations for the independent case established so far to a certain linear framework.

Throughout this section, let $\mathcal{W} = \{P_\gamma | \gamma > 0\}$ be a n-dimensional scaling class with $n \geq 3$ and $P_1(\mathbb{R}^n_+) = 1$. Instead of requiring $X_1,\dots,X_n$ to be independent under $P_1$, let be given a regular n×n-matrix A with nonnegative entries and let $Z = (Z_1,\dots,Z_n)$ be given by $Z = XA$ with $X = (X_1,\dots,X_n)$ such that $Z_1,\dots,Z_n$ are independent and nondegenerate under $P_1$. We shall demonstrate how most of the characterizations presented so far may be carried over to this set-up. The following lemma provides a general unified procedure for reducing the situation now under consideration to the independent case dealt with previously. However, the assumptions required above do not enter into this lemma.

LEMMA 3.1: Let $\mathcal{W} = \{P_\gamma | \gamma > 0\}$ be a n-dimensional scaling class. Moreover, let $T: (\mathbb{R}^n, \mathcal{B}^n) \to (\mathbb{R}^n, \mathcal{B}^n)$ be any one-to-one linear transformation. Writing $\tilde{P}_1 = (P_1)_T$, $\tilde{\mathcal{W}} = \{\tilde{P}_\gamma | \gamma > 0)$ with $\tilde{P}_\gamma = (\tilde{P}_1)_{S_{n,\gamma}}$ $(\gamma > 0)$ and $\mathcal{W}_T = \{(P_\gamma)_T | \gamma > 0\}$ it holds true that:

(I) $T \circ S_{n,\gamma} = S_{n,\gamma} \circ T \qquad (\gamma > 0)$ ;

(II) $\tilde{\mathcal{W}} = \mathcal{W}_T$, i.e. $\tilde{P}_\gamma = (P_\gamma)_T \qquad (\gamma > 0)$;

(III) if $S: (\mathbb{R}^n, \mathcal{B}^n) \to (\mathbb{R}, \mathcal{B})$ is invariantly sufficient w.r.t. $\tilde{\mathcal{W}}$, then $S \circ T$ is invariantly sufficient w.r.t. $\mathcal{W}$.

We point out that one gets dual statements by interchanging the roles of $P_1$ and $\tilde{P}_1$ and replacing T by its inverse $T^*: \mathbb{R}^n \to \mathbb{R}^n$ .

PROOF: Evidently, claim (I) is an immediate consequence of the linearity of T. With the aid of (I) we obtain

$$(P_\gamma)_T = ((P_1)_{S_{n,\gamma}})_T = ((P_1)_T)_{S_{n,\gamma}} = \tilde{P}_\gamma \qquad (\gamma > 0) ,$$

which proves (II). To verify claim (III), let $S: (\mathbb{R}^n, \mathcal{B}^n) \to (\mathbb{R}, \mathcal{B})$ be invariantly sufficient w.r.t. $\mathcal{W}$ and fix some $G \in \mathcal{B}^n_S$. Then, using (I) it follows that we have $T(G) \in \mathcal{B}^n_S$. In view of the invariant sufficiency of S w.r.t. $\tilde{\mathcal{W}}$ there exists a function $h = h_{T(G)}: (\mathbb{R}, \mathcal{B}) \to (\mathbb{R}, \mathcal{B})$ such that

(i) $$\int_{S^{-1}(D)} h \circ S \, d\tilde{P}_\gamma = \int_{S^{-1}(D)} 1_{T(G)} \, d\tilde{P} \qquad (D \in \mathcal{B};\ \gamma > 0) .$$

By means of (II) it follows that

(ii) $$\int_{(S \circ T)^{-1}(D)} h \circ (S \circ T) \, dP_\gamma = \int_{S^{-1}(D)} h \circ S \, d\tilde{P}_\gamma \qquad (D \in \mathcal{B};\ \gamma > 0)$$

and

(iii) $$\int_{(S \circ T)^{-1}(D)} 1_G \, dP_\gamma = \int_{S^{-1}(D)} 1_G \circ T^* \, d\tilde{P}_\gamma =$$

$$= \int_{S^{-1}(D)} 1_{T(G)} \, d\tilde{P}_\gamma \qquad (D \in \mathcal{B};\ \gamma > 0) .$$

Since $G \in \mathcal{B}^n_S$ was chosen arbitrarily, (i), (ii), and (iii) imply the invariant sufficiency of $S \circ T$ w.r.t. $\mathcal{W}$. □

In the following theorem we summarize extensions of previously stated characterizations which may be carried out with the help of Lemma 3.1.

THEOREM 3.2: If under the linear framework introduced above it holds $\mu_k = E_1(Z_k) < \infty$ and $\sigma_k^2 = V_1(Z_k) < \infty$ $(k = 1, \ldots, n)$, then with

$U = \sum_{j=1}^{n} d_j X_j$ where $d_j = \sum_{k=1}^{n} a_{jk}\mu_k/\sigma_k^2$ and with $\tilde{d} = \sum_{k=1}^{n} \mu_k^2/\sigma_k^2$ the following statements are (pairwise) equivalent:

(I) $Z_1,\ldots,Z_n$ have gamma distributions;

(II) U is invariantly sufficient (w.r.t. $\mathcal{W}$);

(III) $\tilde{U} = U/\tilde{d}$ is the MVU estimator for $\gamma$;

(IV) $\tilde{U} = U/\tilde{d}$ is admissible among all unbiased estimators for $\gamma$;

(V) $\tilde{U} = U/\tilde{d}$ is the MVU equivariant estimator for $\gamma$;

(VI) $\hat{U} = U/(1+\tilde{d})$ is absolutely admissible;

(VII) $\hat{U} = U/(1+\tilde{d})$ is the Pitman estimator for $\gamma$.

As before statements (II) to (VII) may be reformulated by saying that there exists some linear statistic obeying the corresponding property.

PROOF: Exemplarily we demonstrate how Lemma 3.1 is to be applied by verifying the equivalence of (I) and (VII). The validity of the other implications may be proved similarly.
We show that $\hat{U}$ is the Pitman estimator for $\gamma$ w.r.t. $\mathcal{W}$ iff

$$\hat{U}\circ Z^* = \frac{1}{1+\tilde{d}} \sum_{j=1}^{n} d_j \sum_{k=1}^{n} a^*_{kj} X_k =$$

$$= \frac{1}{1+\tilde{d}} \sum_{k=1}^{n} X_k \sum_{m=1}^{n} \mu_m/\sigma_m^2 \sum_{m=1}^{n} a^*_{kj} a_{jm} =$$

$$= \frac{1}{1+\tilde{d}} \sum_{k=1}^{n} X_k \mu_k/\sigma_k^2$$

($A^{-1} = (a^*_{jk})$ denotes the inverse of A) is the Pitman estimator for $\gamma$ w.r.t. $\tilde{\mathcal{W}} = \mathcal{W}_Z$ (see Lemma 3.1). Since $Z_1,\ldots,Z_n$ are independent under $P_1$, it follows that $X_1,\ldots,X_n$ are independent under $\tilde{P}_1 = (P_1)_Z$. Thus, the scaling class $\tilde{\mathcal{W}}$ fulfills the assumptions of Theorem 2.4 and it then follows that $\hat{U}$ is the Pitman estimator for $\gamma$ w.r.t. $\mathcal{W}$ iff $X_1,\ldots,X_n$ have gamma distributions under $\tilde{P}_1$, i.e. iff $Z_1,\ldots,Z_n$ have gamma distributions under $P_1$.
First we note that for a statistic S being equivariant w.r.t. $\tilde{\mathcal{W}}$ ($\mathcal{W}$) the statistic $S\circ Z$ ($S\circ Z^*$) is equivariant w.r.t. $\mathcal{W}$ ($\tilde{\mathcal{W}}$).
Now, if $\hat{U}$ is the Pitman estimator for $\gamma$ w.r.t. $\mathcal{W}$, then $\hat{U}\circ Z^* \in L^2(P_1)$ holds and by means of Lemma 1.5 we obtain for any statistic $S \in L^2(\tilde{P}_1)$

being equivariant w.r.t. $\widetilde{\mathcal{W}}$

$$\int (\hat{U} \circ Z^*) \cdot S \, d\widetilde{P}_1 = \int \hat{U} \cdot (S \circ Z) dP_1 = \int S \circ Z \, dP_1 = \int S \, d\widetilde{P}_1$$

which again by Lemma 1.5 implies that $\hat{U} \circ Z^*$ is the Pitman estimator for $\gamma$ w.r.t. $\widetilde{\mathcal{W}}$. The opposite conclusion may be shown analogously. □

We point out that Theorem 2.1 may be extended to the linear framework considered in this section, correspondingly. Under the assumptions and with the notation of Theorem 3.2 the extension of Theorem 2.1 is quite the same like Theorem 2.1. The proof may be carried out by means of Lemma 3.1 in an obvious way.

<u>EXAMPLE 3.3:</u> The processes chosen for illustration are fundamental in time series analysis and are not covered by the ordinary independent case.

First, fix $1 \leq p \leq n$ and let X with XA = Z be an autoregressive process of order p where $Z = (Z_1, \ldots, Z_n)$ fulfills the assumptions encountered at the beginning of this section. Then A is given by

$$A = \begin{bmatrix} a_o & a_1 & \cdot & \cdot & a_p & & & 0 \\ & a_o & a_1 & \cdot & \cdot & a_p & & \\ & & \cdot & \cdot & \cdot & \cdot & \cdot & \\ & & & \cdot & \cdot & \cdot & \cdot & a_p \\ & & & & \cdot & \cdot & \cdot & \cdot \\ & 0 & & & & \cdot & \cdot & \cdot \\ & & & & & & a_o & a_1 \\ & & & & & & & a_o \end{bmatrix} \tag{3.1}$$

with $a_o, a_p \neq 0$. Clearly, A is regular. Thus, the characterizations summarized in Theorem 3.2 can be applied to autoregressive processes of order p, if the structural constants $a_o, a_1, \ldots, a_p$ all are nonnegative.

For the application of the characterizations of this section to moving average processes of order q (i.e. X = ZB where B is of the form (3.1) with $a_o, \ldots, a_p$ being replaced by $b_o, \ldots, b_q$) and to (p,q)-ARMAs (i.e. XD = ZB where B and D both are of the form (3.1)) we only mention that they both may be represented as autoregressive processes (of order $< n$) where the structural constants may be determined recursively as the solutions of some specific difference equations (cf. e.g. [14]).

Finally, we indicate that a quite analogous procedure as presented here for scaling classes may be applied to extend characterizations of the normal distribution in translation classes from the independent case to the linear set-up under consideration. In that way the proofs of Theorem 7.8.1 in [21] and in section II.1 of [29] for autoregressive processes as well as the proofs in section 2 of [8] are simplified and shortened substantially.

The only fact one has to take care of, additionally, is that (I) of Lemma 3.1 remains valid for translation classes $\mathcal{W} = \{P_\gamma \mid \gamma \in \mathbb{R}\}$ and arbitrary translations $T_\gamma = \gamma + X$ ($\gamma \in \mathbb{R}$) solely provided that the linear transformation T is performed by a matrix A with $a_{.k} = \sum_{j=1}^{n} a_{jk} = 1$ ($k = 1,\dots,n$). Clearly, in this case for the inverse $A^{-1} = (a^*_{jk})$ it holds $a^*_{.k} = 1$ ($k = 1,\dots,n$), too. The general case $Z = XA$ with $a_{.k} \neq 0$ ($k = 1,\dots,n$), cf. (2.3) in [8], is easily reduced to $a_{.k} = 1$ ($k = 1,\dots,n$) by considering the matrix $\tilde{A} = (\tilde{a}_{jk})$ with the entries $\tilde{a}_{jk} = a_{jk}/a_{.k}$ ($j,k = 1,\dots,n$) and $\tilde{Z} = X\tilde{A}$. Evidently, the required assumptions as well as the normality are not influenced by this normalization.

## REFERENCES


[1] BARTFAI,P.:*Characterizations by sufficient statistics.* Colloquia Mathematica Societatis János Bolyai 21; Analytic Function Methods in Probability Theory, Debrecen, ed. by B.Gyires (1980),15-19

[2] BASU,D.:*On statistics independent of a complete sufficient statistic.* Sankhya 15 (1955),377-380

[3] BASU,D.:*On statistics independent of sufficient statistics.* Sankhya 20 (1958),223-226

[4] BERK,R.H.:*A note on sufficiency and invariance.* Ann. Math. Statist. 43 (1972),647-650

[5] BONDESSON,L.:*Characterizations of the gamma distribution.* Theory Probab. Appl. 18 (1973),367-369

[6] BONDESSON,L.:*A note on sufficiency and independence.* Preprint, University of Lund, Lund (1977)

[7] EATON,M.L.:*Multivariate Statistics, a Vector Space Approach.* Wiley, New York (1983)

[8] EBERL,W.,jr.:*Invariantly sufficient equivariant statistics and characterizations of normality in translation classes.* Ann. Statist. 11 (1983),330-336

[9] EBERL,W.,jr.:*A note on characterization by sufficiency of the sample mean*. Publ. Math. Debrecen 30 (1983),89-91

[10] EBERL,W.,jr.:*A functional equation for Mellin-Stieltjes transforms of gamma distributions*. J. Math. Anal. Appl. 16 (1986),538-552

[11] EBERL,W.,jr.:*A characterization of the gamma distribution by optimal estimates*. Theory Probab. Appl.30 (1986),855-860 (English translation of Характеризация гамма-распределения оптимальными оценками. Теор. вероятност. и применен. 30 (1985),804-810)

[12] EBERL,W., O.MOESCHLIN:*Mathematische Statistik*. De Gruyter, Berlin (1982)

[13] FINDEISEN,P.:*A simple proof of a classical theorem which characterizes the gamma distribution*. Ann. Statist. 6 (1978), 1165-1167

[14] FULLER,W.A.:*Introduction to Statistical Time Series*. Wiley, New York (1976)

[15] GHURYE,S.G.:*Note on sufficient statistics and two-stage procedures*. Ann. Math. Statist. 29 (1958),155-166

[16] HALL,W., R.A.WIJSMAN, J.K.GHOSH:*The relationship between sufficiency and invariance with applications in sequential analysis*. Ann. Math. Statist. 36 (1965),575-614

[17] HODGES,J.L., E.L.LEHMANN:*Some applications to the Cramér-Rao inequality*. Proc. 2nd Berk. Symp. Math. Statist. Probab. 1, Univ. California, Berkeley and Los Angelos (1951), 13-22

[18] KAGAN,A.M.:*Some analytical aspects of estimation theory*. Selected Statistical Papers 1, Mathematical Centre Tracts 26, Amsterdam (1968),7-14

[19] KAGAN,A.M.:*Estimation theory for families with location and scale parameters, and for exponential families*. Proc. Steklov Inst. Math. 104 (1968),21-103 (English translation of Теория оценивания для семейств с параметрами сдвига, масштаба и экспонентных. Труды мат. инст. Стеклов. 104 (1968),19-87

[20] KAGAN,A.M.:*Partial sufficiency of linear forms*. Stability Problems for Stochastic Models (Proceedings of the 8th International Seminar held in Uzhgorod,USSR,Sep.23-29, 1984), Lecture Notes in Mathematics, Springer, Berlin (1985),60-80

[21] KAGAN,A.M., YU.V.LINNIK, C.R.RAO:*Characterization Problems in Mathematical Statistics*. Wiley, New York (1973) (English translation of Характеризационные задачи математической статистики. Наука, Москва (1972))

[22] KAGAN,A.M., A.L.RUKHIN:*On the estimation of a scale parameter*. Theory Probab. Appl. 12 (1968),672-678 (English translation of К теории оценивания параметра масштаба. Теор. вероятност. и применен. 12 (1967),735-741

[23] KAGAN,F.M.:*Some theorems on the characterization of the gamma distribution and those similar to it.* Selected Transl. in Math. Statist. Probab. 11 (1973),219-234 (English translation of Некоторые теоремы о характеризации гамма-распределения и близких к нему. Литовск. мат. сб. 8 (1968), 265-278)

[24] KAGAN,F.M.: Условия оптимальности некоторых оценок для семейств с параметром масштаба. ДАН УзбССР 6 (1968),3-5 (in Russian)

[25] KELKER,D., T.K.MATTHES:*A sufficient statistics characterization of the normal distribution.* Ann. Math. Statist. 41 (1970), 1086-1090

[26] KHATRI,C.G., C.R.RAO:*Some characterizations of the gamma distribution.* Sankhya, Ser.A 30 (1968),157-166

[27] LEHMANN,E.L., H.SCHEFFÉ:*Completeness, similar regions and unbiased estimation - part* I. Sankhya 10 (1950),305-340

[28] LUCACS,E.:*A characterization of the gamma distribution.* Ann. Math. Statist. 26 (1955),319-324

[29] PHAM, NGOC PHUC:*Autoregressziós típusú Gauss-folyamatok néhány jellemzési problémájáról.* Alkalmazott Matematikai Lapok 5 (1979),303-374 (in Hungarian)

[30] PITMAN,E.J.G.:*The estimation of location and scale parameters of a continuous population of any given form.* Biometrika 30 (1938),391-421

[31] RAO,C.R.:*Some theorems on minimum variance estimation.* Sankhya 12 (1952),27-42

# Efficient Sequential Estimation for an Exponential Class of Processes

Jürgen Franz and Wolfgang Winkler
Sektion Mathematik
Technische Universität Dresden
DDR-8027 Dresden, GDR

SUMMARY: In this paper a general exponential class of random processes is introduced on the basis of a special exponential form of the likelihood function. Many widely used models for Markov processes are of this type. Martingale properties are proved for the corresponding score process. Sequential estimation procedures based on a finite stopping time $\tau$ are considered. A Cramer-Rao inequality is given in the sequential case and the efficiency of sequential estimators is discussed. As an application special results are given for Poisson branching processes.

## 1. EXPONENTIAL CLASS OF RANDOM PROCESSES

Let $X = \{X(t),\ t \geq 0\}$ be a time-continuous Markov process defined on $[\Omega, \mathfrak{A}, \mathfrak{P}]$ with paths in $D^m[0, \infty)$, i.e. X is m-dimensional with values in $[R^m, \mathfrak{B}^m]$ and almost all paths are right-continuous functions having limits from the left. A filtration in $\mathfrak{A}$ is denoted by $\mathfrak{F} = \{\mathfrak{F}_t,\ t \geq 0\}$. We assume that $\mathfrak{F}_t = \sigma\{X(s),\ s \leq t\}$ is the $\sigma$-field generated by the random variables $X(s)$, $0 \leq s \leq t$, and $\mathfrak{A} = \mathfrak{F}_\infty = \sigma\{\mathfrak{F}_t,\ t \geq 0\}$. A random variable $\tau$ defined on $[\Omega, \mathfrak{F}_\infty]$ with values in $[0, \infty]$ is called a $\mathfrak{F}$-stopping time, if $\{\omega: \tau(\omega) \leq t\} \in \mathfrak{F}_t$ holds for every $t \geq 0$. The $\sigma$-field of the $\tau$-past of the process X is then denoted by $\mathfrak{F}_\tau = \{F \in \mathfrak{F}_\infty: F \cap \{\tau \leq t\} \in \mathfrak{F}_t,\ t \geq 0\}$.

The family of probability distributions $\mathfrak{P}$ is assumed to be parametrized, $\mathfrak{P} = \{P_\theta,\ \theta \in \Theta\}$, with unknown parameter $\theta \in \Theta$, where $\Theta$ is an open k-dimensional interval in $R^k$. The measures restricted to $\mathfrak{F}_t$ are denoted by $P_{\theta,t} = P_\theta | \mathfrak{F}_t$. We suppose that the following condition holds

C 1: For every $\theta_1, \theta_2 \in \Theta$, $\theta_1 \neq \theta_2$, the probability measures $P_{\theta_1,t}$ and $P_{\theta_2,t}$ are equivalent for all $t \geq 0$.

Contributions to Stochastics.
Ed. by W. Sendler
© Physica-Verlag Heidelberg 1987

Then Radon-Nikodym-derivatives exist and for fixed $\Theta_0 \in \Theta$ the likelihood function of the process X observed up to t is given by

$$L_\Theta(t) = \frac{dP_{\Theta,t}}{dP_{\Theta_0,t}}, \quad \Theta \in \Theta, \quad 0 \leqq t < +\infty. \tag{1}$$

In the following we assume $0 \in \Theta$ and put $\Theta_0 = 0$.

Now we introduce a general exponential class of random processes. For this purpose we consider processes $M = \{M(t),\ t \geqq 0\}$, where $M(t) = (M_1(t),\dots,M_k(t))^T$, and $V_\Theta = \{V_\Theta(t),\ t \geqq 0\}$, $\Theta \in \Theta$, satisfying the following conditions:

C 2: For each $\Theta \in \Theta$, M is a $(\mathfrak{F}, P_\Theta)$-submartingale (i.e. M is a $\mathfrak{F}$-adapted submartingale with respect to $P_\Theta$) with paths in $D^k[0, \infty)$.

C 3: For each $\Theta \in \Theta$, $V_\Theta$ is a non-decreasing $\mathfrak{F}$-adapted process with paths in $D^1[0, \infty)$. Moreover, the $V_\Theta(t)$ are twice continuously differentiable with respect to the components of $\Theta$ and $V_\Theta(0) = 0$ a.s. $[P_\Theta]$, $V_0(t) = 0$ a.s. $[P_0]$ for all $t \geqq 0$.

Definition: The process X belongs to the (M,V)-exponential family, if there exist processes M and $V_\Theta$ satisfying the conditions C 2 and C 3, respectively, such that the likelihood function of X can be represented in the form

$$L_\Theta(t) = \exp\{\Theta^T M(t) - V_\Theta(t)\}, \quad \Theta \in \Theta, \quad 0 \leqq t < +\infty. \tag{2}$$

The set of all processes X belonging to a (M,V)-exponential family is called the exponential class.

Together with the likelihood function we consider the score function

$$U_\Theta(t) = \frac{d}{d\Theta}(\ln L_\Theta(t)) = M(t) - \dot{V}_\Theta(t). \tag{3}$$

Here, differentiation of a function $f(\Theta)$ is denoted by $\frac{d}{d\Theta} f(\Theta) = \dot{f}(\Theta)$, i.e. $\dot{f}(\Theta) = (\frac{\partial f}{\partial \Theta_1}, \dots, \frac{\partial f}{\partial \Theta_k})^T$, and $\ddot{f}(\Theta) = (\frac{\partial^2 f(\Theta)}{\partial \Theta_i \partial \Theta_j})$ is the matrix of the second derivatives. The score function leads to maximum likelihood estimates which are solutions of the likelihood equation $M(t) = \dot{V}_\Theta(t)$. It is known that the likelihood process $L_\Theta = \{L_\Theta(t),\ t \geqq 0\}$ is a $(\mathfrak{F}, P_0)$-martingale with mean one. By a remark of Feigin [3], the score function $U_\Theta$ forms a zero-mean martingale, if it is possible to interchange integration and differentiation with respect to $\Theta$. In the following theorems we state martingale properties under the condition that the likelihood function is smooth and certain moments related to the processes M and $V_\Theta$ exist.

Theorem 1: Let X be a process of a (M,V)-exponential family. Suppose that

(i) For every $\Theta \in \Theta$ there exist right-sided and left-sided neighbourhoods in which $L_\Theta(t)$ is a convex or concave function of $\Theta$.

(ii) For every $\Theta \in \Theta$ , $t \geqq 0$ it holds $E_\Theta \left|\frac{\partial}{\partial \Theta_i} V_\Theta(t)\right| < +\infty$, $i = 1,2,\ldots,k$.

Then $U_\Theta = \{U_\Theta(t),\ t \geqq 0\}$ is a $(\mathfrak{F}, P_\Theta)$-martingale with mean zero.

Proof: In order to prove the martingale property we restrict our consideration to the first component $U_\Theta^{(1)}(t) = \frac{\partial}{\partial \Theta_1} \ln L_\Theta(t)$ of $U_\Theta(t)$. We assume $\Theta_1 \in (\Theta_{1,u}, \Theta_{1,o})$. Let $\Theta_{1,n} \in (\Theta_{1,u}, \Theta_{1,o})$ be a sequence tending to $\Theta_1$ as $n \to \infty$. Using the notation $\Theta(\Theta_{1,\circ}) = (\Theta_{1,\circ}, \Theta_2, \ldots, \Theta_k)^T$ and $D_n(\Theta_1) = (\Theta_{1,n} - \Theta_1)^{-1} (L_{\Theta(\Theta_{1,n})}(t) - L_\Theta(t))$ it holds $\lim_{n \to \infty} D_n(\Theta_1) = \frac{\partial}{\partial \Theta_1} L_\Theta(t)$. We consider the case that $L_\Theta(t)$ is convex in a right-sided interval $[\Theta_1, \Theta_{1,o})$ and a sequence $\Theta_{1,n} \downarrow \Theta_1$. Then, we have

$$\frac{\partial}{\partial \Theta_1} L_\Theta(t) \leqq D_n(\Theta_1) \leqq \frac{\partial}{\partial \Theta_1} L_{\Theta(\Theta_{1,*})}(t),$$

where $\Theta_{1,*} \in [\Theta_1, \Theta_{1,o})$ and from the Lebesgue-Fatou lemma it follows

$$\frac{\partial}{\partial \Theta_1} \int_F L_\Theta(t) dP_{0,t} = \int_F \frac{\partial}{\partial \Theta_1} L_\Theta(t)\, dP_{0,t}\ ,\quad F \in \mathfrak{F}_t. \tag{4}$$

All other cases lead to the same conclusion. Now, let $s \leqq t$ and $F \in \mathfrak{F}_s$. Using (4) we obtain

$$\int_F U_\Theta^{(1)}(t) dP_{\Theta,t} = \int_F \frac{\partial}{\partial \Theta_1} L_\Theta(t) dP_{0,t} = \frac{\partial}{\partial \Theta_1} \int_F L_\Theta(t) dP_{0,t} = \frac{\partial}{\partial \Theta_1} P_{\Theta,t}(F)$$

$$= \frac{\partial}{\partial \Theta_1} P_{\Theta,s}(F) = \int_F \frac{\partial}{\partial \Theta_1} L_\Theta(s) dP_{0,s} = \int_F U_\Theta^{(1)}(s) dP_{\Theta,s}$$

and hence, the martingale property of $U_\Theta$ is valid. Putting $F = \Omega$ yields $E_\Theta\, U_\Theta(t) = 0$.

In a similar way one can state the martingale property of the process $J_\Theta = \{J_\Theta(t),\ t \geqq 0\}$ with the random matrices $J_\Theta(t) = U_\Theta(t) U_\Theta^T(t) - \ddot{V}_\Theta(t)$:

Theorem 2: Let X be a process of a (M,V)-exponential family. Suppose that

(i) For every $\Theta \in \Theta$ there exist right-sided and left-sided neighbourhoods in which $L_\Theta(t) U_\Theta^{(i)}(t)$ is a convex or concave function of $\Theta$, $i = 1,2,\ldots,k$.

(ii) For every $\Theta \in \Theta$, $t \geqq 0$ it holds $E_\Theta M_i^2(t) < +\infty$,

$$E_\theta\left(\frac{\partial}{\partial\theta_i} V_\theta(t)\right)^2 < +\infty \text{ and } E_\theta\left|\frac{\partial^2}{\partial\theta_i\partial\theta_j} V_\theta(t)\right| < +\infty;\ i,j = 1,2,\ldots,k.$$

Then $J_\theta$ is a $(\mathfrak{F}, P_\theta)$-martingale with mean zero.

As a conclusion of these theorems it follows that $U_\theta$ is a square integrable martingale. Moreover, the following moment relations can be easely obtained

$$E_\theta(M(t)) = E_\theta(\dot{V}_\theta(t)),\ E_\theta(U_\theta(t)U_\theta^T(t)) = E_\theta(\ddot{V}_\theta(t)). \tag{5}$$

We remark that the matrix $\ddot{V}_\theta(t)$ is the observed Fisher information and $i_\theta(t) = E_\theta(\ddot{V}_\theta(t))$ is the usual (expected) Fisher information.

In the case that the submartingale M has an unique compensator $K_\theta$, then it holds $K_\theta(t) = \dot{V}_\theta(t)$ and the martingale $M-K_\theta$ is the score process $U_\theta$. A sufficient condition for the existence of $K_\theta$ is, for instance, that M is a positive submartingale. Obviously, under the conditions of the theorems the compensator of M is $\dot{V}_\theta$ and the compensator of $U_\theta^2 = \{U_\theta(t)U_\theta^T(t),\ t \geqq o\}$ is given by $\ddot{V}_\theta$.

Many widely used models for stochastic processes belong to the exponential class of Markov processes. In particular, if X is a homogeneous process with independent increments with the likelihood representation $L_\theta(t) = \exp\{\theta^T X(t) - \gamma(\theta)t\}$, we have $M = X$, $V_\theta(t) = \gamma(\theta)\cdot t$ (see [7]) and $\dot{\gamma}(\theta)t$ is the deterministic compensator of M. In the case of one-dimensional homogeneous Poisson processes with parameter $\lambda \in (0, \infty)$ we have $L_\theta(t) = \exp\{(\ln\lambda)X(t) - (\lambda-1)t\}$ and with $\theta = \ln\lambda$, $\gamma(\theta) = e^\theta - 1$, $\theta \in R^1$, we get an exponential family of Poisson processes. In the case of one-dimensional Wiener processes with trend $\theta t$ the likelihood function is given by $L_\theta(t) = \exp\{\theta X(t) - \frac{1}{2}\theta^2 t\}$. For $\gamma(\theta) = \frac{1}{2}\theta^2$, $\theta \in R^1$, we obtain an exponential family of Wiener processes. If we consider, more generally, a diffusion process X satisfying the differential equation

$$dX(t) = a(X(t), \theta)dt + b(X(t))dW(t),\ t \geqq o$$

where $\theta \in R^k$, $a(X(t), \theta) = \sum_{i=1}^{k} \theta_i\alpha_i(X(t))$ and W denotes a standard Wiener process then the likelihood function is given by $L_\theta(t) = \exp\{\theta^T M(t) - \frac{1}{2}\theta^T I(t)\theta\}$, where

$$M_i(t) = \int_o^t b^{-2}(X(u))\alpha_i(X(t))dX(u)$$

and $I(t) = (I_{ij}(t))$, $i,j = 1,2,\ldots,k$, with

$$I_{ij}(t) = \int_o^t b^{-2}(X(u))\alpha_i(X(u))\alpha_j(X(u))du$$

(see [5]). In this case we have $V_\Theta(t) = \frac{1}{2}\,\Theta^T I(t)\Theta$, $\Theta \in R^k$ and hence, a (M,V)-exponential family of Markov diffusion processes. The compensator of M is given by $\dot{V}_\Theta(t) = I(t)\cdot\Theta$. Also some subclasses of Markov jump processes as homogeneous birth-death processes and branching processes can be described as exponential families. An example for branching processes is considered in section 3.

## 2. EFFICIENCY OF SEQUENTIAL ESTIMATION PROCEDURES

In view of sequential estimation we consider a stopping time $\tau$ regarded as random observation time of the process. If $\tau$ is finite with probability one, the sequential likelihood function is given by $L_\Theta(\tau) = \exp\{\Theta^T M(\tau) - V_\Theta(\tau)\}$ according to a result of Döhler [2]. The score function is then $U_\Theta(\tau) = M(\tau) - \dot{V}_\Theta(\tau)$ and, under similar conditions as in theorem 1 and 2, the following relations hold

$$E_\Theta(M(\tau)) = E_\Theta(\dot{V}_\Theta(\tau)),\ E_\Theta(U_\Theta(\tau)U_\Theta^T(\tau)) = E_\Theta(\ddot{V}_\Theta(\tau)) . \tag{6}$$

Moreover, if $Y(\tau)$ is an unbiased estimator of a certain differentiable parameter function $h(\Theta)$, i.e. $E_\Theta Y(\tau) = h(\Theta)$, and assuming $E_\Theta Y^2(\tau) < +\infty$, it can be obtained that

$$E_\Theta(U_\Theta(\tau)Y(\tau)) = \dot{h}(\Theta) \tag{7}$$

holds. A sequential procedure for estimating $h(\Theta)$ will be denoted by $\delta = (\tau,\ Y(\tau))$.

C 4: For a sequential procedure $\delta$ it is assumed that
(i) $h(\Theta)$ is a nonconstant differentiable function in $\Theta \in \Theta$,
(ii) $\tau$ is a finite stopping time,
(iii) the estimator $Y(\tau)$ of $h(\Theta)$ is unbiased satisfying $E_\Theta Y^2(\tau) < +\infty$.

Theorem 3: Let X be a process of a (M,V)-exponential family and $\delta = (\tau,\ Y(\tau))$ a sequential procedure for estimating $h(\Theta)$. Under the condition that all required moments exist and the matrix $E_\Theta(\ddot{V}_\Theta(\tau))$ is non-singular we have

$$\mathrm{var}_\Theta(Y(\tau) \geq (\dot{h}(\Theta))^T(E_\Theta\ddot{V}_\Theta(\tau))^{-1}\dot{h}(\Theta),\ \Theta \in \Theta . \tag{8}$$

The sign of equality holds at $\Theta = \Theta^*$ iff

$$Y(\tau) = h(\Theta^*) + (\dot{h}(\Theta^*))^T(E_{\Theta^*}\ddot{V}_{\Theta^*}(\tau))^{-1}U_{\Theta^*}(\tau). \tag{9}$$

The inequality (8) of Cramér-Rao-Wolfowitz type follows immediately from the fact that $(\dot{h}(\Theta))^T(E_\Theta\ddot{V}_\Theta(\tau))^{-1}U_\Theta(\tau) - (Y(\tau)-h(\Theta))$ has mean zero and its variance is non-negative.

A sequential procedure $\delta$ is called efficient at $\theta^*$, if in (8) equality holds for $\theta = \theta^*$. If the equality is fulfilled for all $\theta \in \Theta^* \subsetneq \Theta$, the procedure is called $\Theta^*$-efficient. Motivated by the examples mentioned above it is quite natural to restrict our consideration to (M,V)-exponential families with $V_\theta(t) = \gamma(\theta)\cdot S(t)$, where $\gamma$ is a suitable parameter function and $S = \{S(t),\ t \geqq o\}$ is a one-dimensional process such that condition C 3 will be satisfied. In this case the pair $(S(\tau),\ M(\tau))$ forms a sufficient statistic with respect to observations of X up to $\tau$. This follows directly from the likelihood representation $L_\theta(\tau) = \exp\{\theta^T M(\tau) - \gamma(\theta)S(\tau)\}$. Efficient sequential estimators depend on the sufficient statistic and therefore, we consider sequential estimators of the form $Y(\tau) = Y(S(\tau),\ M(\tau))$. Efficient sequential procedures $\delta$ can be characterized by

Theorem 4: Let X be a process of a (M,V)-exponential family with $V_\theta(t) = \gamma(\theta)S(t)$ and let $\delta = (\tau,\ Y(\tau))$ be an $\Theta^*$-efficient sequential procedure. Suppose that the conditions of theorem 3 are fulfilled. Then there exist constants $c_0, c_1, \ldots, c_k$ non-vanishing simultaneously and $z \neq 0$ such that

$$c_0 S(\tau) + c^T M(\tau) = z \qquad \text{a.s. } [P_\theta],\ \theta \in \Theta^*, \tag{10}$$

where $c^T = (c_1,\ldots,c_k)$.

The proof is based on equation (9) and can be carried out in analogy to [6], theorem 5. In the following section we give an application to homogeneous Markov branching processes. In particular, the efficiently estimable parameter functions and the corresponding efficient sequential estimators are obtained.

## 3. APPLICATION TO BRANCHING PROCESSES

Let $X = \{X(t),\ t \geqq o\}$ be a one-dimensional time-homogeneous Markov branching process with split parameter $\lambda > o$, state space $\{0,1,2,\ldots\}$ and with offspring probabilities $p_i$, $i = 0,2,3,\ldots$ (we put $p_1 = o$ as usual). We assume that the offspring mean $m = \sum_{i=o}^{\infty} i p_i$ is finite. According to Athreya and Keiding [1] the likelihood function is given by

$$C_0(t)\ \lambda^{N(t)} e^{-\lambda S(t)} \prod_{i=1}^{N(t)} p_{X_i - X_{i-1}+1}\ ,$$

where $N(t)$ is the total number of splits, $X_i = X(t_i)$ with splitting points $t_i$ and

$$S(t) = \sum_{i=1}^{N(t)} X_{i-1} T_i + X_{N(t)}(t - \sum_{i=1}^{N(t)} T_i) = \int_0^t X(u)du$$

is the total life time of the population in $[0, t]$, $T_i$ waiting time in state $X_{i-1}$. $C_0(t)$ denotes a random factor independent of $\Theta$. We now specify the offspring distribution to be the following Poissondistribution:

$$p_{x_i-x_{i-1}+1} = P\{X_i=x_i/X_{i-1}=x_{i-1}\} = \frac{\mu^{x_i-x_{i-1}}}{(x_i-x_{i-1})!} e^{-\mu}$$

for $x_i-x_{i-1} > 0$ and $p_0 = e^{-\mu}$. We choose for the unknown parameter $\Theta^T = (\Theta_1, \Theta_2) = (\ln \lambda - \mu + 1, \ln \mu)$ and we put

$$M^T(t) = (M_1(t), M_2(t)) = (N(t), \sum_{i=1}^{N(t)} \max(0, X_i-X_{i-1})).$$

It follows that the likelihood function (1) becomes

$$L_\Theta(t) = \exp\{\Theta^T M(t) - \gamma(\Theta)S(t)\},$$

where $\gamma(\Theta) = \exp\{\Theta_1 + e^{\Theta_2} - 1\} - 1$, $\Theta \in R^2$. So we have an exponential family of Poisson branching processes.
The maximum likelihood estimators can be obtained from the equation $M(t) = \dot{\gamma}(\Theta)S(t)$. In the case $M_1(t) = N(t) > 0$, $t > 0$, the solution is given by

$$\hat{\Theta}_1 = \ln(\frac{N(t)}{S(t)}) - \frac{M_2(t)}{N(t)} + 1, \quad \hat{\Theta}_2 = \ln \frac{M_2(t)}{N(t)} .$$

Asymptotic properties of these estimators are discussed in [1].

Now we turn to the sequential case. Based on theorem 3 we consider efficient sequential procedures $\delta = (\tau, Y(\tau))$, where $Y(\tau) = Y(S(\tau), N(\tau), M_2(\tau))$. From theorem 4 we obtain that there exist constants $c_0, c_1, c_2$ with $c_0^2 + c_1^2 + c_2^2 > 0$ and $z \neq 0$ such that

$$c_0S(\tau) + c_1N(\tau) + c_2M_2(\tau) = z \quad \text{a.s. } [P_\Theta]. \tag{11}$$

Using this linear relation between the components of the sufficient statistic $(S(\tau), N(\tau), M_2(\tau))$ and the corresponding representation (9) of the sequential estimator in the considered special case the following assertion can be proved similar to theorem 5 in [4]:
<u>Theorem 5:</u> Let $X$ be the considered Poisson branching process and let $\delta = (\tau, Y(\tau))$ an efficient sequential procedure. Then, the efficiently estimable parameter functions are of the form

$$h(\Theta) = \frac{\alpha_0 + \alpha_1\gamma(\Theta) + \alpha_2 e^{\Theta^2}(\gamma(\Theta) + 1)}{c_0 + c_1(\gamma(\Theta)+1) + c_2 e^{\Theta^2}(\gamma(\Theta) + 1)},$$

where $\alpha_0$, $\alpha_1$, $\alpha_2$ are arbitrary constants. The corresponding efficient estimators are

(i) in the case $c_0 = 0$: $Y(\tau) = k_0 S(\tau) + k_1 M_2(\tau) + k_2$,
where $zk_0 = \alpha_0 - \frac{c_2}{c_1}\alpha_1$, $zk_1 = \alpha_2$, $c_1 k_2 = \alpha_1$,

(ii) in the case $c_0 \neq 0$: $Y(\tau) = k_0 N(\tau) + k_1 M_2(\tau) + k_2$,
where $zk_0 = \alpha_1 - \frac{c_2}{c_0}\alpha_0$, $zk_1 = \alpha_2$, $c_0 k_2 = \alpha_0$.

Finally, we remark that not all coefficients $c_i$ in (11) will lead to a finite stopping time $\tau$. However, it can be shown that $\tau$ is finite, if

$$c_0 + (\gamma(\Theta)+1)(c_1 + c_2 e^{\Theta^2}) > 0$$

and one of the following two conditions is valid:

(i) $c_0 > 0$, $c_1 \leqq 0$, $c_2 \leqq 0$ and $z > 0$,

(ii) $c_0 = 0$, $c_1 > 0$, $c_2 = 0$ and $\frac{z}{c_1}$ integer-valued.

Moreover, we have $E_\Theta S(\tau) = [c_0 + (\gamma(\Theta)+1)(c_1 + c_2 e^{\Theta^2})]^{-1} z$.

## REFERENCES


[1] Athreya, K.B.; Keiding, N. (1975): Estimation theory for continuous-time branching processes. Preprint No. 6. Inst. of Math. Statistics Univ. Copenhagen.

[2] Döhler, R. (1981): Dominierbarkeit und Suffizienz in der Sequentialanalyse. Math. Operationsforsch. Statist., ser. statist. 12, 101-134.

[3] Feigin, P.D. (1976): Maximum likelihood estimation for continuous time stochastic processes. J. Appl. Prob. 13, 712-736

[4] Franz, J. (1982): Sequential estimation and asymptotic properties in birth-and-death processes. Math. Operationsforsch. Statist., ser. statistics 13, 2; 231-244.

[5] Sørensen, M. (1983): On maximum likelihood estimation in randomly stopped diffusion-type processes. Internat. Statist. Review 51, 93-110.

[6] Winkler, W.; Franz, J. (1979): Sequential estimation problems for the exponential class of processes with independent increments. Scand. J. Statist. 6, 129-139.

[7] Winkler, W.; Franz, J.; Küchler, I. (1982): Sequential statistical procedures for processes of the exponential class with independent increments. Math. Operationsforsch. Statist., ser. statistics 13/1, 105-119.

# On Convergence in Law of Random Elements in Certain Function Spaces

Peter Gaenssler and Wilhelm Schneemeier
Mathematisches Institut
Universität München
D-8000 München, FRG

SUMMARY: The aim of the present paper is to popularize the applicability of a model for convergence in law of random elements in certain (non-separable) function spaces being at first especially appropriate for simplifying the presentation of known functional limit theorems for univariate empirical processes (like the uniform one) and which at the same time allows for a straightforward generalization in handling also empirical processes based on multivariate observations up to empirical processes based on random data in arbitrary sample spaces and being indexed by certain classes of sets or functions, respectively.

INTRODUCTION:

Let $\xi_i$, $i \geq 1$, be i.i.d. $U(0,1)$ r.v.'s defined on some basic p-space $(\Omega, A, \mathbb{P})$; let $\mu_n$ be the *empirical measure* based on $\xi_1, \ldots, \xi_n$, i.e.

$$\mu_n(B) := n^{-1} \sum_{i=1}^{n} 1_B(\xi_i) \quad \text{for any Borel set } B \subset [0,1],$$

and let $\alpha_n = (\alpha_n(t))_{t \in [0,1]}$ be the corresponding *uniform empirical process*, defined by

$$\alpha_n(t) := n^{1/2}(\mu_n([0,t]) - t), \ t \in [0,1].$$

Then, according to *Donsker's functional limit theorem* (*FLT*):

$$\alpha_n \xrightarrow{L} B^0 \quad \text{as} \quad n \to \infty,$$

where $B^0 = (B^0(t))_{t \in [0,1]}$ denotes the Brownian bridge, being a centered

---

Key words and phrases: Convergence in law, functional limit theorems, empirical processes, pseudo metric parameter spaces

AMS 1980 subject classification: Primary 60 B 10, 60 G 07
Secondary 60 B 05

Contributions to Stochastics.
Ed. by W. Sendler
© Physica-Verlag Heidelberg 1987

Gaussian process having all its sample paths in the space $C[0,1]$ of all realvalued continuous functions on $[0,1]$. Here $\alpha_n \overset{L}{\to} B^o$ means *weak convergence* (in the sense of [1]) *of the laws* $L(\alpha_n)$ to $L(B^o)$, each being considered as *Borel measures* on the space $D = D[0,1]$ of all right continuous functions on the unit interval $[0,1]$ that have left hand limits at all points $t \in (0,1]$, endowed with the *Skorokhod topology*, where $L(B^o)(C[0,1]) = 1$.

If, *instead*, $D$ is equipped with the $\sigma$-algebra $B_b(D)$ generated by the open balls w.r.t. the *supremum metric* $\rho$ in $D = D[0,1]$,

then, since (cf. (37) in [4])

$$B_b(D) = \sigma(\{\pi_t : t \in [0,1]\}),$$

$\alpha_n$ as well as $B^o$ can be viewed as random elements (r.e.'s) in $(D, B_b(D))$, i.e. as $A,B_b(D)$-measurable functions from $\Omega$ into $D$, with $L(\alpha_n)$ and $L(B^o)$ being well defined on $B_b(D)$ and where $L(B^o)(C[0,1]) = 1$.

Furthermore, Donsker's FLT can be formulated equivalently (cf. [4], Lemma 18 and (28)) as

$$\alpha_n \overset{L_b}{\to} B^o \quad \text{as} \quad n \to \infty$$

which means that

$$\lim_{n\to\infty} \int f(\alpha_n)\,d\mathbb{P} = \int f(B^o)\,d\mathbb{P} \quad \text{for all} \quad f \in U_b^b(D),$$

where

$$U_b^b(D) := \{g : D \to \mathbb{R} : g \text{ bounded, uniformly } \rho\text{-continuous and } B_b(D)\text{-measurable}\},$$

and where $\mathbb{P}(B^o \in C[0,1]) = 1$.

Thus, given any other sequence $(\beta_n)_{n\in\mathbb{N}}$ of r.e.'s in $(D, B_b(D))$ and a limiting r.e. $\beta_o$ in $(D, B_b(D))$, we may adopt the following *definition of convergence in law*:

$$\beta_n \overset{L_b}{\to} \beta_o \quad \text{iff (i)} \quad \mathbb{P}(\beta_o \in C[0,1]) = 1 \quad \text{and}$$
$$\text{(ii)} \quad \lim_{n\to\infty} \int f(\beta_n)\,d\mathbb{P} = \int f(\beta_o)\,d\mathbb{P} \quad \text{for all} \quad f \in U_b^b(D).$$

With this definition of $L_b$*-convergence* and the corresponding concept of *relative* $L_b$*-sequential compactness* it was shown in [5], Theorem 1.17, that the following theorem holds true:

<u>THEOREM 1</u>. Given a sequence $(\beta_n)_{n\in\mathbb{N}}$ of random elements in $(D, B_b(D))$, the following statements are equivalent:

(i) $(\beta_n)_{n\in\mathbb{N}}$ fulfills the following two conditions (a) and (b):

(a) For any sequence $(f_m)_{m\in\mathbb{N}}$ in $U_b^b(D)$ with $f_m \downarrow 0$ as $m \to \infty$ one has $\limsup_{n\to\infty} \int f_m(\beta_n)\,d\mathbb{P} \to 0$ as $m \to \infty$

(b) $\liminf_{n\to\infty} \int f(\beta_n)\,d\mathbb{P} \geq 1$ for all $f \in U_b^b(D)$ with $f \geq 1_{C[0,1]}$.

(ii) $(\beta_n)_{n\in\mathbb{N}}$ is *relatively* $L_b$*-sequential compact* (i.e. for any subsequence $(\beta_{n'})$ of $(\beta_n)$ there exists a further subsequence $(\beta_{n''})$ of $(\beta_{n'})$ and a limiting random element $\beta_0$ in $(D, \mathcal{B}_b(D))$ such that $\beta_{n''} \overset{L_b}{\to} \beta_0$ as $n'' \to \infty$.

(iii) (A) $\lim_{\delta\downarrow 0} \limsup_{n\to\infty} \mathbb{P}(w_{\beta_n}(\delta) \geq \varepsilon) = 0$ for each $\varepsilon > 0$, and

(B) $\lim_{K\to\infty} \limsup_{n\to\infty} \mathbb{P}(|\beta_n(0)| \geq K) = 0$,

where $w_x(\delta) := \sup\{|x(t_1) - x(t_2)| : t_i \in [0,1] \text{ s.t. } |t_1 - t_2| \leq \delta\}$ for any $x \in D[0,1]$ and $\delta > 0$.

## THE GENERAL MODEL AND RESULT:

Concerning now e.g.

*empirical* $\mathcal{C}$*-processes based on a sequence* $(\xi_i)_{i\in\mathbb{N}}$ *of i.i.d. random elements* $\xi_i$ *in an arbitrary sample space* $(X, \mathcal{X})$

i.e. the random functions $\alpha_n = (\alpha_n(C))_{C\in\mathcal{C}}$ indexed by $\mathcal{C} \subset \mathcal{X}$ and being defined by

$\alpha_n(C) := n^{1/2}(\mu_n(C) - \mu(C))$, $C \in \mathcal{C}$, where $\mu = L(\xi_i)$ denotes the law of the $\xi_i$'s defined on the $\sigma$-algebra $\mathcal{X}$ in $X$ and where $\mu_n$ is again the empirical measure based on $\xi_1,\ldots,\xi_n$ (i.e. $\mu_n(B) := n^{-1} \sum_{i=1}^{n} 1_B(\xi_i)$, $B \in \mathcal{X}$),

the question arises whether the above approach

fitting on the special situation of the uniform empirical process with $\mathcal{C} = \{[0,t] : t \in [0,1]\}$

can be generalized.

FLT's are known for empirical $\mathcal{C}$-processes in case $X = \mathbb{R}^k$, $k \geq 1$, with $\mathcal{C} = \{(-\infty,\underline{t}] : \underline{t} \in \mathbb{R}^k\}$ being the class of all lower left orthants, or, more generally (and under additional measurability assumptions),

for empirical $\mathcal{C}$-processes in case of arbitrary sample spaces $(X, \mathcal{X})$ with $\mathcal{C} \subset \mathcal{X}$ being a so-called *Vapnik-Chervonenkis class*, which means that $\mathcal{C}$ has the following characteristic property (in common with the class of all lower left orthants in $X = \mathbb{R}^k$):

There exists an $s \in \mathbb{N}$ such that for any $F \subset X$ with cardinality

$|F| = s$ one has $|\{C \cap F : C \in \mathcal{C}\}| < 2^s$ .

In these cases $(\mathcal{C}, d_\mu)$ with $d_\mu(C_1, C_2) := \mu(C_1 \Delta C_2)$ for $C_i \in \mathcal{C}$ , $i = 1,2,$ turns out to be (cf. [2], Lemma (7.13)) a

pseudo-metric parameter space being
totally bounded for $d_\mu$ .

Concerning the even more general case of

*empirical F-processes*

with $F$ being a certain class of functions, more general than
$F = \{1_C : C \in \mathcal{C}\}$ with $\mathcal{C} \subset X$ ,

one also works in connection with corresponding FLT's under the assumption that

$(F, d_{\mu,F})$ is totally bounded for the pseudometric $d_{\mu,F}$ defined by
$d^2_{\mu,F}(f_1, f_2) := \int (f_1 - f_2)^2 d\mu - [\int (f_1 - f_2) d\mu]^2$ for $f_i \in F$, $i = 1,2$;

(cf. Theorem 4.1.1 in [3] and Theorem 2.12 in [7]).

So, the appropriate setting in generalizing our special approach mentioned in the introduction is

to start with a pseudo-metric parameter space $T = (T,d)$ assumed to be totally bounded for $d$ ;

next,

to replace $C[0,1]$

by the function space

$S_o := \{x : T \to \mathbb{R} : \|x\|_T := \sup_{t \in T} |x(t)| < \infty$ and x uniformly d-continuous$\}$

and

to replace $D = D[0,1]$

by some $S$ with

$S_o \subset S \subset \ell^\infty(T)$ , where $\ell^\infty(T)$ denotes the space of all bounded realvalued functions on $T$ equipped with the supremum norm $\|\cdot\|_T$ .

In doing this our special approach generalizes in a rather straightforward way:

Given any $S$ such that $S_o \subset S \subset \ell^\infty(T)$ , let $\mathcal{B}_b(S)$ be the $\sigma$-algebra generated by the open $\|\cdot\|_T$-balls in $S$ , assuming here (for simplicity) that the following measurability condition (M) holds true:

(M): $\mathcal{B}_b(S) \subset \mathcal{B} := \sigma(\{\pi_t : t \in T\})$,
where $\sigma(\{\pi_t : t \in T\})$ denotes the $\sigma$-algebra in $S$ generated by the coordinate projections $\pi_t : S \to \mathbb{R}$, defined by $\pi_t(x) := x(t)$ for $x \in S$ .

Then, given any sequence $(\beta_n)_{n\in\mathbb{N}}$ of random elements $\beta_n$ in $(S,\mathcal{B})$ (i.e. $\mathcal{A},\mathcal{B}$-measurable functions $\beta_n : \Omega \to S$, defined on some basic p-space $(\Omega,\mathcal{A},\mathbb{P})$) and a random element $\beta_o$ in $(S, \mathcal{B}_b(S))$, defined on the same p-space $(\Omega,\mathcal{A},\mathbb{P})$,

$(\beta_n)_{n\in\mathbb{N}}$ *is said to converge in law to* $\beta_o$ $(\beta_n \overset{L_b}{\to} \beta_o)$

iff (i) $\mathbb{P}(\beta_o \in S_o) = 1$

and (ii) $\lim\limits_{n\to\infty} \int f(\beta_n)d\mathbb{P} = \int f(\beta_o)d\mathbb{P}$ for all $f \in U_b^b(S)$ ,

where

$U_b^b(S) := \{g : S \to \mathbb{R} : g$ bounded, uniformly $\|\cdot\|_T$-continuous and $\mathcal{B}_b(S)$-measurable$\}$ .

Denoting (as in Theorem 1) with $\omega_x(\cdot)$ the oscillation modulus defined for any $x \in \ell^\infty(T)$ and $\delta > 0$ by

$$\omega_x(\delta) := \sup\{|x(t_1) - x(t_2)| : t_i \in T \text{ s.t. } d(t_1,t_2) \leq \delta\} ,$$

the following criterion generalizes Theorem 1:

<u>THEOREM 1*</u>. Given a sequence $(\beta_n)_{n\in\mathbb{N}}$ of random elements in $(S,\mathcal{B})$, the following four statements are equivalent:

(i) $(\beta_n)_{n\in\mathbb{N}}$ fulfills the conditions (a) and (b) of (i) in Theorem 1 with $D$ replaced by $S$ and with $C[0,1]$ replaced by $S_o$ .

(ii) $(\beta_n)_{n\in\mathbb{N}}$ is relatively $L_b$-sequentially compact.

(iii) (A) $\lim\limits_{\delta\downarrow 0} \limsup\limits_{n\to\infty} \mathbb{P}^*(\omega_{\beta_n}(\delta) \geq \varepsilon) = 0$ for each $\varepsilon > 0$

(where $\mathbb{P}^*$ denotes the outer measure pertaining to $\mathbb{P}$), and

(B) $\lim\limits_{K\to\infty} \limsup\limits_{n\to\infty} \mathbb{P}(|\beta_n(t)| \geq K) = 0$ for all $t \in T$ .

(iv) (A) as in (iii), and

(B') $\lim\limits_{K\to\infty} \limsup\limits_{n\to\infty} \mathbb{P}(\|\beta_n\|_T \geq K) = 0$ .

This theorem especially implies Theorem (1.2) in [2]; cf. Theorem B in [4] and p. 118 there, according to which it follows that under (A) the other condition (B') is automatically fulfilled in case of empirical $\mathcal{C}$-processes $\beta_n = (\beta_n(C))_{C\in\mathcal{C}}$ indexed by classes $\mathcal{C}$ of sets being totally bounded for $d_\mu$ ; the same holds true for empirical $\mathcal{F}$-processes and $d_{\mu,\mathcal{F}}$ instead of $d_\mu$ .

Of course, to apply Theorem 1* in connection with empirical $\mathcal{C}$- (or $\mathcal{F}$-) processes mentioned before, the main problem for obtaining a FLT, i.e. the statement that $\beta_n \overset{L_b}{\to} \beta_o$ , is to verify (iii)(A) (cf. [6] for practicable sufficient conditions), since usually convergence of the finite dimensional distributions (fidis) pertaining to $\beta_n$ follows from classical results, thus yielding that the law of any accumulation point $\beta_o$ of $(\beta_n)_{n\in\mathbb{N}}$ (w.r.t. $L_b$-convergence) is uniquely determined by its fidis and hence also on $\mathcal{B}_b(S)$ according to (M).

The crucial step in proving Theorem $1^*$ is the part verifying that (iv) implies (i):
But this can be done (avoiding the concept of $\delta$-tightness still used in [4], Prop. $B_2$, p. 117) along the same straightforward lines used in [5] to prove Theorem 1.
The only crucial point is to use a substitute for the following simple fact in the classical situation described in the introduction (cf. (**) on p. 67 in [5]):

> Whenever $x \in D[0,1]$ satisfies $w_x(\delta) < \varepsilon$ for some $\delta > 0$ and $\varepsilon > 0$, then $x \in C[0,1]^{2\varepsilon}$, where $C[0,1]^{2\varepsilon}$ denotes the $2\varepsilon$-hull of $C[0,1]$ w.r.t. the supremum metric.

In general, this is achieved with the following

*Approximation Lemma*

which shows at the same time that $(S_o, \|\cdot\|_T)$ is a separable space iff $(T,d)$ is totally bounded for d.

<u>LEMMA</u>. Let $(T,d)$ be a pseudo-metric space assumed to be totally bounded for the pseudo-metric d and let $S_o$ be the space of all realvalued, bounded, and uniformly d-continuous functions on T. Then there exists a countable subset $S_1$ of $S_o$ such that for any $x \in \ell^\infty(T)$ with $w_x(\delta) \leq \varepsilon$ for some $\delta > 0$ and $\varepsilon > 0$ there exists an $y \in S_1$ with $\|x - y\|_T \leq 5\varepsilon$.

As to the proof of this lemma one starts noticing the fact that, by total boundedness of $(T,d)$, there exists for each $n \in \mathbb{N}$ a finite subset $T_n$ of T having the following property:

For each $t \in T$ there exists an $s \in T_n$ such that $d(t,s) < n^{-1}$.

Then, denoting by $\mathbb{Q}$ and $\mathbb{Q}_+$, respectively, the set of all rational and nonnegative rational numbers, respectively, one shows that

$$S_1 := \bigcup_{n\in\mathbb{N}} \bigcup_{q\in\mathbb{Q}_+} \{t \to \sup\{x(s) - q\cdot d(t,s) : s \in T_n\} : x \in \mathbb{Q}^{T_n}\}$$

has the properties stated in the lemma.

For further details, also concerning the proof of Theorem $1^*$, we refer to the forthcoming paper [6].

# R E F E R E N C E S

[1] Billingsley, P. (1968). Convergence of Probability Measures. Wiley, New York.

[2] Dudley, R.M. (1978). Central limit theorems for empirical measures. Ann. Probability 6, 899-929; Correction, ibid. 7, 909-911.

[3] Dudley, R.M. (1984). A Course on Empirical Processes. Springer Lecture Notes in Mathematics, Vol. 1097, 142 p.

[4] Gaenssler, P. (1983). Empirical Processes. IMS Lecture Notes-Monograph Series, Vol. 3, 179 p.

[5] Gaenssler, P. - Haeusler, E. - Schneemeier, W. (1984). Selected Topics on Empirical Processes. In: Proceedings of the Third Prague Symposium on Asymptotic Statistics. Ed. by P. Mandl and M. Huskowá. Elsevier Science Publishers B.V., Amsterdam - New York - Oxford; p. 57-91.

[6] Gaenssler, P. - Schneemeier, W. (1986). On Functional Limit Theorems for a Class of Stochastic Processes indexed by pseudo-metric Parameter Spaces (with Applications to Empirical Processes).
Submitted for publication in the Proceedings of the VI. International Conference on Probability in Banach Spaces, June 16-21, 1986, at Sandbjerg, Denmark.

[7] Giné, E. - Zinn, J. (1984). Some limit theorems for empirical processes. Ann. Probability 12, 929-989.

# An Application of the Mixture Theory to the Decomposition Problem of Characteristic Functions

Béla Gyires
Department of Mathematics
Kossuth L. University
H-4010 Debrecen, Hungary

SUMMARY: In more papers author dealt with a new foundation of the mixture theory of probability distribution functions. In this paper a part of the obtained results is used to give necessary and sufficient condition in order to decide whether a given characteristic function is a factor of another one, or not.

## 1. INTRODUCTION

A function $F(x)$, $x \in R$ is called to be a probability distribution function, if it satisfies the following conditions: a) $F(x)$ is non-decreasing. b) $F(x)$ is right continuous. c) $F(\infty)=1$, $F(-\infty)=0$. $F(x)$ has at most a denumerable number of discontinuities which are all of the first kind. Moreover $F(x)$ is differentiable almost everywhere. The following important decomposition holds:

$$F(x) = F_a(x) + F_j(x) + F_s(x) ,$$

where all these functions on the right are non-decreasing. $F_j(x)$ is a jump function. The discontinuities of $F_j(x)$ are identical with those of $F(x)$. $F_a(x)$ is strictly increasing and absolutely continuous with respect to the Lebesgue measure. $F_a'(x)$ exists almost everywhere. $F_s(x)$ is a singular function, i.e. is continuous, strictly increasing, and $F_s'(x)=0$ almost everywhere. If $F(x)=F_j(x)$, i.e. $F(x)$ is absolutely continuous with respect to the Lebesgue measure, then $f(x)=F'(x)$ is say to be the density function (with respect to the Lebesgue measure). If $F(x)=F_j(x)$ then $F(x)$ is say to be a discret probability distribution function with

Key words: Mixture of probability distribution functions. Decomposition of characteristic functions. AMS subject classifications: 60B15, 60E05.

Contributions to Stochastics.
Ed. by W. Sendler
© Physica-Verlag Heidelberg 1987

finite or denumerable number of discontinuities.

Let us introduce some notations. Let $E$ be the set of the probability distribution functions. Let $E_a$ be the set of the absolutely continuous probability distribution functions with respect to the Lebesgue measure. Let $E_j$ be the set of the discrete probability distribution functions. Let $a<b$ be two real numbers, where $a=-\infty$, $b=\infty$ are permitted too. Denote by $E(a,b)$ the set of the probability distribution functions, which are strictly monotone increasing in $[a,b]$, continuous in the whole real line, and they have values zero and one at the point $a$ and $b$, respectively. The inverse of $F \in E(a,b)$ is denoted by $F^{-1}$.

$F \in E_j$ is say to be degenerate if it has one discontinuity only.

We say

$$G(z,x)\ , \quad z \in R\ , \quad x \in R$$

to be a family of probability distribution functions with parameter $x$, if the following properties are satisfied: a) For each value of $x$ $G(z,x) \in E$ in $z$. b) $G(z,x)$ is a measurable function in $x$.

Let $H \in E$. Then

$$F(z) = \int_{-\infty}^{\infty} G(z,x)dH(x) \in E \tag{1.1}$$

([4], p.199), which is called the mixture of the family of the distribution functions $G(z,x)$ with weight function $H$. An improtant question of the mixture theory of probability distribution functions is the following: Let $F \in E$, and $G(z,x) \in E$ with parameter $x$ be given. What is the necessary and sufficient condition of having a $H \in E$, which satisfies equation (1.1)? This problem was solved by the author ([1]) in whole generality under the assumption that $H \in E_j$. Moreover ([2]) in the special case if $F \in E(a,b)$, $G(z,x) \in E(a,b)$ in $z$ with parameter $x$, and $H \in E_a$ with square integrable density function.

## 2. THE AIM OF THE PAPER

In this paper we deal with the case only if

$$G(z,x) = G(z-x)\ , \quad x \in R\ , \quad z \in R\ , \tag{2.1}$$

where $G \in E$. Obviously that (2.1) is a family of probability distribution functions. By (1.1) we have

$$F(z) = \int_{-\infty}^{\infty} G(z-x)dH(x)\ , \quad z \in R\ , \tag{2.2}$$

i.e. on the right is the convolution of $G \in E$ and $H \in E$.
As in general it is usual, the Fourier transform

$$f(t) = \int_{-\infty}^{\infty} e^{itx}\, dF(x) \ , \quad t \in R \qquad (2.3)$$

of $F \in E$ is said to be the characteristic function of $F$.
It is well-known if $f(t)$, $g(t)$, $h(t)$ are the characteristic functions of $F \in E$, $G \in E$ and $H \in E$, respectively, then by (2.2) we get

$$f(t) = g(t)h(t) \ , \quad t \in R \ . \qquad (2.4)$$

The characteristic function $f(t)$ is said to be decomposable, if it can be written in the form (2.4), where $g(t)$ and $h(t)$ are both characteristic functions of non-degenerate probability distribution functions. We say that $g(t)$ and $h(t)$ are factors of $f(t)$.
By the uniqueness relation between probability distribution function and its characteristic function we get that the mixture problem of (2.2) type is equivalent to the decomposition problem of their characteristic functions. The later question has a rich literature. In this connection it is enough to refer to the large-scale monography [3] of Yu. V. *LINNIK*.
Characteristic function $f(t)$ is said to be non-degenerate, if $F \in E$ is non-degenerate in the representation (2.3).
It seems the following asking in question is new. Let $f(t)$ and $g(t)$ be non-degenerate characteristic functions. What is the necessary and sufficient condition in order that $g(t)$ let be a factor of $f(t)$, i.e. that (2.4) let be satisfied by a non-degenerate characteristic function $h(t)$? By other words, if the non-degenerate $F \in E$ and $G \in E$ are given, does a non-degenerate $H \in E$ exist such that equation (2.2) let be satisfied? As we mention already in the Introduction, this problem was solved by the author in two special cases. In the following we deal with the characteristic functional form of these theorems if the question is the mixture problem of (2.2) type.

## 3. THE CASE OF DISCRETE WEIGHT FUNCTION

In this section we deal with the mixture problem of (2.2) type if $F \in E$ and $G \in E$ are given, and the unknown $H \in E_j$ has a finite number of discontinuities.
Denote by $(R,\mathcal{B})$ the measurable space with Borel $\sigma$-field $\mathcal{B}$ generated by the the set of the real numbers $R$. Denote by $\omega(F)$ the

measure on $\mathcal{B}$ generated by $F \in E$. Let $F \in E$, $G_j \in E$ $(j=1,\dots,n)$ be given, where $n \geqq 2$ is an integer. Let the measures $\omega(F)$, $\omega(G_j)$ $(j=1,\dots,n)$ on $\mathcal{B}$ dominated by the $\sigma$-finite measure $\lambda$ on $\mathcal{B}$, and let $\varphi$, $\gamma_j$ $(j=1,\dots,n)$ be the corresponding density functions with respect to $\lambda$. We introduce the quantities ([1], Definition 2.2.)

$$(G_j,G_k)_F = \int_{-\infty}^{\infty} \frac{\gamma_j \gamma_k}{\varphi} \lambda(dx) \tag{3.1}$$

$$(j,k=1,\dots,n),$$

and assume that quantities (3.1) are finite. Let us introduce the matrix

$$\Gamma = ((G_j,G_k)_F)_1^n ,$$

which is positive definite or semidefinite, and one is positive definite if and only if $G_j \in E$ $(j=1,\dots,n)$ are linearly independent ([1], Theorem 3.3). We say that the positive definite matrix $\Gamma$ is sign-preserving if all row sums of $\Gamma^{-1}$ are positive.

Lemma 3.1. Let $G \in E$, and let the different real numbers $x_j$ $(j=1,\dots,n)$ be given. Then

$$G_j(z) = G(z-x_j) \in E \quad (j=1,\dots,n) \tag{3.2}$$

are linearly independent.

Proof. Suppose there are numbers $\lambda_j$ $(j=1,\dots,n)$ such that

$$\sum_{j=1}^{n} \lambda_j G_j(z) = 0 , \quad z \in R ,$$

i.e.

$$g(t) \sum_{j=1}^{n} \lambda_j e^{itx_j} = 0 , \quad t \in R ,$$

where $g(t)$ is the characteristic function of $G \in E$. Since $g(t) \neq 0$ in a neighbourhood of zero, we have the equality

$$\sum_{j=1}^{n} \lambda_j e^{itx_j} = 0 , \quad t \in R \tag{3.3}$$

in this neighbourhood. Let $\alpha > 0$ be so small that numbers $k\alpha$ $(k=0,1,\dots,n-1)$ let be in this neighbourhood. Substituting these into (3.3) we get the homogeneous linear equation system

$$\sum_{j=1}^{n} \lambda_j e^{ik\alpha x_j} = 0 \quad (k=0,1,\dots,n-1) \tag{3.4}$$

for $\lambda_j$ $(j=1,\dots,n)$. The determinant of this equation system is the determinant of the Vandermonde matrix generated by $e^{i\alpha x_k}$ $(k=1,\dots,n)$. Since numbers $x_k$ $(k=1,\dots,n)$ are different, this Vandermonde matrix is regular, i.e. equation system (3.4) has the only solution $\lambda_j = 0$ $(j=1,\dots,n)$.

Let $F \in E$, $G \in E$, and let the different real numbers $x_j$ $(j=1,\dots,n)$ with $n \geqq 2$ be given. Assume that measures $\omega(F)$ and $\omega(G)$ on $\mathcal{B}$ are dominated by the $\sigma$-finite measure $\lambda$ on $\mathcal{B}$. Denote by $\varphi$ and $\gamma$ the corresponding density function with respect to $\lambda$. If we use notation (3.2) again, we get

$$(G_j,G_k)_F = \int_{-\infty}^{\infty} \frac{\gamma(x-x_j)\gamma(x-x_k)}{\varphi(x)} \lambda(dx) \qquad (3.5)$$

$$(j,k=1,\dots,n)$$

by (3.1).

Using Lemma 3.1. we get the following statement by Theorem 4.9. of paper [1].

<u>Theorem 3.1.</u> Let the different real number $x_j$ $(j=1,\dots,n)$ with $n \geqq 2$ be given. Let $f(t)$ and $g(t)$ the characteristic functions of $F \in E$, and $G \in E$, respectively. Suppose that quantities (3.5) are finite. Then $g(t)$ is a factor of $f(t)$ if any only if matrix $\Gamma$ with entries (3.5) is sign-preserving, and the sum of the entries of $\Gamma^{-1}$ equal to one. In this case

$$f(t) = g(t)h(t) ,$$

where

$$h(t) = \sum_{j=1}^{n} a_j e^{ix_j t} ,$$

and $a_j$ is the sum of the entries of the jth row of $\Gamma^{-1}$.

## 4. THE CASE OF THE ABSOLUTELY CONTINUOUS WEIGHT FUNCTIONS

In this section we deal with the solution of (2.2) by given $F \in E(a,b)$, $G \in E(a,b)$, supposing that the unknown weight function $H \in E_a$, and its density function is square integrable.

If $F \in E(a,b)$, $G \in E(a,b)$, we get that $G(F^{-1}(z)) \in E_a$ $(z [0,1])$ by Theorems 2.4. and 2.5. of paper [1]. Moreover if $(G_1,G_2)_F$ is finite, where $G_j \in E(a,b)$ $(j=1,2)$, then

$$(G_1,G_2)_F = \int_0^1 [\frac{d}{dz} G_1(F^{-1}(z))][\frac{d}{dz} G_2(F^{-1}(z))]dz > 0$$

holds. Let us introduce quantity

$$K(x,y) = (G(z-x) \ , \ G(z-y))_F =$$

$$= \int_0^1 [\frac{d}{dz}G(F^{-1}(z)-x)][\frac{d}{dz} G(F^{-1}(z)-y)]dz \qquad (4.1)$$

for $x,y \in R$, which is continuous in $x$ and $y$. Suppose that

$$\int_{-\infty}^{\infty}\int_{-\infty}^{\infty} K^2(x,y)dxdy < \infty \ ,$$

i.e. (4.1) is a continuous Hilbert-Schmidt kernel.
It is well-known that the eigenvalues of kernel (4.1) are positive numbers. Let $\{\lambda_k\}$ be the set of these enumerated in an increasing way, and let the suitable orthonormal eigenfunctions be the elements of the sequence $\{\varphi_k(x)\}$. Let

$$\alpha_k = \int_{-\infty}^{\infty} \varphi_k(x)dx \qquad (k=1,2,\dots).$$

By Theorem 3.2. of paper [2] we get the following statement.
Theorem 4.1. Let $f(t)$, $g(t)$ be the characteristic functions of $F \in E(a,b)$, and $G \in E(a,b)$, respectively. Let (4.1) be a Hilbert-Schmidt kernel. Then $g(t)$ is a factor of $f(t)$ if and only if the conditions

$$\sum_{k=1}^{\infty} \alpha_k^2 \lambda_k = 1,$$

$$\sum_{k=1}^{\infty} \alpha_k \lambda_k \varphi_k(x) \geqq 0 \ , \qquad x \in R \qquad (4.2)$$

are statisfied. In this case

$$f(t) = g(t)h(t) \ ,$$

where

$$h(t) = \sum_{k=1}^{\infty} \alpha_k \lambda_k \int_{-\infty}^{\infty} e^{itx} \varphi_k(x)dx \ , \qquad t \in R$$

is the characteristic function of the square integrable density function on the left of (4.2).

REFERENCES

[1] Gyires, B. *Contribution to the theory of linear combination of probability distribution functions*. Stud. Sci. Math. Hung. 16(1981), 297-324.

[2] Gyires, B. *The mixture of probability distribution functions by absolutely continuous weight functions*. Acta Sci. Math. 48 (1985), 173-186.

[3] Linnik, Yu. V. *Decomposition of probability distributions*. Oliver & Boyd Ltd., Edinburgh & London, 1964.

[4] Lukacs, E. *Characteristic functions*. Charles Griffin & Co. Limited, London, 1960.

# Construction of Minimax-Tests for Bounded Families of Distribution-Functions

Robert Hafner
Institut für Angewandte Statistik
Universität Linz
A-4045 Linz/Auhof, Austria

SUMMARY: For the composite testing problem $H_0 : F \in \mathcal{F}_0$ against $H_1 : F \in \mathcal{F}_1$, where $\mathcal{F}_i$ i=0,1 are families of distribution-functions on $(\mathbb{R}, \mathcal{B})$ defind by: $\mathcal{F}_i = \{F : \underline{F}_i \leq F \leq \bar{F}_i\}$ i = 0,1, least favourable pairs of distributions and the family of minimax-tests are constructed.

## 1. INTRODUCTION

Consider the composite testing problem $H_0 : F \in \mathcal{F}_0$ against $H_1 : F \in \mathcal{F}_1$, where $\mathcal{F}_0$ and $\mathcal{F}_1$ are families of distribution-functions on $(\mathbb{R}, \mathcal{B})$ defined by:

$$\mathcal{F}_i = \{F : \underline{F}_i \leq F \leq \bar{F}_1\} \qquad i = 0,1, \qquad (1)$$

the bounds $\underline{F}_i$, $\bar{F}_i$ also denoting distribution-functions on $(\mathbb{R}, \mathcal{B})$.

The family of minimax-tests for this testing problem is found by constructing a least favourable pair $(F_0^*, F_1^*)$ of distributions.
The pair $(F_0^*, F_1^*)$ is least favourable in the sense, that

a) $F_0^* \in \mathcal{F}_0$, $F_1^* \in \mathcal{F}_1$,

b) the Neyman-Pearson-tests for the simple testing problem $H_0 : F = F_0^*$ against $H_1 : F = F_1^*$ are minimax-tests for the above composite testing problem, i.e. for every Neyman-Pearson-test $\phi$ we have:

$$E(\phi|F_0^*) \geq E(\phi|F) \qquad \forall F \in \mathcal{F}_0, \qquad (2)$$
$$E(\phi|F_1^*) \leq E(\phi|F) \qquad \forall F \in \mathcal{F}_1.$$

KEYWORDS: Minimax-tests.Robust-tests.Least favourable pairs of distributions.

Contributions to Stochastics.
Ed. by W. Sendler
© Physica-Verlag Heidelberg 1987

This problem is a generalization of the problem to construct a least favourable pair of distributions for $H_0 : P \in \mathcal{P}_0$ against $H : P \in \mathcal{P}_1$, where $\mathcal{P}_i$ are local-variation-neughbourhoods of some central distributions $P_i$ on $(\mathbb{R}, \mathcal{B})$, i.e. $\mathcal{P}_i = \{P : P(B) \leq P_i(B^\varepsilon)\ \forall B \in \mathcal{B}\}$ $i = 0,1$ with $B^\varepsilon = \{x : d(x,B) \leq \varepsilon\}$.

This follows from the fact, that if F, $F_i$ denote the distribution-functions of P,$P_i$ respectively, $P \in \mathcal{P}_i$ iff $\underline{F}_i(x) = F_i(x-\varepsilon) \leq F(x) \leq F_i(x+\varepsilon) = \bar{F}_i(x)$.

The latter problem has been treated in [1] and [2]. It is shown in this paper that the method used in [1] und [2] can be used to solve the more general problem posed above.
Indeed by application of the results given in [3] and [4] for contamination neighbourhoods it will be shown how to constract least favourable pairs of distributions for the testing problem $H_0 : P \in \mathcal{P}_0'$ against $H_1 : P \in \mathcal{P}_1'$ where $\mathcal{P}_i'$ are defined by:

$$\mathcal{P}_i' = \{P : P(B) \leq v_i(B) + \delta_i\} \qquad 0 \leq \delta_i \leq 1, \quad i = 0,1 \tag{3}$$
$$\text{with } v_i(B) = \sup_{F \in \mathcal{F}_i} P_F(B) \quad \forall B \in \mathcal{B}$$

The families $\mathcal{P}_i'$ obviously are generalizations of the Prokhorov-neighbourhoods $\mathcal{P}_{i/Pr} = \{P : P(B) \leq P_i(B^{\varepsilon_i}) + \delta_i\}$ $i = 0,1$ of the central distributions $P_i$, for which least favourable pairs of distributions have been constructed in [1].

With this extension of the results given in [1], [2], [3], [4] and with the results of [5] , [6] and [7] it is now possible to construct robust tests for a large variety of data contamination.
For the statistical relevance of the diverse neighbourhoods in modelling data contamination see [8].

## 2. THE PRE-RISK-FUNCTION FOR SIMPLE ALTERNATIVES

Consider the simple testing problem $H_0 : F = F_0$, $H_1 : F = F_1$ on $(\mathbb{R}, \mathcal{B})$. The pre-risk-function $s(F_0,F_1)$ shall be defind in the following way:

$$s(F_0,F_1) = (a(x,\gamma);b(x,\gamma): -\infty \leq x \leq \infty,\ 0 \leq \gamma \leq 1),$$
$$a(x,\gamma) = (1-\gamma)F_0(x-)+\gamma F_0(x),$$
$$b(x,\gamma) = (1-\gamma)F_1(x-)+\gamma F_1(x). \tag{4}$$

Thus $s(F_0,F_1)$ is nothing else than the P-P-plot of the two distributions $F_0$ and $F_1$ (see [9]).
The definition given here differs a little from that used in [1] and [2], but it is more comfortable to work with.
The graph of $s(F_0, F_1)$, of which (4) is a parameter-description, is a

continuous monotonically increasing curve connecting the points (0,0) and (1,1). If $F_0(x) - F_0(x-) > 0$ and/or $F_1(x) - F_1(x-) > 0$ it contains line segments connecting the points $(F_0(x-), F_1(x-))$ and $(F_0(x), F_1(x))$.
If in x both $dF_i(x)/d\,x = f_i(x)$ $i = 0,1$ exist, the slope of $s(F_0,F_1)$ in $(F_0(x), F_1(x))$ is $f_1(x)/f_0(x)$.

As to the relations between the pre-risk-function and the risk-function of a simple testing problem, see [1] and [2]. We do not however need these relations in the present paper.

Let us now consider the compound testing problem $H_0$: $F \in \mathcal{F}_0$, $H_1$: $F \in \mathcal{F}_1$, $\mathcal{F}_i$ $i = 0,1$ beeing defined by (1).
The family of pre-risk-functions $(s(F_0,F_1) : F_0 \in \mathcal{F}_0,\ F_1 \in \mathcal{F}_1)$ lies between the two extreme curves $\underline{s} = s(\bar{F}_0, \underline{F}_1)$ and $\bar{s} = s(\underline{F}_0, \bar{F}_1)$ (see Fig. 1).

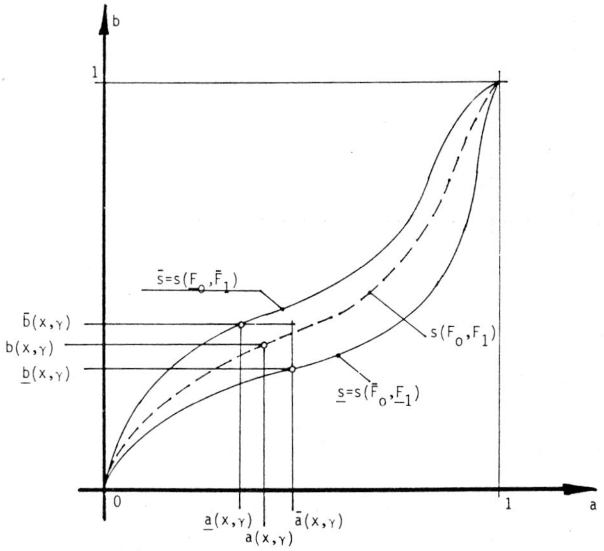


Figure 1: Relative position of $\underline{s} = s(\bar{F}_0,\underline{F}_1)$, $\bar{s} = s(\underline{F}_0,\bar{F}_1)$ and $s(F_0,F_1)$.

To be precise: if $\underline{s} = (\bar{a}(x,\gamma); \underline{b}(x,\gamma))$, $\bar{s} = (\underline{a}(x,\gamma); \bar{b}(x,\gamma)$ and $s(F_0,F_1) =$ $= (a(x,\gamma); b(x,\gamma)$ then:

$$\underline{a}(x,\gamma) \leq a(x,\gamma) \leq \bar{a}(x,\gamma) \ ; \ \underline{b}(x,\gamma) \leq b(x,\gamma) \leq \bar{b}(x,\gamma). \qquad (5)$$

On the other hand, if $s = (a(x,\gamma); b(x,\gamma))$ is the parameter description of a curve connecting the points (0,0), (1,1), where $a(x,1)$ and $b(x,1)$ are destribution functions, $a(x,\gamma) = (1-\gamma)a(x,0) + \gamma\, a(x,1)$ and $b(x,\gamma) =$ $= (1-\gamma)b(x,0) + \gamma\, b(x,1)$ for $0 \leq \gamma \leq 1$ and the unequalities (5) hold, then $a(x,1) = F_0 \in \mathcal{F}_0$, $b(x,1) = F_1 \in \mathcal{F}_1$ and $s = s(F_0,F_1)$.

Especially if $\bar{F}_i$ $\underline{F}_i$ i = 0,1 are continuous we have $\underline{s} = (\bar{F}_0(x), \underline{F}_1(x))$, $\bar{s} = (\underline{F}_0(x), \bar{F}_1(x))$ and if s is an arbitrary continuous monotonic curve connecting (0,0) and (1,1) between $\underline{s}$ and $\bar{s}$, s is the pre-risk-function of a pair $(F_0, F_1) \in \mathcal{P}_0 \times \mathcal{P}_1$. This is seen from Fig. 2, showing how s =(a(x),b(x)) can be parametrized to obey (5). Since obviously this parametrization is not unique there are infinitely many pairs $(F_0, F_1) \in \mathcal{P}_0 \times \mathcal{P}_1$ with pre-risk-function s.

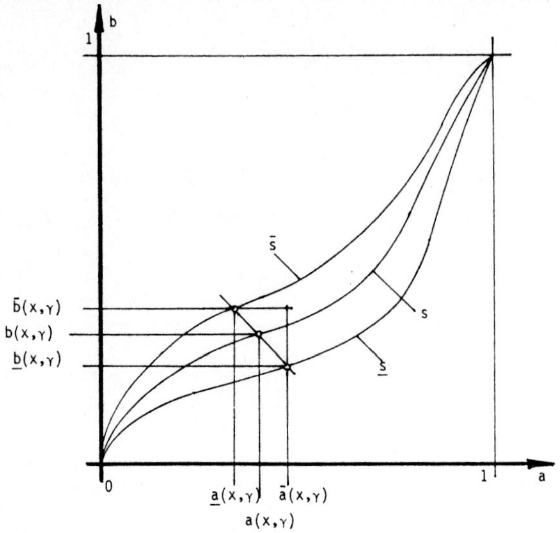


Figure 2: Parametrization of a curve s between $\underline{s}$ and $\bar{s}$

## 3. THE CONSTRUCTION OF LEAST FAVOURABLE PAIRS OF DISTRIBUTIONS

In order to make the presentation more transparent we assume that $\bar{F}_i$ $\underline{F}_i$ i=0,1 are continuous. In addition we postulate the following regularity conditions:

(I) the densities $\bar{f}_i$ $\underline{f}_i$ = $d\bar{F}_i/d\lambda$ $d\underline{F}_i/d\lambda$, i = 0,1 ($\lambda$ denoting Lebesgue-measure) are piecewise continuous,

(II) the measures $\bar{F}_i$ $\underline{F}_i$ i = 0,1 are equivalent and their common carrier is a (possibly infinite) interval [A,B].

The extension to more general cases offers however no serious difficulties.

We denote by $s^*$ the shortest connection of (0,0) and (1,1), subject to the condition, that $s^*$ lies between $\underline{s}$ and $\bar{s}$. (see Fig. 3), and we parametrize $s^*$ as indicated in Fig. 2. $s^*$ is thus the pre-risk-function of a pair $(F_0^*, F_1^*) \in \mathcal{P}_0 \times \mathcal{P}_1$, and $F_0^*$ , $F_1^*$ are themselves continuous with carrier-contained in [A,B] and piecewise continuous derivatives $f_0^*$ and $f_1^*$ . This follows from the fact, that $s^*$ consists of segments of $\underline{s}$ and $\bar{s}$ and straight line segments connecting them. Especially if $s^* = \underline{s}$ for $x_1 \leq x \leq x_2$ then by the parametrization used, $f_0^* = \bar{f}_0$, $f_1^* = \underline{f}_1$ on $[x_1,x_2]$ and simmilarly for $s^* = \bar{s}$. If on the other hand $s^*$ is a straight line segment with slope c for $x_3 \leq x \leq x_4$, then $f_1^* / f_0^* = c$ for $x \in [x_3,x_4]$.

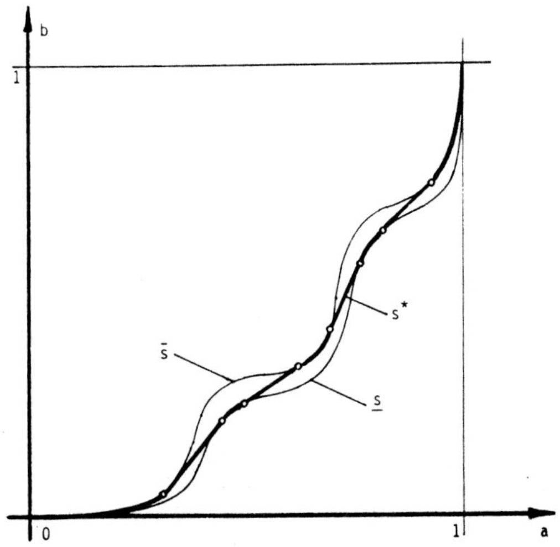


Figure 3: Definition of $s^*$

In order to show, that the pair $(F_0^*,F_1^*)$ is least favourable we define the sets $A_t^* = \{x : f_1^*(x) > t\, f_0^*(x)\}$ and $A_t^{*+} = \{x : f_1^*(x) \geq f_0^*(x)\}$, which because of regularity-condition I are both of the form:

$$A_t^{*(+)} = \bigcup_j \langle u_j, v_j \rangle \quad \text{with } v_j < u_{j+1} \tag{6}$$

LEMMA 3.1. For arbitrary t the points $(a^*(u_j),b^*(u_j))$ lie on $\bar{s}$ and the points $(a^*(v_j), b^*(v_j))$ lie on $\underline{s}$.

<u>PROOF:</u> Since $s^*$is linear in the strip between $\underline{s}$ and $\bar{s}$ with constant slope $f_1^*/f_0^*$, it is obvious that $(a^*(u_j), b^*(u_j))$ and $(a^*(v_j), b^*(v_j))$ must be boundary-points on $\underline{s}$ or $\bar{s}$. Since the parametrization of $s^*$ concides with that of $\underline{s}$ or $\bar{s}$ on boundary-points we necessarily have $(a^*(u_j), b^*(u_j)) = (\underline{a}(u_j), \bar{b}(u_j))$ if $(a^*(u_j), b^*(u_j))$ lies on $\bar{s}$ or $(a^*(u_j), b^*(u_j)) = (\bar{a}(u_j),\underline{b}(u_j))$ if $(a^*(u_j), b^*(u_j))$ lies on $\underline{s}$ and simmilarly for $v_j$.

We prove the proposition for $A_t^*$ and $(a^*(u_j),b^*(u_j))$. By definition of $A_t^*$ the slope of $s^*$ is $>t$ for $x \in (u_j,v_j)$ and $\leq t$ for $x \in (v_{j-1},u_j)$, so that

$s^*$ lies above the line $b - b^*(u_j) = t(a-a^*(u_j))$ in the neighbourhood of $(a^*(u_j), b^*(u_j))$, and strictly so for $x \in (u_j,v_j)$ (see Fig. 4)

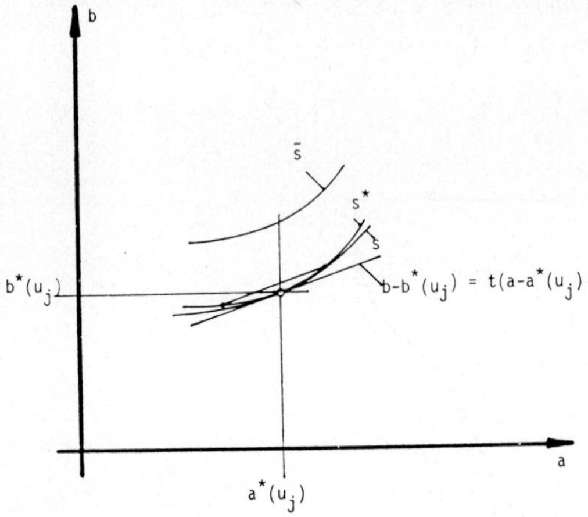


Figure 4: Local form of $s^*$ in the neighbourhood of $(a^*(u_j), b^*(u_j))$

If now $(a^*(u_j), b^*(u_j))$ where a point of $\underline{s}$, and $\bar{s}$ where strictly above $\underline{s}$ in the neighbourhood of this point, then obviously $s^*$ would not be the shortest admissible connection of (0,0) and (1,1) since $s^*$ could be shortened strictly by cutting away an arc in the neighbourhood of $(a^*(u_j), b^*(u_j))$ (see Fig. 4). The proof for $A_t^{*+}$ and for $v_j$ is similar. ∎

Denoting by $P_F$ the probability-measure with distribution function F we now prove:

LEMMA 3.2. For arbitrary t the following inequalities hold:

$$P_{F_0^*}(A_t^{*(+)}) \geq P_F(A_t^{*(+)}) \qquad \forall F \in \mathcal{P}_0$$

$$P_{F_1^*}(A_t^{*(+)}) \leq P_F(A_t^{*(+)}) \qquad \forall F \in \mathcal{P}_1$$

PROOF: This follows immediately from Lemma 3.1 since

$$P_{F_0^*}(A_t^{*(+)}) = \sum_j (F_0^*(v_j) - F_0^*(u_j)) = \sum_j (a^*(v_j) - a^*(u_j)) =$$

$$= \sum_j (\bar{a}(v_j) - \underline{a}(u_j)) = \sum_j (\bar{F}_0(v_j) - \underline{F}_0(u_j)) \geq \sum_j (F(v_j) - F(u_j)) \; \forall F \in \mathcal{P}_0,$$

and

$$P_{F_1^*}(A_t^{*(+)}) = \sum_j (F_1^*(v_j) - F_1^*(u_j)) = \sum_j (b^*(v_j) - b^*(u_j)) =$$

$$= \sum_j (\underline{b}(v_j) - \bar{b}(u_j)) = \sum_j (\underline{F}_1(v_j) - \bar{F}_1(u_j)) \leq \sum_j (F(v_j) - F(u_j)) \; \forall F \in \mathcal{P}_1 \; ∎$$

THEOREM 3.1: The pair ($F_0^*$ , $F_1^*$ ) is least favourable.

PROOF: Since ($F_0^*$, $F_1^*$) $\varepsilon \mathcal{F}_0 \times \mathcal{F}_1$ by construction, it remains to show, that the Neyman-Pearson-tests for the simple testing problem $H_0$:F = $F_0^*$ , $H_1$:F = $F_1^*$ are minimax for the composite problem $H_0$:F $\varepsilon \mathcal{F}_0$, $H_1$:F $\varepsilon \mathcal{F}_1$. But since every Neyman-Pearson-test $\phi$ is of the form:
$\phi = (1-\gamma)A_t^* + \gamma A_t^{*+}$ this is an immediate consequence of Lemma 3.2. ∎

## 4. LEAST FAVOURABLE PAIRS FOR GENERALIZED PROKHOROV-NEIGHBOURHOODS

We consider now the composite testing problem $H_0$:P $\varepsilon \mathcal{P}_0'$, $H_1$:P $\varepsilon \mathcal{P}_1'$ with $\mathcal{P}_i'$ i = 0,1 defined by (3).
As in [1] the construction of a least favourable pair of distributions can be accomplished in two steps.

STEP I: We construct ($F_0^*$,$F_1^*$) as described in section 3.

STEP II: We construct a least favourable pair ($P_0^*$,$P_1^*$) for the composite testing problem $H_0$:P $\varepsilon \mathcal{P}_0''$, $H_1$:P $\varepsilon \mathcal{P}_1''$ with

$$\mathcal{P}_i'' = \{P : P(B) \leq P_{F_i^*}(B) + \delta_i, \forall B \varepsilon \mathcal{B}\} \qquad i = 0,1 \qquad (7)$$

A method for this construction is shown in [3] and [4], the $\mathcal{P}_i''$ being special contamination-neighbourhoods of $P_{F_i^*}$.
The pair ($P_0^*$,$P_1^*$) is then least favourable for the composite testing problem formulated at the beginning of this section.
To see this note first, that ($P_0^*$,$P_1^*$) by construction is admissible.
Further, if we denote the critical sets of the Neyman-Pearson-tests for the pairs ($P_{F_0^*}$, $P_{F_1^*}$) and ($P_0^*$,$P_1^*$) by $A_t^*$ and $B_t^*$ respectively, it has been shown in [3], [4], that:

$$P_0^*(B_t^*) = P_{F_0^*}(B_t^*) + \delta_0, \quad P_1^*(B_t^{*C}) = P_{F_1^*}(B_t^{*C}) + \delta_1 \quad \forall t \geq 0$$

and that ($B_t^*$:t ⩾ 0) c ($A_t^*$:t ⩾ 0), i.e. every $B_t^*$ is an $A_t^*$.

Thus lemma 3.2 shows:

$$P_0^*(B_t^*) \geq P_F(B_t^*) + \delta_0 \quad \forall t \geq 0, \forall F \varepsilon \mathcal{F}_0,$$
$$P_1^*(B_t^*) \leq P_F(B_t^*) + \delta_1 \quad \forall t \geq 0, \forall F \varepsilon \mathcal{F}_1,$$

i.e. the family of Neyman-Pearson-tests $\phi_t = 1_{B_t^*}$ is minimax.
The minimax-property of the randomized Neyman-Pearson-tests is shown along the same lines.

REFERENCES

[1] Hafner,R. (1982). Simple Construction of Least Favourable Pairs of Distributions and of Robust Tests for Prokhorov-Neighbourhoods. Math.Operationsforsch.Statist., Ser.Statistics, 13,33-46.

[2] Österreicher,F. The Construction of Least Favourable Distributions is Traceable to a Minimum Parameter Problem. (Unpublished Mansucript).

[3] Hafner,R. (1982). Construction of Least Favourable Pairs of Distributions and of Robust Tests for Contamination Neighbourhoods. Math.Operationsforsch.Statist., Ser.Statistics, 13, 47-56.

[4] Österreicher,F. (1978). On the Construction of Least Favourable Pairs of Distributions. Z.Wahrscheinlichkeitstheorie verw. Gebiete, 43, 49-55.

[5] Hafner,R. (to appear). Konstruktion von Minimax-Tests für dominierte Familien von Wahrscheinlichkeitsdichten.

[6] Hafner,R. (to appear). Construction of Least Favourable Pairs of Distributions and of Minimax-Tests for Contamination versus Bounded Density Alternatives.

[7] Kusnetzov,V.P. (1983). Minimax Tests for Bounded Families of Distribution Densities. Theory Prob.Appl., 27, 299-309.

[8] Rieder,H.: Robuste Tests. Diplomarbeit, Freiburg.

[9] Wilk,M.B. & Gnanadesikan,R. (1968). Probability Plotting Methods for the Analysis of Data. Biometrika, 55, 1-17.

# An Asymptotic $\chi^2$-Test for Variance Components

Joachim Hartung
Fachbereich Statistik
Universität Dortmund
D-4600 Dortmund, FRG

Bernard Voet
Institut für Statistik
Universität Düsseldorf
D-4000 Düsseldorf, FRG

SUMMARY: In this paper we investigate an asymptotic $\chi^2$-test of a general linear hypothesis of variance components. First a formulation of the model is given and some basic results on exact $\chi^2$- and F-tests are presented. Then, repeated variance component models are introduced and an asymptotic $\chi^2$-test for a general linear hypothesis of variance components is derived. As an example the two way nested classification model with random effects is considered and some tests are stated explicitly.

## 1. INTRODUCTION

The problem of estimating variance components by unbiased resp. by minimum biased quadratic estimators has been considered by a number of authors, see for instance [6],[4] and also the discussions and references given for example by [8] p. 406-408 and [5] p. 196-202. But testing of variance components has been developed only for a limited number and a few types of hypotheses, cf. for instance [7],[3] chapter 15 and [2]. There the F-test of no factor effect resp. $\chi^2$-test for some special linear combinations of variance components are investigated and it is shown that these tests are already uniformly most powerful unbiased tests.
Here we develope an asymptotic $\chi^2$-test for variance components.
First a formulation of the variance component model is given and some basic results on exact $\chi^2$- and F-tests are presented. Further we introduce repeated variance component models and investigate an asymptotic $\chi^2$-test for a general linear hypothesis $K\alpha = d$, $K \in \mathbb{R}^{q\times m}$, $d \in \mathbb{R}^q$, of variance components $\alpha = (\alpha_1,\dots,\alpha_m)'$. As an example the two way nested classification is considered and some asymptotic $\chi^2$-tests are stated explicitly. In order to study the finite properties of this asymptotic

---

AMS 1980 subject classification: 62 J 10
Key words and phrases: Variance component models, commutative quadratic subspaces, repeated models, $\chi^2$- and F-tests, asymptotic $\chi^2$-tests.

Contributions to Stochastics.
Ed. by W. Sendler
© Physica-Verlag Heidelberg 1987

$\chi^2$-test, we compare the finite distribution of the test statistic with the known asymptotic distribution using a short Monte Carlo simulation study.

Throughout this paper let Sym denote the Hilbert space of all real symmetric (nxn) matrices, where the inner product is given by trace AB =: tr(AB). For a real (nxn) matrix A, we denote by $A^+$ the Moore-Penrose-Inverse, by R(A) the range, by rk(A) the rank and by A' the transposed of A. Further we define by $\otimes$ the usual Kronecker product of two matrices and by $I_r$ the (rxr) identity matrix. Other notation will be introduced as needed.

## 2. FORMULATION OF THE MODEL

In this paper we consider a variance component model of the form

$$Y = X\beta + e \tag{2.1}$$

where Y is an (nx1) vector of observable random variables, X is a known (nxp) matrix, $\beta$ a (px1) vector of unknown fixed effect parameters and e is an (nx1) vector of (random) errors with expectation E(e) = 0 and variance-covariance matrix

$$E(ee') = Cov(Y) = \sum_{i=1}^{m} \alpha_i U_i \ . \tag{2.2}$$

Here the m symmetric positive semidefinite (nxn) matrices $U_i$ are known and the (unknown) parameter $\alpha = (\alpha_1,\dots,\alpha_m)'$ varies in $\Omega$, a subset of $\mathbb{R}^m_{(+)}$, the non-negative orthant of $\mathbb{R}^m$.

The problem we consider is testing of the variance components $\alpha$ such that the tests are invariant under the group G of mean value translations:

$$G = \{Y \rightarrow Y + X\beta \mid \beta \in \mathbb{R}^p\} \ . \tag{2.3}$$

Now a maximal invariant linear statistic Z with respect to G is given by, cf. [10],

$$Z = Proj_{R(X)^\perp} Y = (I - XX^+)Y \ . \tag{2.4}$$

From (2.1) and (2.4) we get the reduced (by invariance) linear model

$$Z = (I - XX^+)\ Y \tag{2.5}$$

where $E(Z) = 0$ and $Cov(Z) = \sum_{i=1}^{m} \alpha_i V_i$

with $V_i = (I - XX^+)U_i(I - XX^+)$, $i=1,\dots,m$ .

For this model we make the following three assumptions.

Assumption 1: $Z$ has an n-dimensional normal distribution with mean vector 0 and covariance matrix $\sum_{i=1}^{m} \alpha_i V_i$ .

Assumption 2: $\Psi := \text{span}[V_1,\ldots,V_m]$ forms an m-dimensional commutative quadratic subspace of Sym, i.e. $\Psi$ is a subspace and $A,B \in \Psi$ implies $A^2 \in \Psi$ and $AB = BA$.

By lemma 6 of [9] there exist m pairwise orthogonal projection matrices $P_1,\ldots,P_m$ that form a basis for $\Psi$. So there is a regular matrix

$$\Phi = ((\lambda_{ij})_{i,j=1,\ldots,m}) \tag{2.6}$$

such that

$$V_i = \sum_{j=1}^{m} \lambda_{ij} P_j \ , \quad \text{for } i=1,\ldots,m \ . \tag{2.7}$$

Assumption 3:
$$\tau_j := \sum_{i=1}^{m} \alpha_i \lambda_{ij} > 0 \text{ for } j=1,\ldots,m \text{ and } \alpha \in \Omega \ . \tag{2.8}$$

These three assumptions are fulfilled in the balanced variance component models (ANOVA models) if as usual for the overall variance $\sigma_e^2 > 0$ is assumed. Further if we define

$$f_i = rk(P_i),\ T_i = Z'P_iZ\,/\,f_i \qquad \text{for } i=1,\ldots,m \tag{2.9}$$

and $T = (T_1,\ldots,T_m)'$, then under the above assumptions there holds, cf. [9] and [4],

(i) $E(T) = \Phi'\alpha = \tau$ with $\tau = (\tau_1,\ldots,\tau_m)'$,

(ii) $Cov(T) = 2\cdot \text{diag}(\tau_1^2/f_1,\ldots,\tau_m^2/f_m) =: D(\alpha)$ , (2.10)

(iii) $T_1,\ldots,T_m$ are statistically independent and form a complete sufficient statistic,

(iv) $f_iT_i/\tau_i$ has a central $\chi^2$-distribution with $f_i$ degrees of freedom, $i=1,\ldots,m$ .

Since in ANOVA models the $T_i, i=1,\ldots,m$, considered in (2.10) are the so called 'mean sum of squares', the above assumptions guarantee that the 'sum of squares' $f_iT_i$ have a central $\chi^2$-distribution and are statistically independent. Especially assumption 3 implies that the expected value of the 'sum of squares' is greater than zero. In the usual ANOVA-tables all these properties are implicitly given.

## 3. CLASSICAL $\chi^2$- AND F-TESTS FOR VARIANCE COMPONENTS

From (2.10) we can derive the classical F- and $\chi^2$-tests for special linear combinations of variance components.

If for a fixed $j_0 \in \{1,\dots,m\}$ there holds $\tau_{j_0} = d$, $d > 0$, then from (2.10) (iv) we have that $f_{j_0} T_{j_0}$ has a $d \cdot \chi^2$-distribution with $f_{j_0}$ degrees of freedom. Therefore we reject the hypothesis $H : \tau_{j_0} \leq d$ at level $\gamma \in (0,1)$ if

$$f_{j_0} T_{j_0} > d \cdot \chi^2_{f_{j_0},\gamma} \tag{3.1}$$

where $\chi^2_{f_{j_0},\gamma}$ is the $\gamma$-fractile of the central $\chi^2$-distribution with $f_{j_0}$ degrees of freedom. Hypotheses of the kind $\tau_{j_0} \neq d$ resp. $\tau_{j_0} \geq d$ can be handled analogously.

Another test which can be derived from (2.10) is the well known F-test. If the assumption $\tau_{j_0}/\tau_{\ell_0} = d$, $d > 0$, for fixed $\ell_0, j_0 \in \{1,\dots,m\}$ is fulfilled, then from (2.10) (iii) and (iv) we have that $T_{j_0}/T_{\ell_0}$ has a $d\, F_{f_{j_0},f_{\ell_0}}$-distribution, where

$F_{f_{j_0},f_{\ell_0}}$ denotes the central F-distribution with $f_{j_0}$ and $f_{\ell_0}$ degrees of freedom.

So we reject the hypothesis $H : \tau_{j_0}/\tau_{\ell_0} \leq d$ at level $\gamma \in (0,1)$ if

$$T_{j_0}/T_{\ell_0} > d\, F_{f_{j_0},f_{\ell_0},\gamma} \tag{3.2}$$

where $F_{f_{j_0},f_{\ell_0},\gamma}$ denotes the $\gamma$-fractil of the central $F_{f_{j_0},f_{\ell_0}}$-distribution.

It may be noted that the F-test of no factor effect is equivalent to the hypothesis $\tau_{\ell_0} = \tau_{j_0}$ resp. $\tau_{j_0}/\tau_{\ell_0} = 1$ for a suitable chosen pair $\ell_0$, $j_0 \in \{1,\dots,m\}$, cf. also the following example.
The optimum nature of these $\chi^2$- and F-tests is for example shown in [2] and for some models these exact tests are considered explicitly in [3].

For example let us discuss the two way nested classification model with random effects extensively considered in [4]. For this model, i.e.

$$Y_{ijk} = \mu + a_i + b_{ij} + e_{ijk}\,, \qquad \begin{array}{l} i=1,\dots, r > 1 \\ j=1,\dots, s > 1 \\ k=1,\dots, t > 1 \end{array} \tag{3.3}$$

we get

$$\Phi = \begin{bmatrix} st & 0 & 0 \\ t & t & 0 \\ 1 & 1 & 1 \end{bmatrix} \tag{3.4}$$

and with the usual dot notation

$$\begin{aligned} T_1 &= st \sum_{i=1}^{r} (\overline{Y}_{i\cdot\cdot} - \overline{Y}_{\cdot\cdot\cdot})^2/(r-1) \\ T_2 &= t \sum_{i=1}^{r} \sum_{j=1}^{s} (\overline{Y}_{ij\cdot} - \overline{Y}_{i\cdot\cdot})^2/r(s-1) \\ T_3 &= \sum_{i=1}^{r} \sum_{j=1}^{s} \sum_{k=1}^{t} (Y_{ijk} - \overline{Y}_{ij\cdot})^2/rs(t-1) \ . \end{aligned} \tag{3.5}$$

Let $\sigma_a^2, \sigma_b^2$ and $\sigma_e^2$ denote the variance components corresponding to the random effects $a_i$, $b_{ij}$ and $e_{ijk}$, respectively, then $\tau_1 = \sigma_a^2 st + \sigma_b^2 t + \sigma_e^2, \tau_2 = \sigma_b^2 t + \sigma_e^2$ and $\tau_3 = \sigma_e^2$. So only a few types of hypotheses are testable using the above $\chi^2$- and F-tests, cf. (3.1) and (3.2). For example the test of no factor effect rejects the hypothesis $H : \sigma_a^2 = 0$ at level $\gamma \in (0,1)$ if

$$T_1/T_2 > F_{r-1,r(s-1),\gamma} \ . \tag{3.6}$$

However, no exact tests for example are known for the following linear hypotheses ($d_i > 0$, $i=1,2$, given) $\sigma_a^2 = d_1$ resp. $\sigma_b^2 = d_2$ or $\sigma_a^2 = d_1$ and $\sigma_b^2 = d_2$ or $\sigma_a^2 = d_1 \sigma_b^2$, which are special linear hypotheses of the form $H : K\alpha = d$ with $\alpha = (\sigma_a^2, \sigma_b^2, \sigma_e^2)'$, $K \in \mathbb{R}^{q \times 3}$, $q \leq 3$ and $d \in \mathbb{R}^q$.

In the following section we develope an asymptotic $\chi^2$-test for a general linear hypothesis $H : K\alpha = d$ with $K \in \mathbb{R}^{q \times m}$, $rk(K) = q \leq m$, $d \in \mathbb{R}^q$ and $\alpha = (\alpha_1, \ldots, \alpha_m)'$.

## 4. AN ASYMPTOTIC $\chi^2$-TEST FOR VARIANCE COMPONENTS

If the invariant model (2.5) is n-times statistically independently repeated, we get the following model

$$Z^N = 1_N \otimes Z, \qquad \text{where } 1_N = (1,\ldots,1)' \in \mathbb{R}^N \tag{4.1}$$

with $E(Z^N) = 0$ and $Cov(Z^N) = \sum_{i=1}^{m} \alpha_i (I_N \otimes V_i)$.

It should be noted that $span[I_N \otimes V_1, \ldots, I_N \otimes V_m]$ forms an m-dimensional commutative quadratic subspace of Sym and the m pairwise orthogonal projection matrices are given by

$$P_i^N = I_N \otimes P_i \ , \quad \text{for } i,\ldots,m \ ,$$

with (4.2)

$$rk(P_i^N) = N \cdot f_i \ .$$

Hence the three assumptions of section 2 and so the statements under (2.10) in a corresponding way are fulfilled in the repeated model if they are fulfilled in the non-repeated model; note that the basis transformation matrix in the repeated model is also $\Phi$, as in the basic model.

Let us define, cf. also (2.9),

$$T_i^N = Z^{N'} P_i^N Z^N / Nf_i \ , \qquad i=1,\ldots,m$$

and

$$T^N = (T_1^N,\ldots,T_m^N)' \ , \quad \text{with } E(T^N) = \Phi'\alpha = \tau \ , \tag{4.3}$$

$$D^N(\alpha) = 2\text{diag}(\tau_1^2/Nf_1,\ldots,\tau_m^2/Nf_m) = D(\alpha)/N \ .$$

Here we repeat the reduced (by invariance) linear model and therefore the mean value is allowed to vary at each stage of the replication. However, one also can replicate the original model with non-zero mean value and then reduce the grand model for all N replications with respect to mean value translations. If this kind of replications is chosen, we can develope an asymptotic $\chi^2$-test in this model in a complete analogous way. The essential difference of these two kinds of replication lies in the number of the degrees of freedom. If we repeat the reduced model we 'lose' at each stage of the N replications the degrees of freedom for reducation (= rk(X)), and in the once reduced grand model we 'lose' only once the degrees of freedom for reduction.

Now to develope an asymptotic $\chi^2$-test for variance components we need the likelihood function of $T_i^N$, i=1,...,m. We have, cf. (2.10) (iv), that $Nf_i T_i^N/\tau_i$ has a central $\chi^2$-distribution with $Nf_i$ degrees of freedom, i=1,...,m, and so with (2.10) (iii) the likelihood function of $T_1^N,\ldots,T_m^N$ is given by

$$L^N(\alpha) = \prod_{i=1}^{m} C_i^N \, (Nf_i/\tau_i)^{Nf_i/2} \, T_i^{N\,(Nf_i-2)/2} \exp(-Nf_i T_i^N / 2\cdot\tau_i) \tag{4.4}$$

where

$$(C_i^N)^{-1} = 2^{Nf_i/2} \, \Gamma(Nf_i/2), \quad i=1,\ldots,m \ ,$$

and

$\Gamma$ denotes the usual gamma function.

The log-likelihood function $l^N(\alpha)$ is

$$l^N(\alpha) = \log L^N(\alpha)$$
$$= \sum_{i=1}^{m} \left[\log C_i^N + Nf_i \log(Nf_i/\tau_i)/2 + (Nf_i-2)\log(T_i^N)/2 - Nf_i T_i^N/2\tau_i\right] . \qquad (4.5)$$

For the first and second derivatives of $l^N$ with respect to $\alpha$ we get

$$\frac{\partial l^N(\alpha)}{\partial\alpha} = \Phi\, D^N(\alpha)^{-1}(T^N - \Phi'\alpha) , \qquad (4.6)$$

$$\frac{\partial^2 l^N(\alpha)}{\partial\alpha^2} = -\Phi\, D^N(\alpha)^{-1}\, \Phi' . \qquad (4.7)$$

Since $\Phi$ and $D^N(\alpha)$ are regular matrices, we get from (4.6) and (4.7) the following maximum likelihood estimator for $\alpha$, cf. also [4],

$$\tilde{\alpha}^N = \Phi'^{-1}\, T^N . \qquad (4.8)$$

Now for a general linear hypothesis $H : K\alpha = d$ let $\bar{\alpha}^N \in \mathbb{R}^m$ denote a solution of

$$\underset{\text{subject to:}K\alpha=d}{\text{maximize}} \quad l^N(\alpha) . \qquad (4.9)$$

The following theorem states that under the present assumptions $\tilde{\alpha}^N$ is asymptotically normal.

<u>Theorem 1</u>

Under the present assumptions there holds:

$\sqrt{N}(\tilde{\alpha}^N-\alpha)$ has asymptotically a normal distribution with mean value 0 and variance-covariance matrix $(\Phi D(\alpha)^{-1}\Phi')^{-1}$.

This theorem holds under more general assumptions than considered here, cf. for instance [1]. However, under the commutative quadratic subspace condition we get a more explicit and easy representation of the limiting covariance matrix and are able to prove some results about asymptotic tests.

In the next theorem we consider the likelihood ratio statistic

$$\Lambda^N = \max_{\alpha\in\mathbb{R}^m} L(\alpha) \Big/ \max_{K\alpha=d} L(\alpha) . \qquad (4.10)$$

Taking logarithms and using (4.8),(4.9), the statistic may be written as

$$2 \log \Lambda^N = 2(l^N(\tilde{\alpha}^N) - l^N(\bar{\alpha}^N)) , \qquad (4.11)$$

and we have the following theorem.

Theorem 2

Under the general linear hypothesis $K\alpha = d$ there holds

$$2(1^N(\tilde{\alpha}^N) - 1^N(\bar{\alpha}^N)) \underset{(a)}{\sim} \chi^2_{rk(K)}$$

i.e., $2 \log \Lambda^N$ is asymptotically central $\chi^2$-distributed with rk(K) degrees of freedom.

This theorem mainly can be established by using a Taylor expansion of $1^N(\tilde{\alpha}^N)$ at $\bar{\alpha}^N$ and the assertion of theorem 1.

Now let $Q(T^N) = 2(1^N(\tilde{\alpha}^N) - 1^N(\bar{\alpha}^N))$, then by (4.5) and (4.8) we have

$$Q(T^N) = \sum_{j=1}^{m} Nf_j\left(T_j^N \Big/ \sum_{i=1}^{m} \bar{\alpha}_i^N\lambda_{ij} - \log(T_j^N \Big/ \sum_{i=1}^{m} \bar{\alpha}_i^N\lambda_{ij}) - 1\right) \tag{4.12}$$

and under the hypothesis $K\alpha = d$ we get, cf. theorem 2,

$$Q(T^N) \underset{(a)}{\sim} \chi^2_{rk(K)}. \tag{4.13}$$

Therefore the hypothesis $K\alpha = d$ is rejected at the (asymptotic) level $\gamma \in (0,1)$ if

$$Q(T^N) > \chi^2_{rk(K),\gamma} . \tag{4.14}$$

As an example let us consider the repeated two way nested classification model with random effects, cf. (3.3).

If we repeat the reduced model (3.3), we get, cf. (4.2) and (4.3), with the usual dot notation

$$\begin{aligned} T_1^N &= st \sum_{l=1}^{N} \sum_{i=1}^{r} \left(\bar{Z}_{li\cdot\cdot}^N - \bar{Z}_{\cdot\cdot\cdot\cdot}^N\right)^2 \Big/ N(r-1) \\ T_2^N &= t \sum_{l=1}^{N} \sum_{i=1}^{r} \sum_{j=1}^{s} \left(\bar{Z}_{lij\cdot}^N - \bar{Z}_{li\cdot\cdot}^N\right)^2 \Big/ Nr(s-1) \\ T_3^N &= \sum_{l=1}^{N} \sum_{i=1}^{r} \sum_{j=1}^{s} \sum_{k=1}^{t} \left(Z_{lijk}^N - \bar{Z}_{lij\cdot}^N\right)^2 \Big/ Nrs(t-1) . \end{aligned} \tag{4.15}$$

Now from (4.14) we can derive asymptotic tests for the general linear hypothesis $H : K\alpha = d, K \in \mathbb{R}^{q\times 3}, rk(K) = q, d \in \mathbb{R}^q$.

Putting for example $K = I_3$, $d_i \geq 0$, i=1,2,3, we can test $H : (\sigma_a^2,\sigma_b^2,\sigma_e^2)' = (d_1,d_2,d_3)'$, and H is rejected at the (asymptotic) level $\gamma \in (0,1)$ if $Q(T^N) > \chi^2_{3,\gamma}$, where

$$Q(T^N) = N(r-1)\left[T_1^N/(d_1st+d_2t+d_3)-\log(T_1^N/(d_1st+d_2t+d_3))-1\right]$$
$$+ Nr(s-1)\left[T_2^N/(d_2t+d_3)-\log(T_2^N/(d_2t+d_3))-1\right] \qquad (4.16)$$
$$+ Nrs(t-1)\left[T_3^N/d_3-\log(T_3^N/d_3)-1\right].$$

Further testable hypotheses are for example $H : \sigma_a^2 = d,\ d > 0$ or $H : \sigma_a^2 = u \cdot \sigma_b^2, u>0$. It may be noted, that for this hypotheses no exact test is known.

In order to study the finite distribution of $Q(T^N)$, cf. (4.16), under the hypothesis we use a short Monte Carlo simulation study.

From theorem 2 we have that $Q(T^N)$ is asymptotically $\chi_3^2$-distributed as $N \to \infty$ and under the special hypothesis $H : (\sigma_a^2,\sigma_b^2,\sigma_e^2)' = (d_1,d_2,d_3)'$ there holds, cf. (2.10)(iv),

$N(r-1)T_1^N/(d_1st+d_2t+d_3)$ is $\chi_{N(r-1)}^2$-distributed.

Therefore pseudorandom uniform variables were generated using a congruence method and then transformed to pseudorandom $\chi^2$-variables with $N(r-1)$, $Nr(s-1)$ and $Nrs(t-1)$ degrees of freedom.

For different values of r,s,t and N the following tables present the $\gamma$-fractile $\chi_{3,\gamma}^2$ of the $\chi_3^2$-distribution and the empirical $\gamma$-fractiles $q_{3,\gamma,N}$ of

$$\tilde{Q}(T^N) = U_1 - N(r-1)\log(U_1)+N(r-1)(\log(N(r-1))-1)$$
$$+ U_2 - Nr(s-1)\log(U_2)+Nr(s-1)(\log(Nr(s-1))-1)$$
$$+ U_3 - Nrs(t-1)\log(U_3)+Nrs(t-1)(\log(Nrs(t-1))-1),$$

where $U_i$, $i=1,2,3$ stands for a $\chi^2$-distributed random variable with $N(r-1)$, $Nr(s-1)$ and $Nrs(t-1)$ degrees of freedom.

Note that under the considered hypothesis $Q(T^N)$ has the same distribution as $\tilde{Q}(T^N)$, cf. (2.10). The results of this simulation study show that under the hypothesis the finite distribution of $Q(T^N)$, cf. (4.16), and the asymptotic distribution of $Q(T^N)$ (here a $\chi_3^2$-distribution) are essentially the same already for this small values of r,s,t and N.

So for testing the hypothesis $H : K\alpha = d$ one can use the statistic $Q(T^N)$ with the percentage points of the $\chi_{rk(K)}^2$-distribution.

Table 1 Exact and empirical $\gamma$-fractiles for N=1,3

| $\gamma$ | $\chi^2_{3,\gamma}$ | r=s=t=2 $q_{3,\gamma,1}$ | r=s=t=2 $q_{3,\gamma,3}$ | r=s=t=4 $q_{3,\gamma,1}$ | r=s=t=4 $q_{3,\gamma,3}$ |
|---|---|---|---|---|---|
| 0.990 | 0.115 | 0.129 | 0.114 | 0.122 | 0.130 |
| 0.975 | 0.216 | 0.256 | 0.232 | 0.230 | 0.223 |
| 0.950 | 0.352 | 0.415 | 0.370 | 0.369 | 0.371 |
| 0.900 | 0.584 | 0.697 | 0.615 | 0.611 | 0.607 |
| 0.750 | 1.213 | 1.432 | 1.269 | 1.278 | 1.239 |
| 0.500 | 2.366 | 2.760 | 2.526 | 2.486 | 2.449 |
| 0.250 | 4.108 | 4.755 | 4.364 | 4.343 | 4.227 |
| 0.100 | 6.251 | 7.214 | 6.675 | 6.590 | 6.326 |
| 0.050 | 7.815 | 8.965 | 8.255 | 8.204 | 7 893 |
| 0.025 | 9.348 | 10.65 | 9.796 | 9.678 | 9.447 |
| 0.010 | 11.34 | 13.17 | 11.99 | 11.56 | 11.34 |
| 0.005 | 12.84 | 15.11 | 13.34 | 12.98 | 12.92 |

Table 2 Exact and empirical $\gamma$-fractiles for N=1

| $\gamma$ | $\chi^2_{3,\gamma}$ | r=3,s=4,t=5 $q_{3,\gamma,1}$ | r=4,s=3,t=5 $q_{3,\gamma,1}$ | r=5,s=4,t=3 $q_{3,\gamma,1}$ | r=3,s=2,t=4 $q_{3,\gamma,1}$ |
|---|---|---|---|---|---|
| 0.990 | 0.115 | 0.132 | 0.122 | 0.122 | 0.127 |
| 0.975 | 0.216 | 0.240 | 0.235 | 0.224 | 0.240 |
| 0.950 | 0.352 | 0.379 | 0.360 | 0.369 | 0.394 |
| 0.900 | 0.584 | 0.616 | 0.608 | 0.625 | 0.637 |
| 0.750 | 1.216 | 1.282 | 1.268 | 1.248 | 1.347 |
| 0.500 | 2.366 | 2.478 | 2.464 | 2.432 | 2.591 |
| 0.250 | 4.108 | 4.284 | 4.315 | 4.214 | 4.452 |
| 0.100 | 6.251 | 6.495 | 6.561 | 6.471 | 6.790 |
| 0.050 | 7.815 | 8.224 | 8.196 | 8.178 | 8.452 |
| 0.025 | 9.348 | 10.01 | 9.773 | 9.909 | 10.26 |
| 0.010 | 11.34 | 12.11 | 12.00 | 11.83 | 12.00 |
| 0.005 | 12.84 | 13.53 | 13.72 | 13.19 | 14.06 |

## REFERENCES

[1] ANDERSON, T.W. (1973). Asymptotically Efficient Estimation of Covariance Matrices with Linear Structure. Ann. Statist. 1, p. 135-141.

[2] ELBASSIOUNI, M.Y./SEELY, J. (1980). Optimal Tests for Certain Functions of the Parameters in a Covariance Matrix with Linear Structure. Sankhya 42, p. 64-77.

[3] GRAYBILL, F.A. (1976). Theory and Application of the Linear Model. Duxbury Press, North Scituate, Massachusetts.

[4] HARTUNG, J. (1981). Nonnegative Minimum Biased Invariant Estimation in Variance Component Models. Ann. Statist. 9, p. 278-292.

[5] LEHMANN, E.L. (1983). Theory of Point Estimation. Wiley, New York.

[6] RAO, C.R. (1972). Estimation of Variance and Covariance Components in Linear Models. J. Amer. Statist. Assoc. 67, p. 112-115.

[7] SCHEFFÉ, H. (1959). The Analysis of Variance. Wiley, New York.

[8] SEARLE, S.R. (1971). Linear Models. Wiley, New York.

[9] SEELY, J. (1971). Quadratic Subspaces and Completeness. Ann. Math. Statist. 42, p. 710-721.

[10] SEELY, J. (1972). Completeness for a Family of Multivariate Normal Distribution. Ann. Math. Statist. 43, p. 1644-1647.

# Probabilities on Groups: Submonogeneous Embedding and Semi-Stability

Wilfried Hazod
Fachbereich Mathematik
Universität Dortmund
D-4600 Dortmund, FRG

Summary: There are different generalizations of the concept of stable and semistable probability distributions on the real line. In this paper we show that two natural generalizations coincide for large classes of Lie groups. Whereas however for general locally compact groups these concepts are different.

INTRODUCTION

In a sequence of papers the following generalization of semistable probabilities on locally compact groups was introduced:
A continuous convolution semigroup $(\mu_t, t \geq 0)$ of probabilities on a locally compact group G is called (strictly) semistable, if there exist $\tau \in \mathrm{Aut}(G)$, $c \in (0,1) \cup (1,\infty)$, such that

$$\tau(\mu_t) = \mu_{ct}, \quad t > 0. \tag{1}$$

(See e.g. [ 3 - 7, 17 ]).

On the other hand, starting from P. Baldi [ 1 ] and his definition of stable measures on groups, T. Drisch and L. Gallardo were led to the following definition: A measure $\mu$ is B-semistable if there exist $k,l \in \mathbb{N}$, $k \neq l$ and $\tau \in \mathrm{Aut}(G)$ s.t. $\tau(\mu^{*k}) = \mu^{*l}$. (Cf. [ 2 ] for B-stability). It follows (private communication) that in this case there exists a non-cyclic subsemigroup S of rational numbers, a homomorphism $S_+ \ni s \longmapsto \mu_s \in M^1(G)$ and $c > 0$, with $c \cdot s \in S$ for $s \in S$, such that $\mu_{s_0} = \mu$ for some $s_0 \in S_+$ and

$$\tau(\mu_s) = \mu_{cs}, \quad s \in S_+. \tag{2}$$

The question arises under which conditions on G these definitions coincide. I.e. under which conditions is $s \longmapsto \mu_s$ continuous?

We show (cf. § 4) that for certain discrete groups and also for certain connected locally compact groups it is possible to construct

Contributions to Stochastics.
Ed. by W. Sendler
© Physica-Verlag Heidelberg 1987

semigroups $\{\mu_t\}_{t \in S_+}$ which fulfill (2) but not (1), i.e. the semigroups are not continuously embeddable.
In § 1 resp. § 2 we study the case of discrete groups resp. of Lie groups, and give sufficient conditions for the equivalence of (1) and (2).

The situation is rather simple if we suppose the measures to be symmetric. We sketch this briefly in § 3.

NOTATIONS

G denotes a locally compact group, $M^1(G)$ the set of probability measures, * denotes convolution.
Let S be a subgroup of $\mathbb{R}$, $S_+ := S \cap (0,\infty)$, the additive subsemigroup of positive elements. A convolution semigroup is a homomorphism $S_+ \ni s \longmapsto \mu_s \in M^1(G)$ (with respect to convolution). $\{\mu_s\}$ is called a continuous convolution semigroup (abr. c.c.s.) if $S_+ = \mathbb{R}_+$ and if $s \longmapsto \mu_s$ is weakly continuous.

A subsemigroup $S_+ \subseteq \mathbb{R}_+$ is called submonogeneous if there exist sequences $\{r_n\}_{n\geq 1} \subseteq S_+$, $\{k_n\} \subseteq \mathbb{N}$, such that

$$k_n r_{n+1} = r_n, \quad n \in \mathbb{N}$$

and $\bigcup_n \langle \mathbb{N} r_n \rangle = S_+$, where $\langle \mathbb{N}\ r \rangle$ denotes the cyclic semigroup generated by r.
A homomorphic image $\{\mu_s\}_{s \in S_+}$ of a submonogeneous semigroup $S_+$ is called a submonogeneous convolution semigroup.

Submonogeneous subgroups of topological semigroups were studied by K.H. Hofmann [ 9 ], in connection with probabilities on groups by L. Schmetterer and the author [ 13, 14 ] (s. [ 8 ] for a survey). Recently M. McCrudden [ 10 ] published new results on embedding submonogeneous semigroups into c.c.s. The role of submonogeneous semigroups in connection with stability was studied by E. Siebert [ 17 ]. Later on we will use heavily the results of McCrudden and Siebert.

Aut(G) always denotes the group of topological automorphisms of G.

In the following we always assume G to be a locally compact group,
$\tau \in \mathrm{Aut}(G)$,
$c \in (0, 1) \cup (1, \infty)$,

$S_+$ to be a submonogeneous semigroup of $\mathbb{R}_+$, such that $c \cdot s \in S_+$ for $s \in S_+$ and $\{\mu_s\}_{s \in S_+}$ to be a submonogeneous convolution semigroup (of $M^1(G)$) which is semistable, i.e. $\tau(\mu_s) = \mu_{cs}$, $s \in S_+$.

## § 1. DISKRETE GROUPS

We collect some results concerning semistable submonogeneous convolution semigroups on discrete groups G.

The measures $\mu_s$, $s \in S_+$, are now discrete measures, therefore we can apply E. Siebert's results (cf. [ 17 ] § 2 Theorem 2 and remark 3):

1.1 Proposition. Let G, $\{\mu_s\}_{s \in S_+}$, $\tau$, c as above (G discrete). Then there exists a finite $\tau$-invariant subgroup $F \subseteq G$, a homomorphism $S_+ \ni s \longmapsto x_s \in G$, such that $\mu_s = \varepsilon_{x_s} * \omega_F = \omega_F * \varepsilon_{x_s}$, $s \in S_+$. (Here $\omega_F$ is the normalized Haar measure on F).

We call $\{\mu_s\}$ trivial if $\{\mu_s\}_{s \in S_+}$ is finite. In this case $H := \{x_s \cdot F,\ s \in S_+\}$ is a finite group, such that $\text{supp}(\mu_s) \subseteq H$ for $s \in S_+$. Furthermore there exists $k \in \mathbb{N}$, such that $\mu_{ks} = \omega_H$, $s \in S_+$.

$\{\mu_s\}$ is called nontrivial if $\{\mu_s\}$ is infinite, hence G contains an infinite submonogeneous subgroup. We obtain:

1.2 Proposition. Suppose that any abelian subgroup of G is finitely generated. Then any submonogeneous semistable convolution semigroup $\{\mu_s\}$ is trivial.

1.3 Remarks 1. If $\{\mu_s\}_{s \in S_+}$ is semistable and embeddable into a continuous convolution semigroup, then $\mu_s = \omega_F$, $s \in S_+$.
[ For $\mu_s$ is a Poisson semigroup in this case, and semistable Poisson semigroups are trivial. ]

2. $\{\mu_s\}_{s \in S_+}$ is called locally tight if for $r \in S_+$ the set $\{\mu_s : 0 < s \leq r\}$ is relatively compact (cf. McCrudden [ 10 ]). Locally tight semistable convolution semigroups $\{\mu_s\}$ are always trivial.
[ $\mu_s$ has the form $\mu_s = \varepsilon_{x_s} * \omega_F$. Hence local-tightness implies finiteness of $\{x_s F : s \in S_+\}$ ].

3. The hypothesis of 1.2 is fulfilled if G is a finite extension of a polycyclic group. Any subgroup is finitely generated then (cf. e.g. [ 15 ] ). For this interesting class of groups the embedding problem is solved: any infinitely divisible measure is Poisson (McCrudden [11]).

4. Examples of nontrivial semistable convolution semigroups are constructed in §4.

## § 2. LIE GROUPS

Let G be a Lie group, let $\{\mu_s\}_{s \in S}$, $\tau$, c as above.

We combine some recent contributions to the embedding problem with the results mentioned in § 1.

Let $\Gamma$ be the closed subgroup generated by the supports of $\mu_s$, $s \in S_+$. Let $G_o$ resp. $\Gamma_o$ be the connected component of the unit element in G resp. $\Gamma$. Furthermore let $\pi_1 : \Gamma \longrightarrow \Gamma / \Gamma_o$ resp. $\pi : G \longrightarrow G/G_o$ be the canonical homomorphisms and finally we denote $\bar{\mu}_s := \pi(\mu_s)$ resp. $\nu_s := \pi_1(\mu_s)$. Obviously $\{\bar{\mu}_s\}_{s \in S_+}$ and $\{\nu_s\}_{s \in S_+}$ are semistable submonogeneous convolution semigroups on the discrete groups $G/G_o$ resp. $\Gamma/\Gamma_o$, hence the results mentioned in § 1 apply.

2.1 Propositon. Assume that the abelian subgroups of $\Gamma/\Gamma_o$ are finitely generated.

Then (i) $\Gamma$ is a finite extension of $\Gamma_o$,

(ii) $\{\nu_s\}_{s \in S_+}$ and $\{\bar{\mu}_s\}_{s \in S_+}$ are trivial

(iii) $\{\mu_s\}_{s \in S_+}$ is locally tight

( i.e. for any $r \in S_+$ $\{\mu_s : o < s < r\}$ is relatively compact).

Proof: We apply 1.2 to $\{\nu_s\}_{s \in S_+}$. Hence $\nu_s = \varepsilon_{x_s} * \omega_F$, and $H = \{x_s F : s \in S\}$ is a finite subgroup supporting the measures $\nu_s$. Therefore $\Gamma/\Gamma_o \cong H$.

(iii) follows from McCrudden [ 10 ] Theorem 1. □

2.2 Proposition. Assume that all subgroups of $G/G_o$ are finitely generated and assume $\Gamma$ to be solvable. Then again (i), (ii), (iii) of 2.1 hold.

Proof: From McCrudden [ 10 ] Theorem 2 follows that $\{\mu_s\}_{s \in S_+}$

is locally tight. Hence $\{\nu_s\}_{s \in S_+}$ is locally tight then. But 1.3.2 yields the finiteness of $\Gamma / \Gamma_0$ in this case. □

The proof of the following proposition is essentially due to M.McCrudden (private communication):

2.3 Proposition. Let G be connected. Then again (i), (ii), (iii) of 2.1 hold.

Proof: We use the following powerful result (cf. G.D. Mostow [ 12 ] Theorem 1', Remark):

Let G be a connected Lie group, K a closed subgroup. If $K/K_0$ is solvable then $K/K_0$ is polycyclic (and rank $(K/K_0) \leq \dim(G/K)$ ).

Put now again $\pi_1 : \Gamma \longrightarrow \Gamma / \Gamma_0$, $A := \{ x_s, s \in S \}$.
Put $H := \pi_1^{-1}(A)$. H is a closed subgroup of G with finite index in $\Gamma$. $H/H_0$ is submonogeneous, hence abelian. Therefore $H/H_0$ is finitely generated and submonogeneous, hence $H/H_0$ and $\Gamma / \Gamma_0$ are finite. □

2.4 Lemma. Let $(\nu_s)_{s \in S_+}$ be a locally tight submonogeneous convolution semigroup. Then there exist a continuous convolution semigroup $(\rho_t)_{t \in \mathbb{R}_+}$ with $\rho_0 = \omega_K$, a compact subgroup $L \subseteq N(K) \subseteq G$, with LK/K abelian, a homomorphism $S \ni s \longmapsto x(s) \in KL/K$, such that

$$\nu_s = \varepsilon_{x(s)} * \rho_s = \rho_s * \varepsilon_{x(s)}, \quad s \in S_+.$$

Furthermore there exist $k \in \mathbb{N}$, such that $x(k\,s)K$ lies in the connected component of the unit in LK/K, $s \in S_+$. Hence any $x(k\,s)K$ lies on a one parameter subgroup, and hence any $\nu_{ks}$ is continuously embedded.

Proof: If $S = \mathbb{Q}$ the submonogeneous group of rational numbers, the reader is referred to [17 ] or [ 8 ] III. 3.4 for a proof. In this case we have $k = 1$.

The proof in the general case goes along the same steps. We have to observe that the compact abelian subgroup H consisting of limit points of $\{\nu_s\}_{s \longrightarrow 0}$ in $M^1(G)$ is not necessarily connected. This is due to the fact that there may be nontrivial homomorphisms of S into the finite group $H/H_0$. But if we put $k := \mathrm{ord}(H/H_0)$ we obtain that any limit point of $\{\nu_{ks}\}_{s \longrightarrow 0}$ is contained in the connected subgroup $H_0$.

We omit the details. □
Combining the steps 2.1 - 2.4 we obtain the main result:

2.5 Theorem. Let G be a Lie group, $(\mu_s)_{s \in S_+}$ a submonogeneous convolution semigroup which is semistable w.r.t. $(\tau, c)$. Suppose that the hypotheses of 2.1 resp. 2.2 resp. 2.3 hold.

Then there exist a continuous convolution semigroup $(\nu_t)_{t \in \mathbb{R}_+}$ with $\nu_o = \omega_K$, a compact subgroup $L \subseteq N(K)$, a map $S \ni s \mapsto x(s) \in L$, such that $s \mapsto x(s)K \in LK/K$ is a homomorphism, and such that $\mu_s = \varepsilon_{x(s)} * \nu_s = \nu_s * \varepsilon_{x(s)}$, $s \in S_+$. Furthermore there exist $k \in \mathbb{N}$, such that any $\varepsilon_{x(ks)}$ and hence any $\mu_{ks}$ is continuously embeddable.

The continuous convolution semigroup $(\nu_t)_{t \in \mathbb{R}_+}$ is again semistable, but in a wider sense, i.e. there exists a map $\mathbb{R}_+ \ni t \mapsto y(t) \in L$ such that $\tau(\nu_t) = \nu_{ct} * \varepsilon_{y(t)} = \varepsilon_{y(t)} * \nu_{ct}$, $t \in \mathbb{R}_+$.

Proof: The first assertions follow immediately from 2.1 - 2.3 in combination with 2.4.

We have to prove the semistability of $(\nu_t)_{t \in \mathbb{R}_+}$:

Let $s \in S_+$. Then $\nu_s = \mu_s * \varepsilon_{x(s)^{-1}} = \varepsilon_{x(s)^{-1}} * \mu_s$.

Hence $\tau(\nu_s) = \tau(\mu_s) * \varepsilon_{\tau(x(s)^{-1})} = \mu_{cs} * \varepsilon_{\tau(x(s)^{-1})} =$

$$= \nu_{cs} * \varepsilon_{x(cs) \cdot \tau(x(s)^{-1})}.$$

The proof follows if we define $y(s) := x(cs)\tau(x(s)^{-1})$, and if we observe that $S_+$ is dense in $\mathbb{R}_+$ and $t \mapsto \nu_t$ and $t \mapsto \nu_{ct}$ are continuous. □

2.6 Corollary. If in 2.5 we have $\mu_s = \nu_s$, $s \in S_+$, e.g. if the measures $\mu_s$ are positive definite or if G has no non-trivial compact subgroups, then $(\nu_s)_{s \in \mathbb{R}_+}$ is semistable in the strict sense.

## § 3. POSITIVE DEFINITE MEASURES

As it is well known, and asserted in several papers, the things become quite simple if the measures are supposed to be symmetric or positive definite, since in this case Hilbertspace - technics may be applied. We illustrate this in the following example. A measure $\mu$ on a locally compact group G is called positive definite if the convolution operator $T_\mu : f \longrightarrow \mu * f$ operates on $L^2(G)$ as a non-negative definite operator.

3.1 Proposition. Let $(\mu_s)_{s \in S_+}$ be a submonogeneous convolution semigroup, suppose that the measures $(\mu_s)_{s \in S_+}$ are positive definite. Then there exists a continuous convolution semigroup $(\nu_t)_{t \in \mathbb{R}_+}$, such that $\mu_t = \nu_t$, $t \in S_+$.

Indeed a more general result yields:

Assume $\mu \in M^1(G)$. Let $k_n \in \mathbb{N}$, $k_n \uparrow \infty$ and suppose that there exist

$\mu_{(n)} \in M^1(G)$, such that $\mu_{(n)}^{k_n} = \mu$, $n \in \mathbb{N}$.

Assume furthermore that $\mu_{(n)}$ are positive definite, $n \in \mathbb{N}$. Then there exists a continuous convolution semigroup

$(\nu_t)_{t \in \mathbb{R}_+} \subseteq M^1(G)$ with $\mu = \nu_1$, $\mu_{(n)} = \nu_{1/k_n}$.

And $(\nu_t)_{t \in \mathbb{R}_+}$ is uniquely determined by $\mu$.

Proof: We represent the measures by the corresponding convolution operators on $L^2(G)$. Then $T_{\mu_{(n)}}$ is the non-negative definite $k_n$-th root of $T_\mu$.

Hence, if $T_\mu = \int_0^1 x \, d E_x$ is the spectral representation, we have $T_{\mu_{(n)}} = \int_0^1 x^{1/k_n} \, d E_x$, $n \in \mathbb{N}$.

Let $D \subseteq \mathbb{R}_+$ be the additive subsemigroup generated by $\{1/k_n, n \in \mathbb{N}\}$. We obtain a homomorphism $D \ni d \longmapsto \mu_d \in M^1(G)$, such that $\mu_1 = \mu$, $\mu_{(n)} = \mu_{1/k_n}$, $n \in \mathbb{N}$, and $T_{\mu_d} = \int_0^1 x^d \, d E_x$.

Since $\{ T_t := \int_0^1 x^t \, d E_x \}_{t \in \mathbb{R}_+}$ is a strongly continuous group of operators and since D is dense in $\mathbb{R}_+$ by assumption, we obtain the existence of $\mu_t \in M^1(G)$, $t \in \mathbb{R}_+$, with $T_t = T_{\mu_t}$. □

Let now $(\mu_t)_{t \in \mathbb{R}_+}$ be a continuous semigroup consisting of positive definite measures, which is semistable w.r.t. $(\tau, c)$. Let $(E_x)_{x \in [0,1]}$ be the corresponding spectral resolution, such that $T_{\mu_t} = \int_0^1 x^t dE_x$. Semistability is now expressible as a condition on the spectral resolution:

Let $\hat{\tau}$ be the natural action of $\tau$ on the self adjoint operators on $L^2(G)$. Then the relation $\tau\mu_t = \mu_{ct}$, $t \in \mathbb{R}_+$ is equivalent to $\hat{\tau}E_x = E_{x^{1/c}}$, $x \in [0,1]$.

[ This follows immediately from the uniqueness of the spectral representation of $\hat{\tau}\, T_{\mu_t}$ resp. $T_{\mu_{ct}}$, $t > 0$ ].

Furthermore, if $(F_x := E_{e^{-x}})_{x \in \mathbb{R}_+}$ is the spectral resolution of the infinitesimal generator of $(T_{\mu_t})_{t \in \mathbb{R}_+}$, we obtain

$$\hat{\tau}\, F_x = F_{x/c}, \qquad x \in \mathbb{R}_+.$$

Define for $k \in \mathbb{Z}$

$$G_x^{(k)} := \begin{cases} 0 & x < c^{k+1} \\ F_x - F_{c^{k+1}+} & x \in [c^{k+1}, c^k). \\ F_{c^k} - F_{c^{k+1}+} & x > c^k \end{cases}$$

Then we obtain $F_x = \sum_{k \in \mathbb{Z}} G_x^{(k)}$, $x \in \mathbb{R}$, where $G_x^{(0)}$ is concentrated on $[c,1)$ and $G_x^{(k)} = \hat{\tau}^{-k}\, G_x^{(0)}$.

It can be shown that this is closely related to the canonical representation of the Lévy-measure of a semistable convolution semigroup.

## § 4. EXAMPLES

1. We show the existence of a discrete, non finitely generated group, supporting a non continuously embeddable submonogeneous and semistable convolution semigroup:

Let S be an infinite submonogeneous subgroup of $\mathbb{Q}$ (e.g. $S = \mathbb{Q}$), let $c > o$ be an element of S, such that $cs \in S$ for $s \in S$. Define $\tau \in \mathrm{Aut}(S)$ via $\tau(s) := cs$, $s \in S$.

Now we define the convolution semigroup $\mu_s := \varepsilon_s$, $s \in S_+$. Obviously $\{\mu_s\}_{s \in S_+}$ is not continuously embeddable, but on the other hand $\tau(\mu_s) = \mu_{cs}$ by definition.

2. There exist locally compact groups, supporting a locally tight semistable submonogeneous convolution semigroup, which is not continuously embeddable. In this case G is not a Lie group, but G may be choosen connected, abelian and compact. Furthermore $\mathrm{supp}(\mu_s) = G$, $s \in S_+$.

Let S be the submonogeneous group of dyadic numbers, put $c := 2$.
Let K be a compact group containing an element $x_0$ of order 3. Then we define $G := K^{\mathbb{Z}}$ furnished with the product topology. Elements of G are considered as functions $f : \mathbb{Z} \longrightarrow K$. $\tau \in \mathrm{Aut}(G)$ is defined by $(\tau f)(k) := f(k+1)$, $k \in \mathbb{Z}$.
Put further for $n \in \mathbb{Z}$ $\mathbb{Z}_n := \{n, n+1, \ldots\}$ and
$G_n := K^{\mathbb{Z}_n} = \{ f : \mathbb{Z} \longrightarrow K$ such that $f(k) = e$ for $k < n \}$.
$\{G_n\}_{n \in \mathbb{Z}}$ is a sequence of compact subgroups of the compact group G, such that $\bigcup G_n = G$, $\bigcap G_n = \{e\}$, $G_n \supseteq G_{n+1}$, and $\tau G_n = G_{n+1}$, $n \in \mathbb{Z}$.

We define $\eta := \sum_{k \in \mathbb{Z}} 2^{-k} \tau^k \omega_{G_k}$. $\eta$ is a positive measure on $G \setminus \{e\}$, which is bounded on complements of neigbourhoods of $\{e\}$. Hence $\eta$ is a Lévy measure generating a continuous semigroup $(\nu_t)_{t \geq 0}$ of probabilities. From $\tau\eta = 2\eta$ we obtain $\tau\nu_t = \nu_{2t}$, $t \geq 0$.

The group G contains an element $f_0$, such that
$\tau f_0 = f_0^2 = f_0^{-1}$, $f_0^3 = e$.
[ We define $f_0 : \mathbb{Z} \longrightarrow K$ via $f_0(k) = x_0$ if k is even,
$f_0(k) = x_0^2$ if k is odd. ]

Then we define $\mu_{1/2^k} := \varepsilon_{f_0^k} * \nu_{1/2^k}$.

We have $(\mu_{1/2^{k+1}})^{*2} = \mu_{1/2^k}$, $k \in \mathbb{Z}$, hence a submonogeneous convolution semigroup $\{\mu_d\}_{d \in S_+}$ is generated.

But the construction yields

$$\tau(\nu_{1/2^k}) = \varepsilon_{\tau(f_0^k)} * \tau\nu_{1/2^k} = \varepsilon_{f_0^{k-1}} * \nu_{1/2^{k-1}}, \text{ hence } \tau\mu_d = \mu_{2d},\ d \in S_+.$$

Hence, if we put $K = T$, the one-dimensional torus group, we obtain an example of a compact abelian connected group supporting a semi-stable submonogeneous convolution semigroup, which is not continuously embeddable.

Of course the submonogeneous semigroup is continuously embeddable up to a shift in this case. We do not know if there exist examples without this property.

To obtain an example of a submonogeneous semistable semigroup $(\mu_d)$ with an infinite set of accumulation points we consider the following modification: Let $G, \tau, (\nu_t)_{t \in \mathbb{R}_+}$ as above. Let $y_0 \in T$ be an element of infinite order. Then we define $g_0 \in G$:

$$g_0(k) := y_0^{2^k}, \qquad k \in \mathbb{Z}.$$

Obviously $\tau g_0 = g_0^2$. Hence, if we put $g_{2^n} := g_0^{2^n}$ $(= \tau^n g_0)$, $n \in \mathbb{Z}$, an injective homomorphism of the dyadic numbers $d \longmapsto g_d \in G$ is defined. $\mu_d := \nu_d * \varepsilon_{g_d}$ is the desired example.

REFERENCES

[ 1 ] P. Baldi: Lois stables sur les déplacements de $\mathbb{R}^d$. In: Probability measures on groups. Proceedings Oberwolfach (1978). Lecture Notes in Math. 706, 1 - 9. Springer (1979).

[ 2 ] T. Drisch, L. Gallardo: Stable laws on the Heisenberg group. In: Probability measures on groups. Proceedings Oberwolfach (1983). Lecture Notes Math. 1064, 56 - 79 (1984).

[ 3 ] W. Hazod: Stable probabilities on locally compact groups. In: Probability measures on groups. Proceedings Oberwolfach (1981). Lecture Notes Math. 928, 183 - 211 (1982).

[ 4 ] W. Hazod: Remarks on [ semi- ] stable probabilities. In: Probility measures on groups. Proceedings Oberwolfach (1983). Lecture Notes Math. 1064, 182 - 203 (1984).

[ 5 ] W. Hazod: Stable and semistable probabilities on groups and vector spaces. In: Probability theory on vector spaces III. Proceedings Lublin (1983). Lecture Notes Math. 1080, 69 - 89 (1984).

[ 6 ] W. Hazod: Semigroups de convolution [ demi- ] stables et autodécomposables sur les groupes localement compacts. In: Probabilités sur les structures géometriques. Actes des Journées Toulouse (1984). Publ. du Lab. Stat. et Prob. Université de Toulouse, 57 - 85 (1985).

[ 7 ] W. Hazod: Stable probability measures on groups and on vector spaces. A survey. Probability measures on groups VIII. Proceedings Oberwolfach (1985). Lecture Notes in Math. 1210, 304 - 352. Springer: Berlin-Heidelberg-New York (1986).

[ 8 ] H. Heyer: Probability measures on locally compact groups. Ergebnisse der Math. Berlin-Heidelberg-New York. Springer (1977).

[ 9 ] K.H. Hofmann: Topologische Halbgruppen mit dichter submonogener Unterhalbgruppe. Math. Z. 74, 232 - 277 (1960).

[10] M. McCrudden: Local tightness of convolution semigroups over locally compact groups. In: Probability measures on groups, Proceedings Oberwolfach (1981), 304 - 314. Lecture Notes in Math. 928. Springer: Berlin-Heidelberg-New York (1982).

[11] M. McCrudden: Infinitely divisible probabilities on polycyclic groups are Poisson. (Unpublished manuscript).

[12] G.D. Mostow: On the fundamental group of a homogeneous space. Ann. of Math. 66 (1957) 249 - 255.

[13] L. Schmetterer: On Poisson laws and related questions. In: Proceedings of the 6th Berkeley Symposium on Mathematical Statistics and Probability Vol. II, pp. 169 - 185. Berkeley and Los Angeles: University of California Press 1970.

[14] L. Schmetterer, W. Hazod: Über einige mit der Wahrscheinlichkeitstheorie zusammenhängende Probleme der Gruppentheorie. J. reine angew. Math. 262/263, 144 - 152 (1973).

[15] D. Segal: Polycyclic groups. Cambridge Univ. Press (1983).

[16] E. Siebert: Einbettung unendlich teilbarer Wahrscheinlichkeitsmaße auf topologischen Gruppen. Z. Wahrscheinlichkeitstheorie und verw. Gebiete 28, 227 - 247 (1974).

[17] E. Siebert: Semistable convolution semigroups on measurable and topological groups. Ann. Inst. H. Poincaré 20, 147 - 164 (1984).

# On the Use of Predictive Distributions in Portfolio Theory

Rudolf Henn and Peter Kischka
Institut für Statistik und
Mathematische Wirtschaftstheorie
Universität Karlsruhe
D-7500 Karlsruhe, FRG

SUMMARY: In this paper we consider a multiperiod portfolio problem from a Bayesian point of view. Subjective estimates and/or estimation risk enter the predictive distributions which are the foundations for the determination of optimal portfolios. Predictive distributions are adjusted from period to period. It is shown that this adjustment process exhibits some natural requirements for the behaviour of investors. Contrary to other multiperiod models this approach can be seen to be endogeneously defined since transition probabilities are implied by Bayes´ formula.

## Introduction

The predictive or unconditional distribution is a main instrument in Bayesian analysis and portfolio theory is a main application of Bayesian methods in economic theory (see [1], [3]). In this paper we consider the role of predictive distribution in portfolio theory. From an economic point of view the predictive distribution should contain all information on future returns available at the moment when the investor has to make its decision. In general, in the next period additional informations are available and therefore using Bayes´ formules a new predictive distribution has to be applied for decision making.

In portfolio theory new information means that there is at least a new realization of returns. In this paper we will consider a multiperiod model of portfolio theory using the predictive distribution which arise by successive realizations of returns. Other sources of information are neglected. This Bayesian approach can be modelled as a Markovian model; in contrast to other Markovian models in portfolio theory (see e.g. [5]) the transition probabilities are given by the

Contributions to Stochastics.
Ed. by W. Sendler
© Physica-Verlag Heidelberg 1987

model, i.e. there are no additional assumptions to be made whether a singleperiod or a multiperiod model is considered from a Bayesian point of view.

The importance of Bayesian procedures for portfolio theory was first mentioned in Mao/Sãrndal [4]. The applications to single period theory are summarized in Bawa et al. [1]. In Winkler [8] the Bayesian approach for multiperiod models is recommended but the approach is not elaborated.

The learning process of the investor is described in section I. Since we are mainly interested in the distribution of returns we only consider the case without intermediate consumption. The problem therefore is to maximize expected utility of wealth at the end of the K-th period taking into account that information on the returns of the risky assets will grow during the planning period. In section II we briefly sketch the underlying Bayesian decision model and derive the functional equations. The case of normally distributed returns with different prior distributions is examined seperately. Section III contains the derivation of some properties of the predictive distribution induced by the learning process. We show that these distributions e.g. have the property to attribute a higher probability to returns which have realized. Surely, this is a natural requirement in a multiperiod portfolio model.

## I The learning process

We use the following notations

$W_o$ initial wealth

$K$ number of periods

$W_t$ wealth at the end of the period t $(1 \leq t \leq K)$

$R_t = (R_{1t},\dots,R_{nt})$ random variable denoting the return in period t of n securities

$r_t = (r_{1t},\dots,r_{nt})$ realization of $R_t$ $(1 \leq t \leq K)$

$a_t = (a_{1t},\dots,a_{nt})$ portfolio in period t

$a_{jt}$ denotes the fraction of wealth $W_{t-1}$ invested in security t, $\sum_{j=1}^{n} a_{jt} = 1$.

$W_t\,(W_{t-1},a_t,r_t)$ wealth at the end of period t, if at the end of period t-1 the wealth is $W_{t-1}$, portfolio $a_t$ is chosen and the return $r_t$ is realized $(1 \leq t \leq K)$.

It is implicit in this notation that the possibilities to invest are the same for all periods. The investor acts as a price taker and wants to maximize expected

utility of final wealth: $E(U(W_k)) \longrightarrow \max$.
The solution of this maximization can be achieved by successive solutions of the functional equations when the distributions of returns are specified. E.g. in Merton [5] the distribution in period t depends on the state $\gamma_{t-1}$ of the previous period; the functional equations are given by

$$\phi(W_{t-1},\gamma_{t-1}) = \max_{a_t} E_t(\phi_{t+1}(W_t(W_{t-1},a_t,R_t)),\gamma_t) \qquad (1)$$

where $E_t$ denotes the expected value with respect to the distribution determined by $\gamma_{t-1}$. As mentioned above the transition from $\gamma_{t-1}$ to $\gamma_t$ frequently is not specified.
Now we assume that the distribution of returns depends on some unknown parameter $\lambda$ for which the investor has a prior distribution. For fixed $\lambda$ the returns are assumed to be stochastically independent; therefore, if the true state of the world, say $\lambda_0$, would be known returns are random with respect to a stationary distribution and independent of each other. Since this is not the case the investor gets more information on $\lambda$ observing realized returns. At the starting point t=0 of the multiperiod portfolio model the investor has some prior distribution $\xi$ for the parameter expressing his subjective estimates and/or information contained in past returns; $\xi$ can be a posterior distribution with respect to returns $r_{-L},\dots,r_0$ and some initial prior $\varrho$ for $\lambda$. If $\xi$ denotes a density and if $f(\ |\lambda)$ is the density function of returns conditional on $\lambda$ the unconditional distribution of the K future returns at t=0 is given by the density of the predictive distribution.

$$v_1(y_1,\dots,y_K) = \int \prod_{j=1}^{K} f(y_j|\lambda)\xi(\lambda)d\lambda \ ^{1)} \qquad (2)$$

This is the multiperiod equivalent to the density used in [1]. At t=0 the density $v_1$ contains all information on future returns. After t periods there are t realizations $r_1,\dots,r_t$ and by Bayes theorem the posterior distribution is given by

$$\psi_t(\lambda|r_1,\dots,r_t) \propto \xi(\lambda) \prod_{j=1}^{t} f(r_j|\lambda) \ ^{2)} \qquad (3)$$

---

1) We denote by $y_j$ the argument of the density function of returns in period j. The discrete case or the mixed case can be handled by analogy.

2) If the normalizing constant $\int\xi(\lambda)\prod_{j=1}^{t} f(r_j|\lambda)d\lambda$ in the definition of $\psi_t$ equals o we set $\psi_t(\lambda) = \xi(\lambda)$.

From Bayes theorem we have

$$\psi_t(\lambda|r_1,\ldots,r_t) \propto f(r_t|\lambda)\psi_{t-1}(\lambda|r_1,\ldots,r_{t-1})$$

whenever $\int f(r_t|\lambda)\xi(\lambda)d\lambda > 0$. In any case we can write

$$\psi_t = \psi_t(\psi_{t-1}, r_t) \tag{4}$$

i.e. the posterior distribution at time t is a function of the posterior distribution at time t-1 and of the last observation $r_t$. Corresponding to equation (2) at time t the predictive distribution of future returns conditional on past observations has the density

$$v_{t+1}(y_{t+1},\ldots,y_k|r_1,\ldots,r_t) = \int \prod_{j=t+1}^{K} f(y_j|\lambda)\psi_t(\lambda|r_1,\ldots,r_t)d\lambda \tag{5}$$

We can also consider this density conditional on the posterior distribution $\psi_t$

$$v_{t+1}(y_{t+1},\ldots,y_k|\psi_t) = \int \prod_{j=t+1}^{K} f(y_j|\lambda)\psi_t(\lambda)d\lambda \tag{5´}$$

Equations (4) and (5) show the learning process: at time t-1 all informations on the unknown parameter $\lambda$ are contained in $\psi_{t-1}$. The return $R_t$ of the following period is assumed to be distributed with density

$$v_t(y_t|r_1,\ldots,r_{t-1}) = v_t(y_t|\psi_{t-1}) = \int f(y_t|\lambda)\psi_{t-1}(\lambda)d\lambda \text{ [3]} \tag{6}$$

In period t some realization $r_t$ of $R_t$ is observed. $\psi_{t-1}$ is ´improved´ by (4) and the returns $R_{t+1}$ at time t are assumed to be distributed with density $(\psi_t=\psi_t(\psi_{t-1},r_t))$

$$v_{t+1}(y_{t+1}|r_1,\ldots,r_t) = v_{t+1}(y_{t+1}|\psi_t) = \int f(y_{t+1}|\lambda)\psi_t(\lambda)d\lambda \tag{7}$$

In this way the distribution of returns are adjusted from period to period. Future returns are not independent as can be seen from (2), (5) and the distribution of returns are, in general, non stationary as can be seen from (6), (7). If the posterior distributions are defined regularly one can write for (5)

$$v_{t+1}(y_{t+1},\ldots,y_K|r_1,\ldots,r_t) =$$
$$v_{t+1}(y_{t+1}|r_1,\ldots,r_t)\cdot \prod_{j=t+2}^{K} v_j(y_j|r_1,\ldots,r_t,y_{t+1},\ldots,y_{j-1}) \tag{5´´}$$

The density used at time t for the future returns therefore takes into account that there will be additional information.

## II The functional equations

At time t=0 the relevant distribution is given by (2). The problem at the begin-

3) We use the same symbol $v_t$ for the density of the marginal distribution of $R_t$ and for that of the joint distribution of $(R_t,\ldots,R_K)$.

ning therefore is to maximize $E(U(W_K))$ where terminal wealth $W_K$ depends on portfolios $a_1,\dots,a_K$ and E is evaluated with respect to distribution (2). The portfolios $a_1,\dots,a_K$ depend on the wealths $W_0,\dots,W_{K-1}$ achieved at these times. Maximizing $E(U(W_K))$ with respect to (2) means maximization of a weighted average of expected utility conditional on $\lambda$. The mathematical framework for such a problem is given by a Bayesian decision model as described e.g. in Rieder [7]. Since the investor acts as a price taker the distribution of returns doesn´t depend on the portfolios actually chosen; this gives a remarkable simplification compared to the general model. On the other hand there is a special aspect arising from the fact that the relevant state variables are wealths but the posterior distributions have to be computed with the returns[4]. The Bayesian decision model therefore must contain returns as state variables. It can be reduced to a Markovian model with completely known transition law by augmenting the state space with the set of distributions on the parameter set. The functional equations corresponding to the model are given by (see [3], pp 96)

$$\Phi_t(W_{t-1},\psi_{t-1}) = \max_{a_t} \int \Phi_{t+1}(W_t(W_{t-1},a_t,y_t),\ \psi_t(\psi_{t-1},y_t))v_t(y_t|\psi_{t-1})dy_t \tag{8}$$

defined backwards starting with

$$\Phi(W_{K-1},\psi_{K-1}) = \max_{a_K} \int U(W_K(W_{K-1},a_K,y_K))v_K(y_K|\psi_{K-1})dy_K \tag{9}$$

At time K-1 wealth $W_{K-1}$ and the posterior distribution $\psi_{K-1}$ are given. The portfolio defined by (9) is therefore optimal with respect to the learning process described above.

Equivalently the optimal portfolio at time $t<K-1$ is given by (8) using the distribution (7) which contains information available at this time.

By (8) a special case of (1) is given, the transition probabilities of which are derived endogenously by the model. In section III we show that these transition probabilities fulfill some requirements natural for the behaviour of an investor in a multiperiod model. Like in Fama [2] one can show that the derived utility functions (8) are concave with respect to wealth whenever U in concave.

We now want to consider the special case of (8) when returns are assumed to be normally distributed with unknown mean vector $\lambda$ and known covariance matrix $\Phi$; $f(y|\lambda)$ therefore is the density of a $N(\lambda,\Phi)$ distribution. If one considers $\xi$ as the posterior distribution with respect to past observations $r_{-L},\dots,r_0$ and some

---

4) Each $\lambda$ induces a distribution of $W_t$ conditional on $a_t$ and $W_{t-1}$ and therefore posterior probabilities could be computed using realized wealth instead of realized returns. But the vector of returns $r_t$ contains more information than the scalar $W_t$.

´original´ prior distribution $\varrho$ for the parameter $\lambda$ there are two different situations discussed in the literature (see [1]).

i) $\varrho$ is the density of a normal distribution $N(\mu,\eta)$

ii) $\varrho$ is an improper diffuse distribution, $\varrho(\lambda) \equiv c$ $(\lambda \in \mathbb{R}^n)$

In case i) the choice of $\mu$ and $\eta$ reflects subjective estimates whereas in case ii) only estimation risk is encountered by the Bayesian procedure.

In both cases the posterior distribution $\xi$ is the density of a normal distribution (see Raiffa/Schlaifer [6], Zellner [9]). Extensive interpretations of informative conjugate priors (case i)) or diffuse priors for the singelperiod case are given in Bawa et al. [1]. At time t=0 these argumentes hold for the multiperiod portfolio model, too. Since there are no essential differences we only consider case ii); with $m_o = \frac{1}{L+1} \sum_{i=-L}^{o} r_i$ we have for the distribution $\xi$ computed as posterior distribution with resepect to $r_{-L},\ldots,r_0$ and the diffuse prior $\varrho$:

$$\xi(\lambda) \sim N(m_o, \frac{1}{L+1}\Phi).$$

With a normal prior the densities $\psi_t$ (see (3)) as well as the densities $v_t$ (see (5)) are normal (see [9]); we have

$$\psi_t(\lambda|r_1,\ldots,r_t) \sim N(m_t, \frac{1}{L+t+1}\Phi) \tag{10}$$

$$v_{t+1}(y_{t+1}|r_1,\ldots,r_t) \sim N(m_t, \frac{L+t+2}{L+t+1}\Phi) \tag{11}$$

with $m_t = \sum_{i=-L}^{t} \frac{1}{L+t+1} r_i$.

At time t the investor regards the actual sample mean $m_t$ as expected value for next period´s return. Since the posterior distribution of time t only depends on $m_t$ and the time passed we can write $v_{t+1}(y_{t+1}|m_t)$ for the distribution (11). With the mapping

$$Q_t(x,y) := \frac{L+t}{L+t+1} x + \frac{1}{L+t+1} y \tag{12}$$

the functional equations (8), (9) now can be written as

$$\Phi_t(W_{t-1},m_{t-1}) = \max_{a_t} \int \Phi_{t+1}(W_t(W_{t-1},a_t,y_t),Q_t(m_{t-1},y_t))v_t(y_t|m_{t-1})dy_t, \tag{13}$$

$$\Phi_K(W_{K-1},m_{K-1}) = \max_{a_K} \int U(W_K(W_{K-1},a_K,y_K))v_K(y_K|m_{K-1})dy_K. \tag{14}$$

For the case of normally distributed returns and utility functions which exhibit constant absolute risk aversion -i.e. U(W)=-exp(-qW) with some parameter q>0- one can compute the solutions of the functional equations. Especially one can compare two investors both of them assume the returns to have the same normal distribution in the initial period. One investor acts classically, assuming this distribution for the following periods, too, the other investor takes into account that he will learn in the way described above. Then the 'Bayesian investor' will invest less in the risky asset in the initial period (see [3], pp. 122).

## III Some implications of the learning process

At time t-1 the investor has some wealth $W_{t-1}$ and he regards $v_t(y_t|\psi_{t-1})$ as the relevant density of next period's return. Assume that $\bar{r} = r_t$ is the return acutally achieved in period t. The relevant density for period t+1 is given by $v_{t+1}(y_{t+1}|\psi_t(\psi_{t-1},\bar{r}))$ (see (4)). We will first show that by learning in a Bayesian way the investor will attribute a higher probability for period t+1 to $\bar{r}$ than he did for period t; loosely speaking this means $v_t(\bar{r}|\psi_{t-1})<v_{t+1}(\bar{r}|\psi_t(\psi_{t-1},\bar{r}))$.

The second result is an extension for the case of a discrete parameter space $\Lambda$; so e.g. discrete mixtures of normal distributions are included. A weaker form of these results is proved in [3].

Before stating the propositions we have to make some technical preliminaries. For simplicity we assume that densities of random variables are continuous functions. Given $\psi_{t-1}$ we consider the set

$$T_t = \{y|v_t(y|\psi_{t-1}) > 0\}. \tag{15}$$

$T_t$ can be interpreted as the set of returns which occur in period t with positive probability (given $\psi_{t-1}$). The set of returns which contain new information on $\lambda$ with respect to $\psi_{t-1}$ is

$$\mathring{T}_t = \{y \in T_t | f(y|\lambda_1) \neq f(y|\lambda_2) \text{ for some } \lambda_1,\lambda_2 \text{ with } \psi_{t-1}(\lambda_1) > 0, \; \psi_{t-1}(\lambda_2) > 0\} \tag{16}$$

For every $y \in \mathring{T}_t$ the variace $\sigma^2_{t-1}(y)$ of the random variable $\lambda \to f(y|\lambda)$ is positive with respect to the distribution $\psi_{t-1}$.

If the investor observes in period t some return $\bar{r}=r_t \in \mathring{T}_t$ the probability of some set U of returns in period t+1 is

$$\int_U v_{t+1}(y|\psi_t(\psi_{t-1},\bar{r}))dy$$

If U is a small set containing $\bar{r}$ it is natural to assume that this probability is higher than

$$\int_U v_t(y|\psi_{t-1})dy,$$

the probability of the set U in period t. We have

Proposition 1:

Given $\psi_{t-1}$, for $\overline{r}\in\mathring{T}_t$, $\alpha\in[0,1[$ there is some open set $U_\alpha$ containing $\overline{r}$ such that

$$\int_{U_\alpha} v_{t+1}(y|\psi_t(\psi_{t-1},\overline{r}))dy - \int_{U_\alpha} v_t(y|\psi_{t-1})dy > \alpha\sigma^2_{t-1}(\overline{r})(v_t(\overline{r}|\psi_{t-1}))^{-1}\int_{U_\alpha} dy \qquad (17)$$

Especially the difference is positive.

Proof:

With $\psi_t=\psi_t(\psi_{t-1},\overline{r})$ we have $v_{t+1}(\overline{r}|\psi_t)-v_t(\overline{r}|\psi_{t-1}) = \int f(\overline{r}|\lambda)\psi_t(\lambda)-f(\overline{r}|\lambda)\psi_{t-1}(\lambda)d\lambda$.
Because of $\psi_t(\lambda) = f(\overline{r}|\lambda)\psi_{t-1}(\lambda)(v_t(\overline{r}|\psi_{t-1}))^{-1}$ this difference equals
$[\int(f(\overline{r}|\lambda))^2\psi_{t-1}(\lambda)d\lambda - (\int f(\overline{r}|\lambda)\psi_{t-1}(\lambda))d\lambda)^2](v_t(\overline{r}|\psi_{t-1}))^{-1}$.
The expression in the first bracket equals $\sigma^2_{t-1}(\overline{r})$. For $r\in\mathring{T}_t$ this variance is positive. (17) therefore follows from continuouty.

The variance $\sigma^2_{t-1}(\overline{r})$ is a measure of deviation of $\lambda$ for the special return $\overline{r}$. The higher this variance the higher is additional information. Now assume $\Lambda = \{\lambda_1,\dots,\lambda_m\}$; in this case the parameter can be interpreted as different states of economic development (see [4]). We show now that the probability of any return r rises which has the same ranking with respect to $\Lambda$ (see below) as an observed return $\overline{r}$. $\psi_t$ now denotes a discrete distribution.

Proposition 2:

Let $\Lambda = \{\lambda_1,\dots,\lambda_m\}$ and let $\psi_{t-1}$, $\overline{r}\in\mathring{T}_t$ be given. Assume
$f(\overline{r}|\lambda_1) > f(\overline{r}|\lambda_2) \geq \dots \geq f(\overline{r}|\lambda_m)$, $0 < \psi_{t-1}(\lambda_1) < 1$.
Then for any set U with

$$\int_U f(y|\lambda_1)dy > \int f(y|\lambda_2)dy \geq \dots \geq \int_U f(y|\lambda_m)dy$$

we have

$$\int_U v_{t+1}(y|\psi_t(\psi_{t-1},\overline{r}))dy - \int v_t(y|\psi_{t-1})dy > 0. \qquad (18)$$

Proof:

Like in the proof of (17) one can write for (18)

$$[(\sum_{j=1}^m f(\overline{r}|\lambda_j)\psi_{t-1}(\lambda_j)\int_U f(y|\lambda_j)dy) -$$
$$- (\sum_{j=1}^m f(\overline{r}|\lambda_j)\psi_{t-1}(\lambda_j))(\sum_{j=1}^m \int_U (y|\lambda_j)dy\psi_{t-1}(\lambda_j))](v_t(\overline{r}|\psi_{t-1}))^{-1}.$$

The assertion follows from the lemma below with
$\alpha_j = \psi_{t-1}(\lambda_j)$, $\beta_j = \int_U f(y|\lambda_j)dy$, $\delta_j = f(\overline{r}|\lambda_j)$.

Since there is always a set U with $\overline{r}\in U$ fulfilling the assumption of proposition 2 the second assertion of proposition 1 follows from proposition 2.

Lemma:

Let $\alpha_j$, $\beta_j$, $\delta_j \in \mathbb{R}$ $(1 \leq j \leq m)$ be nonnegative real numbers s.t. $0 < \alpha_1 < 1$, $\sum_{j=1}^{m} \alpha_j \leq 1$, $\beta_1 > \beta_2 \geq \beta_3 \geq \ldots \geq \beta_m$, $\delta_1 > \delta_2 \geq \delta_3 \geq \ldots \geq \delta_m$. Then we have

$$\sum_{j=1}^{m} \alpha_j \beta_j \delta_j - \left( \sum_{j=1}^{m} \alpha_j \beta_j \right)\left( \sum_{j=1}^{m} \alpha_j \delta_j \right) > 0. \tag{19}$$

Proof:

The assertion holds for m=1. The proof is by induction. We assume that the asssertion holds for some m-1.

Define $A = \sum_{j=1}^{m} \alpha_j$, $a = \sum_{j=1}^{m-1} \alpha_j$. Then we can write for (19) (all summation from 1 to m-1):

$$\Sigma\alpha_j\beta_j\delta_j + A\beta_m\delta_m - a\beta_m\delta_m - \Sigma\alpha_j\beta_j\Sigma\alpha_j\delta_j - A\delta_m\Sigma\alpha_j\beta_j + a\delta_m\Sigma\alpha_j\beta_j - A\beta_m\Sigma\alpha_j\delta_j + a\beta_m\Sigma\alpha_j\delta_j$$

$$- A^2\beta_m\delta_m + 2aA\beta_m\delta_m - a^2\beta_m\delta_m =: B$$

Because of

$\beta_m\delta_m a - \beta_m\Sigma\alpha_j\delta_j \leq 0$, $\beta_m\delta_m a - \delta_m\Sigma\alpha_j\beta_j \leq 0$, $A \leq 1$ we have

$B \geq \Sigma\,(\alpha_j(\beta_j - \beta_m)(\delta_j - \delta m)) - \Sigma(\alpha_j(\beta_j - \beta_m))\Sigma(\alpha_j(\delta_j - \delta_m))$.

With $\beta_j' := \beta_j - \beta_m$, $\delta_j' := \gamma_j - \delta_m$ $(1 \leq j \leq m-1)$ the assertion follows from the assumption for m-1.

**References**


[1] Bawa, S., J.Brown, W.Klein (1979), Estimation Risk and Optimal Portfolio Choice. Amsterdam - New York - Oxford.

[2] Fama, E. (1970), Multiperiod Consumption - Investment Decisions. The American Economic Review 60.

[3] Kischka, P. (1984), Bestimmung optimaler Portfolios bei Unsicherheit, Mathematical Systems in Economics 97, Königstein.

[4] Mao, J., C.Särndal (1966), A Decision Theory Approach to Portfolio Selection . Management Science 12.

[5] Merton, C. (1982), On the Microeconomic Theory of Investment under Uncertainity, in: Handbook of Mathematical Economics II, K.J.Arrow, M.D. Intriligator (eds.). Amsterdam - New York - Oxford.

[6] Raiffa, H., R. Schlaifer (1961), Applied Statistical Decision Theory. Cambridge (Mass.), London.

[7] Rieder, U. (1975), Bayesian Dynamic Programming, Advanced Applied Probability Theory 17.

[8] Winkler, R.L. (1979), Statistical Decision Theory and Financial Decision Making, in: J.Bichsler (ed.), Handbook of Financial Economics. Amsterdam - New York - Oxford.

[9] Zellner, A. (1971), An Introduction to Bayesian Inference in Econometrics. New York - London.

# A Short Proof for the Contraction-Property of Markov Chains with Attractive States

Ferdinand Österreicher
Institut für Mathematik
Universität Salzburg
A-5020 Salzburg, Austria

SUMMARY: A short proof for the contraction-property of the map $TP = P.\mathbb{P}$ for homogeneous Markov chains with attractive states is presented 1). It enables the application of Banach's fixed point theorem to show the existence of the stable distribution. This approach may serve as an alternative for the classic approach ( cf. [1], chapter v.2 ) presented in postcalculus texts on the basics of Markov chains.

Let $(S,\mathbb{P})$ be a homogeneous Markov chain with finite or countable infinite state space $S$ and transition function $\mathbb{P} = (P(x,y))_{x,y\in S}$, which has an attractive state, i.e.

(*) there exists a state $y_0 \in S$ and a power $\mathbb{P}^{n_0}$ of $\mathbb{P}$, such that $\delta^{n_0}(y_0) = \inf\{P^{n_0}(x,y_0): x\in S\} > 0$.

Furthermore, let $V(S) = \{P = (p(x): x\in S): p(x)\geq 0 \ \forall x\in S, \sum_{x\in S} p(x)=1\}$ be the set of (probability-) distributions on $S$, let $P,Q \in V(S)$ be two initial distributions, $P_n = P.\mathbb{P}^n$ and $Q_n = Q.\mathbb{P}^n$ their iterates and $P^*$ a stable distribution, i.e. an element $P^* \in V(S)$ satisfying $P^* = P^*.\mathbb{P}$. Then the following well-known results hold

(R1) $\|Q_n - P_n\|_1 \leq \kappa(\mathbb{P}^{n_0})^{[n/n_0]}\|Q - P\|_1$

with a coefficient of ergodicity $\kappa(\mathbb{P}^{n_0}) \in [0,1)$,

(R2) there exists a unique stable distribution $P^*$.

---

1) It was motivated by the observation that $\Delta_y(1)$ in [4], §2 can be represented as

$$\Delta_y(1) = \min\{ \sum_{x\in S} P(x,y)(q(x)-p(x))^+, \sum_{x\in S} P(x,y)(q(x)-p(x))^-\} .$$

Contributions to Stochastics.
Ed. by W. Sendler
© Physica-Verlag Heidelberg 1987

We use a simple trick to establish the basic contraction-property of the map $T:V(S) \to V(S)$ defined by $TP = P.\mathbb{P}$ .

<u>LEMMA</u>: $\|(Q - P).\mathbb{P}\|_1 \leq \kappa(\mathbb{P}).\|Q - P\|_1$

with $\kappa(\mathbb{P}) = 1 - \sum_{y \in S} \delta(y)$ and $\delta(y) = \inf\{P(x,y): x \in S\}$ .

<u>PROOF</u>: Let $y \in S$ and choose the sign $s = s(y) \in \{-1,+1\}$ so that $|q_1(y) - p_1(y)| = s(q_1(y) - p_1(y))$. Then

$$|a| - s.a = 2.\max\{-s.a,0\} \geq 0 \quad \text{and} \quad \sum_{x \in S} (q(x) - p(x)) = 0$$

yield

$$|q_1(y) - p_1(y)| = \sum_{x \in S} s(q(x) - p(x))P(x,y) =$$

$$= \sum_{x \in S} |q(x)-p(x)|P(x,y) - \sum_{x \in S} [\,|q(x)-p(x)| - s(q(x)-p(x))\,]P(x,y) \leq$$

$$\leq \sum_{x \in S} |q(x)-p(x)|P(x,y) - \sum_{x \in S} [\,|q(x)-p(x)| - s(q(x)-p(x))\,].\delta(y) =$$

$$= \sum_{x \in S} |q(x) - p(x)|(P(x,y) - \delta(y)) \text{ 2)}.$$

Summing these inequalities over $y \in S$, interchanging the order of summation and taking into account $\sum_{y \in S} P(x,y) = 1$ gives the assertion

(R1) ( with $\kappa(\mathbb{P}^{n_o}) = 1 - \sum_{y \in S} \delta^{n_o}(y)$ ) is an immediate consequence of this Lemma.

(R2): Since $(V(S), \|.\|_1)$ is a complete space and since, according to the Lemma, the map $T^{n_o}$ is a contraction the Banach fixed point theorem ensures (R2).

<u>REMARKS</u>: In the classic approach ( cf. [1], p. 172-174 ) the following weakening of our Lemma

$$\|(Q - P).\mathbb{P}\|_\infty \leq 1 - \delta.\text{card}(\{y \in S: \delta(y) \geq \delta\}), \quad \delta \in (0,1)$$

is used ( cf. footnote 2 ). The former is designed for finite S, where the choice of the norm is irrelevant and (R2) is obvious. Of course,

---

2) and $2 \geq \|Q - P\|_1 = 2.\sum_{x \in S} \max(q(x)-p(x),0) =$

$2.\sum_{x \in S} \max(p(x)-q(x),0) \geq 2.\|Q - P\|_\infty$

the advantage of the $l_1$-norm reveals itself primarily for countable infinite S.

In [2], p. 112-115, for example,

$$\|(Q - P).\mathbb{P}\|_1 \leq (1 - \delta(y_0)).\|Q - P\|_1$$

is first proved for the case when $Q \neq P$ are distributions concentrated in one single state ( hence $\|Q - P\|_1 = 2$ ) and then extended to the general case. However, our Lemma could be proved along the lines of [2] as well.

Slightly more sophisticated arguments based on $sa = \frac{|a|+sa}{2} - \frac{|a|-sa}{2}$ instead of $sa = |a| - [|a|-sa]$ and applied to both sets $\{y \in S,\ s(y)=s\}$, $s \in \{-1,+1\}$ separately yield $\|(Q-P)\mathbb{P}\|_1 \leq \kappa_1(\mathbb{P})\|Q-P\|_1$ with

$$\kappa_1(\mathbb{P}) = 1 - \inf_{x,x' \in S} \sum_{y \in S} \min(P(x,y),P(x',y)) \leq \kappa(\mathbb{P}).$$

This coefficient of ergodicity is best possible, i.e.

$$\kappa_1(\mathbb{P}) = \sup\{ \|(Q-P)\mathbb{P}\|_1/\|Q-P\|_1 : P,Q \in V(S),\ Q \neq P \}$$

( cf. e.g. [5], Lemma 4.3 ). For the case card(S) = 2, which is investigated in Markov's paper [3], $\|(Q-P)\mathbb{P}\|_1 = \kappa(\mathbb{P})\|Q-P\|_1$ and hence $\kappa(\mathbb{P}) = \kappa_1(\mathbb{P})$.

REFERENCES:

[1] Doob, J.L.: Stochastic processes. John Wiley, New York 1953 ( 7th printing, 1967 ).

[2] Jacobs, K.: Markov-Prozesse mit endlichvielen Zuständen. Entry from K. Jacobs (edt.): Selecta Mathematika IV. Springer, Berlin 1972.

[3] Markov, A.A.: Investigations of a notable case of dependent trials (in Russian). Izvestia Akad. Nauk S.P.-B.(6), 1, 61-80 (1907).

[4] Österreicher, F. and M. Thaler: Analysing Markov chains by risk sets. Studia Sci. Math. Hungar. 15, 411-419 (1980).

[5] Seneta, E.: Non-negative matrices and Markov chains. Springer, New York 1981 ( 2nd edition ).

# A Characteristic Property of Atomic Measures with Finitely Many Atoms Resp. Atomless Measures

Detlev Plachky
Institut für Mathematische Statistik
Universität Münster
D-4400 Münster, FRG

SUMMARY: It is proved that a finite measure on a $\sigma$-algebra is atomic with a finite number of atoms if and only if there does not exist a $\{0,1\}$-valued pure charge on the same $\sigma$-algebra whose system of zero sets is larger than the family of zero sets of the finite measure. Furthermore, an example is given which shows that there exists a $\{0,1\}$-valued pure charge whose system of zero sets is not larger than the family of zero sets of any finite measure. Finally it is proved that a finite measure on a $\sigma$-algebra is atomless if and only if the support of the regular Borel measure of the corresponding Stone representation does not contain a $\sigma$-additive measure.

We start from the following simple observation: Let $\mu$ be a finite atomic measure on a $\sigma$-algebra $\mathbb{A}$ of subsets of a set $\Omega$ with finitely many atoms $A_1,\ldots,A_n$. Then it is impossible to find a $\{0,1\}$-valued pure charge $\nu$ on $\mathbb{A}$ (i.e. $\nu$ is purely finitely additive) such that $\mu(N) = 0$ implies $\nu(N) = 0$ for $N \in \mathbb{A}$ (symbol: $\nu \ll \mu$). Indeed, this property characterizes finite measures of the type introduced above, namely:

THEOREM 1. A finite measure $\mu$ on a $\sigma$-algebra $\mathbb{A}$ is atomic with a finite number of atoms if and only if there does not exist a $\{0,1\}$-valued pure charge $\nu$ on $\mathbb{A}$ such that $\nu \ll \mu$ holds.

Proof. If $\mu$ is not atomic, i.e. $\mu$ has a non zero atomless part $\mu_0$, we introduce the $\sigma$-algebra $\mathbb{N}_\mu$ consisting of all $\mu$-zero sets and their complements. Since the convex set of all finitely additive, non negative set functions $\rho$ on $\mathbb{A}$ extending the restriction of $\mu_0$ to $\mathbb{N}_{\mu_0}$ has at least one extremal point $\rho_0$, which must be

---

AMS 1980 subject classification: 28 A 12

Key words and phrases: atomless measure, atomic measure, charge, Choquet's representation theorem, Stone's representation theorem

Contributions to Stochastics.
Ed. by W. Sendler
© Physica-Verlag Heidelberg 1987

$\{0, \mu_0(\Omega)\}$-valued (cf. [2]), $\nu = \rho_0/\rho_0(\Omega)$ is the desired pure charge, if we can show that $\rho_0$ is not $\sigma$-additive. For this purpose, we observe that there exists a countably generated sub-$\sigma$-algebra $\mathbb{A}_0$ of $\mathbb{A}$ such that the restriction $\bar{\mu}_0$ of $\mu_0$ to $\mathbb{A}_0$ is atomless too (cf. [1]). If $\rho_0$ would be $\sigma$-additive, the restriction $\bar{\rho}_0$ of $\rho_0$ to $\mathbb{A}_0$ would be concentrated on an atom $A_0$ of $\mathbb{A}_0$ and the relation $\bar{\rho}_0 << \bar{\mu}_0$ would imply that $A_0$ is an atom of $\bar{\mu}_0$, too. Suppose that $\mu$ is atomic with an infinite number of atoms $A_i$, $i \in \mathbb{N}$, i.e. $\mu(A \Delta \bigcup_{i \in I(A)} A_i) = 0$ for $A \in \mathbb{A}$ where $I(A)$ is a subset of $\mathbb{N}$ with $I(A) = I(A')$ if $\mu(A \Delta A') = 0$ holds. Denoting by $[A]$ for $A \in \mathbb{A}$ the equivalence class consisting of all $A' \in \mathbb{A}$ with $\mu(A \Delta A') = 0$ the mapping defined by $[A] \to I(A)$ is one to one and $[A_1] \cup [A_2] \to I_{A_1 \cup A_2}$ for $A_i \in \mathbb{A}$, $i = 1,2$, holds. Finally, let $\nu'$ denote a $\{0,1\}$-valued charge on the $\sigma$-algebra $\mathbb{A}'$ consisting of all (countable) unions of the atoms $A_i$ and their complements with $\nu'(A_i) = 0$, $i \in \mathbb{N}$ (the existence of such a charge follows for example by choosing any extremal extension $\nu''$ to $\mathbb{A}'$ of $\nu'$ as a charge defined on the algebra consisting of all finite unions of the atoms $A_i$ and their complements by $\nu''(\bigcup_{i \in I} I_{A_i}) = 0$, if $I$ is a finite subset of $\mathbb{N}$ and 1, otherwise), then $\nu$ with $\nu(A) = \nu'(I(A))$, $A \in \mathbb{A}$, is the desired charge.

EXAMPLE. Let $\mathbb{P}(\Omega)$ denote the set of all subsets of a set $\Omega$ with cardinality equal to the first uncountable ordinal. According to a result of Ulam every finite measure on $\mathbb{P}(\Omega)$ is discrete. Therefore every $\{0,1\}$-valued pure charge of the type occuring in the theorem above is concentrated on a countable subset of $\Omega$. However, there exist $\{0,1\}$-valued pure charges vanishing on all countable subsets of $\Omega$: Consider any $\{0,1\}$-valued extension as a charge to $\mathbb{P}(\Omega)$ of the charge defined on the $\sigma$-algebra generated by the singletons of $\Omega$ to be equal to zero on every countable subset of $\Omega$ and to be equal to one for their complements.

The characterization of atomless probability measures $\mu$ on a $\sigma$-algebra $\mathbb{A}$ of subsets of a set $\Omega$ is connected with the Stone representation, which follows from Choquet's representation $\mu(A) = \int_\Delta \rho(A)\rho_\mu(d\rho)$, $A \in \mathbb{A}$, where $\rho_\mu$ is a uniquely determined, regular probability measure defined on the Borel subsets of the compact set $\Delta$ consisting of all $\{0,1\}$-valued, finitely additive set functions $\rho$ on $\mathbb{A}$ with $\rho(\Omega) = 1$, equipped with the weak* topology of the set of all bounded, finitely additive set functions on $\mathbb{A}$. From $\mu(A) = \rho_\mu(\{\rho \in \Delta: \rho(A) = 1\})$, $A \in \mathbb{A}$, follows, that the support $T_{\rho_\mu}$ consists of all $\rho \in \Delta$ with $\rho << \mu$. Furthermore, if $A_0 \in \mathbb{A}$ is a $\mu$-Atom, where $\mu$ is a probability measure on $\mathbb{A}$, the probability measure $\rho_0$ defined by $\rho_0(A) = \mu(A \cap A_0)/\mu(A_0)$, $A \in \mathbb{A}$, is an isolated point of $T_{\rho_\mu}$. Otherwise, there would exist $\rho_1 \in T_{\rho_\mu}$ and $A_1 \in \mathbb{A}$, such that $\rho_1(A_0 \cap A_1) = 1$ and $\rho_0(A_0 \cap A_1^c) = 1$ holds. Furthermore, $\rho_\mu(\{\rho \in \Delta: \rho(A_0 \cap A_i) = 1\}) = \mu(A_0 \cap A_i) > 0$, $i = 1,2$, $A_2 = A_1^c$, is valid because of $\rho_i \in T_{\rho_\mu}$, $i = 1,2$, from which the contradiction $\mu(A_0 \cap A_i) = \mu(A_0)$, $i = 1,2$, follows. Now it is not difficult to prove the following characterization of atomless probability measures:

THEOREM 2. A probability measure $\mu$ on a $\sigma$-algebra is atomless if and only if the support of the corresponding regular Borel measure of Stone's representation of $\mu$ does not contain a $\sigma$-additive measure.

Proof. First of all it will be shown that the support $T_{\rho_\mu}$ of $\rho_\mu$ is equal to $\{\rho\in\Delta: \rho << \mu\}$. For this purpose let $\rho_0 \in T_{\rho_\mu}$ hold. The existence of some $A_0 \in \mathbb{A}$ with $\rho_0(A_0) = 1$ and $\mu(A_0) = 0$ leads to the contradiction $0 = \mu(A_0) = \rho_\mu(\{\rho\in\Delta: \rho(A_0) = 1\}) > 0$. The property $\rho_0 << \mu$ of some $\rho_0 \in \Delta$ implies $\rho_0 \in T_{\rho_\mu}$, since otherwise there would exist an $A_0 \in \mathbb{A}$ such that $\rho_0(A_0) = 1$ and $\rho_\mu(\{\rho\in\Delta: \rho(A_0) = 1\}) =$ hold which implies the contradiction $\mu(A_0) = 0$. If now $\mu$ is atomless the support $T_{\rho_\mu}$ does not contain any $\sigma$-additive measure because of $T_{\rho_\mu} = \{\rho\in\Delta: \rho << \mu\}$. In the case where $T_{\rho_\mu}$ does not contain any $\sigma$-additive measure, $\mu$ must be atomless since otherwise there would exist a $\mu$-atom $A_0 \in \mathbb{A}$, which implies that $\rho_0$ defined by $\rho_0(A) = \mu(A \cap A_0)/\mu(A_0)$, $A \in \mathbb{A}$, belongs to $T_{\rho_\mu}$. □

COROLLARY. If a finite measure on a $\sigma$-algebra is atomless, the support of the corresponding regular Borel measure does not contain any interior point. If the support of the representing regular Borel measure of a finite measure on a $\sigma$-algebra contains an isolated point, the finite measure is not atomless.

Proof. The first part of the corollary follows from the observation that every $\{0,1\}$-valued finitely additive set function $\nu$ is the setwise limit of a net of Dirac measures. One can choose for the directed index set the system of all measurable sets A with $\nu(A) = 1$ directed by inclusion. Furthermore, the net $\mu_A$ of Dirac measures is defined by $\delta_a$ for some $a \in A$. The second part of the corollary is a consequence of the uniqueness of the representing regular Borel measure. □

REMARK. Theorem 2 yields immediately that every finite, atomless measure on a $\sigma$-algebra is the limit of a net of pure charges relative to the weak$^*$ topology. Under the general assumption that $T_{\rho_\mu}$ for a bounded, non-negative, and finitely additive set function $\mu$ on a $\sigma$-algebra $\mathbb{A}$ is a Baire subset of $\Delta$, $\mu$ is a pure charge if and only if the interior of $T_{\rho_\mu}$ is empty. This follows from the characterization of $\mu$ to be a pure charge by the property that the regular Borel measure $\rho_\mu$ corresponding to $\mu$ in the Stone representation is carried by a meager Baire subset of $\Delta$ (cf. [3], p. 323). This last characterization is no longer true for an algebra $\mathbb{A}$ instead of a $\sigma$-algebra, e.g. if in addition $\mathbb{A}$ is countable, $T_{\rho_\mu}$ is a Baire subset of $\Delta$.

REFERENCES

[1] Bhaskara Rao, K. P. S. and M. Bhaskara Rao: Theory of Charges. Academic Press, New York, 1983.

[2] Plachky, D.: Extremal and monogenic additive set functions. Proc. Amer. Math. Soc. 54 (1976), 193 - 196.

[3] Semadeni, Z.: Banach spaces of continuous functions, Vol. I, PWN-Polish Scientific Publishers, 1971.

# Estimation in the Presence of Nuisance Parameters

Ludger Rüschendorf
Institut für Mathematische Statistik
Universität Münster
D-4400 Münster, FRG

Abstract. We consider MVUE and median unbiased estimation in the presence of nuisance parameters. For several different situations sufficient assumptions are formulated guaranteeing that conditionality and marginalization principles lead to optimal estimators. We also discuss some related questions such as the maximality of ancillary statistics, completeness in markovian models, and a generalization of the independence of complete sufficient statistics and ancillary statistics due to Basu.

## 1. Marginalization and Ancillarity

Let $\mathfrak{P} = \{P_{\theta,\eta}; \theta \in \Theta, \eta \in \Gamma\}$ be a class of distributions on $(X,\mathfrak{A})$ and consider $\theta \in \Theta$ as a parameter of interest, $\eta \in \Gamma$ as a nuisance parameter. We can decompose $\mathfrak{P}$ as

$$\mathfrak{P} = \bigcup_{\eta\in\Gamma} \mathfrak{P}_\eta = \bigcup_{\theta\in\Theta} \mathfrak{P}_\theta, \tag{1}$$

where $\mathfrak{P}_\eta = \{P_{\theta,\eta}; \theta \in \Theta\}$, $\mathfrak{P}_\theta = \{P_{\theta,\eta}; \eta \in \Gamma\}$ are the corresponding sections. (In most parts of the paper we could consider the more general assumption that $\theta:\Omega \to \Theta$, $\eta:\Omega \to \Gamma$ are identifiable parameters.)

DEFINITION 1. (Marginalization, Sufficiency) Let $T: (X,\mathfrak{A}) \to (Y,\mathfrak{B})$ be a statistic,

a) T is called sufficient for $\theta$, if T is sufficient for $\mathfrak{P}_\eta$, $\forall \eta \in \Gamma$.

b) T is called complete for $\theta$, if $\mathfrak{P}_\eta^T$ is complete for all $\eta \in \Gamma$ ($\mathfrak{P}_\eta^T$ denoting the set of images of $\mathfrak{P}_\eta$ under T). □

Similar to Definition 1 we define sufficiency and completeness for the nuisance parameter.

---

AMS 1979 subject classification: 62 F 10, 62 A 20

Key words and phrases: Partial ancillarity, partial sufficiency, MVUE, median unbiased estimator, maximal ancillary statistic

---

Contributions to Stochastics.
Ed. by W. Sendler
© Physica-Verlag Heidelberg 1987

<u>DEFINITION</u> 2. (Ancillarity) For a statistic $S: (X,\mathcal{A}) \to (Y,\mathcal{B})$ define

a) S is ancillary for $\theta$, if $P^S_{\theta,\eta} = P^S_\eta$ for all $\theta \in \Theta$, $\eta \in \Gamma$.

b) S is complete ancillary for $\theta$, if S is ancillary for $\theta$ and S is complete for $\eta$.

c) S is weakly ancillary for $\theta$ if for all $(\theta,\eta) \in \Theta$, $\theta_0 \in \Theta$ there exists a $\eta_0 \in \Gamma$ such that $P^S_{\theta,\eta} = P^S_{\theta_0,\eta_0}$. □

The sufficiency (marginalization) principle proposes to base the statistical inference for $\theta$ on the class of functions of T (a statistic sufficient for $\theta$), while the conditionality principle proposes to consider the class of conditional distributions

$$\mathcal{P}^y = \{P^y_{\theta,\eta};\ \theta \in \Theta,\ \eta \in \Gamma\}, \quad y \in Y, \tag{2}$$

where S is an ancillary statistic for $\theta$ and where we assume the conditional distributions $P^y_{\theta,\eta} = P_{\theta,\eta}(\cdot|S=y)$ throughout the paper to exist.

For these principles to lead to optimal statistical methods, one needs some additional conditions.

<u>DEFINITION</u> 3. (Partial sufficiency, partial ancillarity)

a) T is <u>p-sufficient for $\theta$</u>, if
   1. T is sufficient for $\theta$ and
   2. T is ancillary for $\eta$.

b) S is <u>p-ancillary for $\theta$</u>, if
   1. S is sufficient for $\eta$ and
   2. S is ancillary for $\theta$.

c) S is <u>weakly p-ancillary for $\theta$</u>, if
   1. S is sufficient for $\eta$ and
   2. S is weakly ancillary for $\theta$. □

For this history and ample discussion of these notions we refer to Fraser [11], Sandved [22], Andersen [2], Dawid [10], Bahadur [4], Basu [8], Barndorff-Nielsen [5], Godambe [12] and Lehmann [16]. We remark that in Barndorff-Nielsen [5] in a), b) of Definition 3 the terminology S-sufficient and S-ancillary is used.

The various definitions are most easily remembered by the following suggestive factorization properties:

1. S is ancillary for $\theta$, iff

$$P_{\theta,\eta} = P^y_{\theta,\eta} \oplus P^S_\eta, \tag{3}$$

where $P^y_{\theta,\eta} = P_{\theta,\eta}(\cdot|S=y)$ (cf. (2)) and where $\oplus$ denotes the usual product of the stochastic kernel $P^y_{\theta,\eta}$ and the image measure $P^S_\eta$.

2. S is p-ancillary for $\theta$, iff

$$P^y_{\theta,\eta} = P^y_\theta = P_{\theta,\cdot}(\cdot|S=y) \text{ and } P_{\theta,\eta} = P^y_\theta \oplus P^S_\eta. \tag{4}$$

3. Similarly, T is sufficient for $\theta$, iff

$$P_{\theta,\eta} = P^t_\eta \oplus P^T_{\theta,\eta}, \text{ with } P^t_\eta = P_{\theta,\eta}(\cdot|T=t). \tag{5}$$

4. T is p-sufficient for $\theta$, iff

$$P_{\theta,\eta} = P_{\eta}^{t} \oplus P_{\theta}^{T}. \tag{6}$$

So p-sufficient and p-ancillary models concern special cases of mixture models; in the first case the structural parameter $\theta$ is concerned with the mixing distribution (cf. (6)) while the mixed kernels contain the nuisance parameter, in the second case $\theta$ and $\eta$ change their roles.

In the following part we want to apply these notions to the construction of optimal statistical procedures. While there is an ample literature on various aspects of the marginalization and ancillarity principle concerning comparison of ancillaries by the conditional Fisher information, bayesian aspects of this problem as in the Raiffa and Schlaifer result, relations to group structure, relations to the likelihood principle, asymptotic properties of conditional likelihoods etc., there does not seem to be a more systematic treatment of the application to the construction of optimal classical statistical procedures like MVUE, UMP tests and median unbiased estimators. We collect several results of this type and in order to get a good impression of the various possibilities of marginalization and conditionality principles and their interrelations, the following part also contains several relevant results known from the literature resp. slight extensions of them.

## 2. Construction of MVUE

### a) MVUE and Partial Sufficiency

We start with the application of the sufficiency principle to minimum variance unbiased estimation (MVUE). The following sequence of relations 1 - 5 is essentially known from papers of Bondesson [9] and Fraser [11].

1. If $P \in \mathfrak{P}_{\eta}$ and d is MVUE in P w.r.t. $\mathfrak{P}_{\eta}$, then d is MVUE in P w.r.t. $\mathfrak{P}$ (for its expectation).

The proof of 1. is obvious; the corresponding uniform version is due to Bondesson [9].

2. If d is UMVUE w.r.t. $\mathfrak{P}_{\eta}$, $\forall \eta \in \Gamma$, then d is UMVUE in $\mathfrak{P}$.

Relation 2. implies:

3. (cf. Bondesson [9]) If T is sufficient and complete for $\theta$, then $h \circ T$ is UMVUE for each real function h with $h \circ T \in L^2(\mathfrak{P})$.

Some more constructive versions are possible for the case that T is p-sufficient for $\theta$. Let $g\colon \Theta \to \mathbb{R}^1$ and let $D_g$ be the set of all unbiased estimators of g.

4. If T is p-sufficient for $\theta$ and $d \in D_g$, then for any $\eta_0 \in \Gamma$, $d_{\eta_0} = E_{\eta_0}(d|T) \in D_g$ and $d_{\eta_0}$ has lower risk than d.

Under an additional completeness assumption in 4. one gets the final constructive (but also more restrictive) version due to Fraser [11]:

5. If T is p-sufficient and complete for $\theta$ and $d \in D_g$, then $d^* = E_{\eta}(d|T)$ is defined independently from $\eta \in \Gamma$ and $d^*$ is UMVUE for g w.r.t. $\mathfrak{P}$.

EXAMPLE 1. Let $\theta\in\Theta=\mathbb{R}^1$, $\eta=(\mu,\sigma^2)\in\Gamma\subset\mathbb{R}^1\times(0,\infty)$ and $P_{\theta,\eta}=\bigotimes_{i=1}^{n} N(\theta,\sigma^2)\otimes N(\mu,\sigma^2)$ with Lebesgue-density

$$f_{\theta,\eta}(x,y)\sim\exp\{n\frac{\theta\bar{x}+\mu\bar{y}}{\sigma^2}-\frac{1}{2}\sum_{i=1}^{n}(x_i^2+y_i^2)/_{\sigma^2}\}.$$

For $\eta$ fixed $\bar{x}$ is sufficient and complete for $\theta$ and, therefore, $\bar{x}$ is UMVUE for $\theta$. Since $\Gamma$ is an arbitrary subset of $\mathbb{R}^1\times(0,\infty)$ there does not exist a complete sufficient statistic for $\mathfrak{P}$. If, additionally, we have the restriction $\Theta=\mathbb{N}_0$, then all estimates of the form $\sum_{i=-k}^{\infty} c_i \exp\{\frac{in}{\sigma_0^2}\bar{x}\}$ are MVUE for $\theta\in\mathbb{N}$, $\theta\geq k$ and $(\mu,\sigma_0^2)\in\Gamma$. If, especially, $\Gamma\subset\{(\mu,\sigma_0^2);\ \mu\in\mathbb{R}^1\}$, then all estimators of the form $\sum_{i=0}^{\infty} c_i\exp\{\frac{in}{\sigma_0^2}\bar{x}\}$ are UMVUE. □

Some further examples for the application of 1 - 5 can be found in the papers of Bondesson [9] and Fraser [11]. In the next step we consider a statistic which is sufficient for $\eta$.

PROPOSITION 1. Assume that

1. S is complete and sufficient for $\eta$.
2. For $\mathfrak{P}^S$ almost all $y\in Y$ is $d_y^*$ the MVUE for $g=g(\theta)$ in $P_{\theta_0}^y$ (UMVUE for g) w.r.t. $\mathfrak{P}^y$.
3. There exists a w.r.t. $\mathfrak{P}$ a.s. measurable equivalent version $d^*$ of $d_{S(x)}(x)$.

Then $d^*$ is MVUE for g in $P_{\theta_0,\eta}$, $\eta\in\Gamma$, (UMVUE) w.r.t. $\mathfrak{P}$.

Proof. Let $d\in D_g$, i.e. $E_{\theta,\eta}\,d=g(\theta)$, $\theta\in\Theta$, $\eta\in\Gamma$. Since $E_{\theta,\eta}\,d=E_{\theta,\eta}\,E_\theta(d|S=y)=g(\theta)$, $\forall\,\eta\in\Gamma$, we obtain using completeness of S, that $E_\theta(d|S=y)=\int d\,dP_\theta^y=g(\theta)$, $\forall\,\theta\in\Theta$ and $\mathfrak{P}^S$ almost all $y\in Y$. Since $d_y^*$ is MVUE in $P_{\theta_0}^y$ (UMVUE) for g w.r.t. the conditional model $\mathfrak{P}^y$, we obtain that

$$E_{\theta_0,\eta}(d^*)^2=E_{\theta_0,\eta}[\int(d_y^*)^2\,dP_{\theta_0}^y]\leq E_{\theta_0,\eta}[\int d^2 dP_{\theta_0}^y]=E_{\theta_0,\eta}d^2,$$

i.e. $d^*$ is MVUE in $P_{\theta_0,\eta}$ (UMVUE) for g w.r.t. $\mathfrak{P}$. □

Call an estimator <u>d conditional unbiased for</u> $g=g(\theta)$ if $E_\theta(d|S=y)=g(\theta)$ $[\mathfrak{P}^S]$, $\forall\theta\in\Theta$, i.e. if d is unbiased for g w.r.t. $\mathfrak{P}^y$ for $\mathfrak{P}^S$ a.a. $y\in Y$. Call <u>d CMVUE (CUMVUE) for g</u> in $P_{\theta_0,\eta}$, if $d^*$ is conditional unbiased and d is MVUE (UMVUE) for g w.r.t. the conditional model $\mathfrak{P}^y$ in $P_{\theta_0}^y$ for $\mathfrak{P}^S$ a.a. $y\in Y$. Proposition 1 can then be formulated in the following way;

PROPOSITION 2. Let S be complete and sufficient for $\eta$; then

1. Any unbiased estimator d for g is conditional unbiased.
2. If $d^*\in D_g$ is CMVUE for g in $P_{\theta_0,\eta}$ (CUMVUE) then $d^*$ is MVUE for g in $P_{\theta_0,\eta}$ (UMVUE). □

The uniqueness of MVUE implies also the converse of 2 under the assumption of existence of a measurable version of the conditional problem as in Proposition 1.3. The assumption of a complete and sufficient statistic S for $\eta$ was used by Andersen [2] and Pfanzagl [18], pg. 232 - 235, to prove the asymptotic efficiency of the conditional maximum likelihood estimator. Godambe [12] used these conditions to prove that the conditional likelihoods provide optimal estimation equations for the interesting parameter $\theta$. In these papers one can find several examples for which by Proposition 2 now MVUE can be constructed.

EXAMPLE 2. (cf. Godambe [12], p. 156)
Let $(X,\mathbb{A}) = (\mathbb{R}^n,\mathbb{B}^n)$, $P_{\theta,\eta} = \bigotimes_{i=1}^{n} \Pi(\theta^i\eta)$, $\theta,\eta > 0$, $\Pi$ denoting the Poisson distribution.
For $S(x) = \sum_{i=1}^{n} x_i$ holds that

$$P_{\theta,\eta}(x|S=y) = \frac{y!(1-\theta)^y}{(1-\theta^{n+1})^n \prod_{i=1}^{n} x_i!}\,\theta^{\Sigma i x_i}$$

for $x \in \mathbb{N}_0^n$ with $\sum_{i=1}^{n} x_i = y$. Therefore, if $\Gamma = (0,\infty)$, S is complete and sufficient for $\eta$. If $\Theta \subset (0,\infty)$ contains an open intervall, then the conditional model $\mathbb{P}^y$ is a complete exponential family in $\theta$ and $T(x) = \sum_{i=1}^{n} i x_i$ and, therefore, any function $f(T(x),S(x)) \in D_g$ is a UMVUE. If $\Theta$ is smaller, say $\Theta = \mathbb{N}$, then for $\theta_0 \in \mathbb{N}$ and $\theta \in \mathbb{N}$ by a wellknown result of Barankin $(\frac{\theta}{\theta_0})^{\sum_{i=1}^{n} i x_i} f(\sum_{i=1}^{n} x_i)$ is CMVUE in $\theta_0$ if it is conditionally unbiased for $g(\theta)$ and, therefore, by Proposition 2 also MVUE in $\theta_0$. Note that this MVUE property holds for all $\theta_1 \in \mathbb{N}$ such that $\theta_1 \cdot \frac{\theta}{\theta_0} \in \mathbb{N}$.
While in the first case the UMVUE property can be explained from the completeness and sufficiency of (T(x), S(x)) the second case seems to have not a similar easy explanation without refering to the conditional model. □

A connection between 1 - 5 and Proposition 1 is the following. Under some restrictions on the class of distributions (dominatedness and overlapping support, cf. Andersen [1]) it holds true that (T,S) is (minimal) sufficient for $\mathbb{P}$, if T is (minimal) sufficient for $\theta$ and S is (minimal) sufficient for $\eta$. Furthermore, if S is complete and sufficient for $\eta$, then T,S are independent, iff T is ancillary for $\eta$ (extending Basu's theorem, cf. Andersen [1]); so the conditionality principle and the sufficiency principle coincide in this case, since $P^{T|S}_{\theta,\eta} = P^T_{\theta,\eta} = P^T_\theta$.

b) MVUE and Partial Ancillarity

The idea of Propositions 1 and 2 also works to a certain extent, if completeness of S is replaced by ancillarity.

PROPOSITION 3. Let S be weakly p-ancillary for $\theta$ and $g = g(\theta)$.
1. If d is MVUE for g in $P_{\theta_0,\eta}$, then $E_{\theta_0}(d|S=y) = g(\theta_0)$ $[P^S_{\theta_0,\eta}]$.
2. If d* is CMVUE for g in $P_{\theta_0,\eta}$, then d* is MVUE for g in $P_{\theta_0,\eta}$.
3. If d* is CUMVUE for g, then d* is UMVUE for g w.r.t. $\mathbb{P}$.
4. If d* is UMVUE for g and if for $\mathbb{P}^S$ a.a. $y \in Y$, $d^*_y(\theta,\cdot)$ is the CMVUE in $P_{\theta,\eta}$, then the existence of a measurable version of $d^*_{S(x)}(\theta,x)$ for $\theta \in \Theta$ implies that d* is

CUMVUE for g.

Proof.

1. If d is MVUE for g in $P_{\theta_0,\eta}$, then define

$$\tilde{d} = d - E_{\theta_0}(d|S) + g(\theta_0). \tag{7}$$

For $(\theta,\eta) \in \Theta \times \Gamma$ there exists an element $\eta_0 \in \Gamma$ such that $P^S_{\theta,\eta} = P^S_{\theta_0,\eta_0}$ and, therefore, $E_{\theta,\eta}\tilde{d} = E_{\theta,\eta} d - E_{\theta_0,\eta_0} E_{\theta_0}(d|S) + g(\theta_0) = g(\theta)$; i.e. $\tilde{d} \in D_g$. Furthermore, the assumption $P_{\theta_0,\eta}\{E_{\theta_0}(d|S) \neq g(\theta_0)\} > 0$ would imply for the variance in $P_{\theta_0,\eta}$: $V_{\theta_0,\eta}(\tilde{d}) < V_{\theta_0,\eta}(d)$, contradicting the assumption, that d is MVUE in $P_{\theta_0,\eta}$.

2., 3., 4. are immediate consequences of 1. □

So under the assumption of a weakly p-ancillary statistic for $\theta$ we can restrict to the conditional model. The idea of the construction in relation (7) in the case without nuisance parameters is due to Rao [21], p. 54.

The picture of Proposition 3 gains further clearness by the following extension of a result due to Bahadur [4], Theorem 1, concerning the case without nuisance parameters.

PROPOSITION 4. If S is weakly p-ancillary for $\theta$, if T is p-sufficient for $\theta$ and if $d = d \circ T$ is bounded UMVUE, then d and S are stochastically independent w.r.t. $\mathfrak{P}$.

Proof. Let $D^*_b$ denote the set of all bounded UMVUE w.r.t. $\mathfrak{P}$ and $\mathfrak{A}_0 = \mathfrak{A}(D^*_b)$ the generated $\sigma$-algebra. A theorem due to Bahadur [3] states that every element of $L^2(\mathfrak{A}_0,\mathfrak{P})$ is a UMVUE. Therefore, if $h \circ d \in L^2(\mathfrak{P})$, then $h \circ d$ is UMVUE for $\mathfrak{P}$. Since T is ancillary for $\eta$ we obtain $E_{\theta,\eta}\, h \circ d = E_\theta h \circ d$ and, therefore, by the proof of Proposition 3 we get that $E_\theta(h \circ d|S) = E_\theta h \circ d$ $[\mathfrak{P}]$. Since this is true for all functions h such that $h \circ d \in L^2(\mathfrak{P})$, we get that d,S are stochastically independent. □

Therefore, provided a bounded UMVUE exists, there will be no problem which of the weakly p-ancillary statistics to choose for the conditionality principle. Compared to the remarks following Proposition 2 related to Basu's result, the interesting point here is that no completeness assumption is necessary to establish the independence property. The boundedness condition on d can obviously be weakened to the assumption that $d \in L^2(\mathfrak{A}_0,\mathfrak{P})$.

One can now formulate a completely constructive result corresponding to Fraser's result in the sufficiency case.

COROLLARY 1. Assume that either S is complete and sufficient for $\eta$ or that S is weakly p-ancillary for $\theta$.

If $d^*_y$ is MVUE for $g = g(\theta)$ in $P^y_{\theta_0}$ (UMVUE) for $\mathfrak{P}^S$ a.a. $y \in Y$,
and if there exists a measurable version $d^*$ of $d^*_{S(x)}(x)$, (8)
then $d^*$ is MVUE for g in $P_{\theta_0,\eta}$ (UMVUE). □

The construction of UMVUE in Corollary 1 is often rendered possible by the assumption (9) replacing (8):

$$\text{If } T \text{ is complete and sufficient for } \mathfrak{P}^y \text{ for } \mathfrak{P}^S \text{ a.a. } y \in Y \text{ and if } d \in D_g\text{, then define } d_y^*(t) = E_{P^y}(d|T=t). \text{ If } d^*(x) \text{ is a measurable version of } d^*_{S(x)}(T(x)), \text{ then } d^* \text{ is a UMVUE for } g. \tag{9}$$

EXAMPLE 3. Let $S: (X,\mathbb{A}) \to (\mathbb{R}^k, \mathbb{B}^k)$, $T: (X,\mathbb{A}) \to (\mathbb{R}^m, \mathbb{B}^m)$ $h: (X,\mathbb{A}) \to (\mathbb{R}^1, \mathbb{B}^1)$ and consider for $\theta \in \mathbb{R}^m$, $\eta \in \mathbb{R}^k$ an exponential family of the type

$$\frac{dP_{\theta,\eta}}{d\lambda}(x) = c^{-1}(\theta,\eta) \exp\{<\theta,T(x)> + <\eta,S(x)> + h(x)\}, \ \theta \in \Theta \in \mathbb{R}^m, \eta \in T \subset \mathbb{R}^k.$$

Let $\overline{\lambda}(A|y)$ be the conditional measure of $\lambda$ given $S = y$, i.e. $\lambda(A \cap S^{-1}(A)) = \int_A \overline{\lambda}(A|y) d\lambda^S(y)$ and $\gamma(y,\theta) := \int_{\{S=y\}} \exp\{<\theta,T(x)) + h(x)\} d\overline{\lambda}(x|y)$; then

$$\frac{dP_\theta^y}{d\overline{\lambda}(\cdot|y)}(x) = \frac{\exp\{<\theta,T(x)> + h(x)\}}{\gamma(y,\theta)} \quad \text{and} \quad \frac{dP^S_{\theta,\eta}}{d\lambda^S}(y) = c^{-1}(\theta,\eta)\exp(<\eta,y>\}\ \gamma(y,\theta) =: g_{\theta,\eta}(y).$$

For $(\theta_0,\eta_0) \in \Theta \times \Gamma$ holds:

a) If there exists a function $\eta: \Theta \to \Gamma$ such that $g_{\theta,\eta(\theta)} = g_{\theta_0,\eta_0}$, then $\log \gamma(y,\theta) = <a(\theta),y> + b(\theta) + c(y)$.

b) If, conversely, $\log \gamma(y,\theta) = <a(\theta),\eta> + b(\theta) + c(y)$ and $\eta(\theta) = \eta_0 + a(\theta_0) - a(\theta) \in \Gamma$ for $\theta \in \Theta$, then $g_{\theta,\eta(\theta)} = g_{\theta_0,\theta_0}$.

The proof of a), b) was given in the case $k = m = 1$ by Andersen [2] and extends to the more general situation. Andersen [2] considers a series of examples where he proves that the conditional mle is asymptotically efficient for the estimation of $\theta$. In these examples Proposition 3 and Corollary 1 allow to construct MVUE's resp. UMVUE for the finite sample situation, so e.g. in the normal resp. Poisson case considered by Andersen 2 the c.m.l.e. $\overline{\theta}_n = \overline{x}_n - \overline{y}_n$ resp. $\overline{\theta}_n = \log \overline{x}_n - \log \overline{y}_n$ are UMVUE. Moreover we can consider meager subsets of the parameter spaces such as $\Theta = \mathbb{N}^m$, $\Gamma = \mathbb{N}^k$ destroying the completeness properties of S and T and typically will get only MVUE. □

In the dominated case assumption (9) corresponds to densities of the form $f_{\theta,\eta}(x) = C(\theta,\eta)h(T(x),\theta)g(S(x),\theta,\eta)$. Assumption (9) is quite strong since it implies, together with the completeness of S for $\eta$, that (S,T) is complete for $\mathfrak{P}$ as the following lemma shows.

For a Markov kernel Q from $(X,\mathbb{A})$ to $(Y,\mathbb{B})$ and a probability measure P on $\mathbb{A}$ define

$$Q \oplus P(A \times B) = \int_A Q(x,B)\, dP(x), \ A \in \mathbb{A},\ B \in \mathbb{B}. \tag{10}$$

Let $\mathfrak{P}$ be a set of distributions on $(X,\mathbb{A})$ and $\mathfrak{Q}$ be a set of markov kernels from $(X,\mathbb{A})$ to $(Y,\mathbb{B})$.

LEMMA 1. If $\mathfrak{P}$ is (boundedly) complete on $(X,\mathbb{A})$ and if for $\mathfrak{P}$ a.a. $x \in X$ we have that $\mathfrak{Q}^x = \{Q(x,\cdot);\ Q \in \mathfrak{Q}\}$ is (boundedly) complete on $(Y,\mathbb{B})$, then

$$\mathbb{Q}\oplus \mathbb{P} = \{Q\oplus P;\ Q\in \mathbb{Q},\ P\in \mathbb{P}\} \text{ is (boundedly) complete on } (X,\mathbb{A})\oplus(Y,\mathbb{B}) . \tag{11}$$

Proof. Let $h\in L^1(\mathbb{Q}\oplus \mathbb{P})$ (h bounded) satisfy

$$\iint h(x,y)\ Q(x,dy)\ P(dx) = 0,\ \forall\ Q\in \mathbb{Q},\quad P\in \mathbb{P},$$

then by completeness of $\mathbb{P}$ we get

$$\int h(x,y)\ Q(x,dy) = 0\ [\mathbb{P}] ,$$

i.e. for all $x\in N_1^c$ - $N_1$ being a $\mathbb{P}$ null set - we obtain by completeness of $\mathbb{Q}^x$, that $h(x,\cdot) = 0\ [\mathbb{Q}^x]$. Therefore, for $x\in N_1^c$, the function $g(x) = \int h(x,y)\ 1_{A_2}(y)Q(x,dy) = 0$ for all $A_2\in \mathbb{B}$. For any $A_1\in \mathbb{A}$ we then get

$$0 = \int g(x)\ 1_{A_1}(x)dP(x) = \iint h(x,y)\ 1_{A_1}(x)\ 1_{A_2}(y)Q(x,dy)P(dx),$$

implying that $h = 0\ [\mathbb{Q}\oplus \mathbb{P}]$. □

The case of bounded completeness in Lemma 1 is due to Plachky [19], while in the case of product measures Lemma 1 is due to Landers and Rogge [14]. Lemma 1 leads by induction to completeness results for measures constructed as 'Ionescu-Tuclea products', so, especially, for Markov chains.

An interesting question concerning ancillary statistics is to decide whether an ancillary statistic S is maximal. This question was investigated by Basu [7] for the case without nuisance parameters. His man result (Theorem 7 of Basu [7]) states: If S is ancillary, T is boundedly complete and sufficient and $\mathbb{A}(S,T)$ is essentially equivalent to $\mathbb{A}$, then S is essentially maximal ancillary. A hint to a differend kind of solution was given by Lehmann [16] who proposes to consider the completeness of the conditional model. In the case with nuisance parameters this idea leads to the following result.

PROPOSITION 5. If S is p-ancillary for $\theta$ and if the conditional model $\mathbb{P}^y$ is boundedly complete for $\mathbb{P}^S$ a.a. $y\in Y$, then S is essentially maximal in the class of ancillary statistics for $\theta$.

Proof. Let $\tilde{S}$ be ancillary for $\theta$ and let $\mathbb{A}(S)\subset\mathbb{A}(\tilde{S})\ [\mathbb{P}]$. For $M\in \mathbb{A}(\tilde{S})$ and $F\in \mathbb{A}(S)$ holds:

$$P_{\theta,\eta}(M\cap F) = P_{\eta}(M\cap F) = \int_F P_{\theta,\cdot}(M|S=y)dP_{\eta}^S(y),\ \forall\theta\in\Theta,\ \eta\in\Gamma. \tag{12}$$

Therefore, there exists a version $P_{\cdot}(M|S=y)$ of $P_{\theta,\cdot}(M|S=y)$ which is independent of $\theta\in\Theta$, i.e. M is ancillary also w.r.t. the conditional model $\mathbb{P}^y$. For $d = 1_M - P_{\cdot}(M|S)$ holds $\int d\ dP_\theta^y = P_{\theta,\cdot}(M|S=y) - P_{\cdot}(M|S=y) = 0$ for $\mathbb{P}^S$ a.a. $y\in Y$. The completeness of $\mathbb{P}^y$ then implies that $1_M = P_{\cdot}(M|S)$ a.s. w.r.t. $\mathbb{P}$; i.e. $\mathbb{A}(\tilde{S}) = \mathbb{A}(S)\quad [\mathbb{P}]$ . □

3. UMP-Tests and Median Unbiased Estimators

Suppose that $\Theta\subset \mathbb{R}^1$; since median unbiased estimates are closely related to tests at level $\alpha = \frac{1}{2}$ we start with some results on the construction of UMP-tests. Let $\eta\in\Gamma$ again denote a nuisance parameter. The first situation is concerned with the marginalization principle. It extends a result of Fraser [11], p. 117, where the case $|\Theta| = 2$ is considered.

PROPOSITION 6. Let $\alpha \in (0,1)$, $\theta_0 \in \Theta$ and $\Theta_0 = \Theta \cap (-\infty, \theta_0] \neq \emptyset$, $\Theta_1 = \Theta \cap (\theta_0, \infty) \neq \emptyset$. Assume that T is p-sufficient for $\theta$.

a) If there exists a UMP-test $\varphi^* = \varphi^* \circ T$ at level $\alpha$ within the class of all level $\alpha$-tests of the form $\varphi = \varphi \circ T$, then $\varphi^*$ is UMP at level $\alpha$.

b) If $\mathfrak{P}^T = \{P_\theta^T;\ \theta \in \Theta\}$ has a monotone likelihood ratio in the real statistic h, then

$$\varphi^* = 1_{\{h \circ T > c_\alpha\}} + \gamma(\alpha)\, 1_{\{h \circ T = c_\alpha\}} \tag{13}$$

with $E_{\theta_0} \varphi^* = \alpha$ is a UMP-test at level $\alpha$ for $\Theta_0 \times \Gamma : \Theta_1 \times \Gamma$.

Proof. a) Let $\psi \in \phi_\alpha(\mathfrak{P})$ be any level $\alpha$-test and define for fixed $\eta_1 \in \Gamma$, $\tilde{\psi} = E_{\eta_1}(\psi | T)$. Then for $\theta \in \Theta_0$, $\eta \in \Gamma$ holds: $E_{\theta,\eta} \tilde{\psi} = E_{\theta,\eta_1} \tilde{\psi} = E_{\theta,\eta_1} \psi \leq \alpha$, i.e. $\tilde{\psi} \in \phi_\alpha(\mathfrak{P})$ and for $\theta \in \Theta_1$ holds $E_{\theta,\eta_1} \psi = E_{\theta,\eta_1} \tilde{\psi} \leq E_{\theta,\eta_1} \varphi^*$, since $\varphi^*$ is UMP within the class of level $\alpha$-tests of the form $\tau = \tau \circ T$.

b) follows from a). □

In the above mentioned paper Fraser [11] also shows that a maximin test can be found within the class of tests which are functions of T. In this direction Fraser's result was generalized by Hajek [13] and Plachky and Rüschendorf [20]. The assumption that $\Theta \subset \mathbb{R}^1$ is not necessary for part a). For part b) also multivariate hypotheses on $\theta$ can be considered, which are described by real functions in the form $\{g(\theta) \leq g(\theta_0)\}$: $\{g(\theta) > g(\theta_0)\}$ such that the marginal distributions of T have a monotone likelihood ratio in h and the new parameter $g(\theta)$ (examples might be $g(\theta) = \max \theta_i$ or $g(\theta) = \Sigma\, \theta_i^2$).

The next result is concerned with statistics sufficient for $\eta$. This result is a slight generalization of Lehmann [15], p. 134 - 140, who is concerned with exponential families. Let $\phi_{\alpha,u}(\mathfrak{P})$ denote the set of all unbiased level $\alpha$-tests for $\Theta_0 \times \Gamma : \Theta_1 \times \Gamma$ and $\phi_{\alpha,s}(\mathfrak{P})$ be the set of all level $\alpha$-tests, which are similar on the boundary $\{\theta_0\} \times \Gamma$ at level $\alpha$. Let for a class Q of distributions N(Q) denote the null sets of Q.

PROPOSITION 7. Let $\alpha$, $\Theta = \Theta_0 + \Theta_1$ be as in Proposition 6 and assume that $\phi_{\alpha,u}(\mathfrak{P}) \subset \phi_{\alpha,s}(\ )$. Let, furthermore, S be complete and sufficient for $\eta$ and $N(\mathfrak{P}^S_{\theta_0}) \subset N(P^S_{\theta,\eta})$, $\forall \theta \in \Theta_1$, $\eta \in \Gamma$.

a) If $\varphi^* \in \phi_\alpha(\mathfrak{P})$ is a UMP level $\alpha$-test for $\{P^y_{\theta_0}\} : \{P^y_\theta;\ \theta \in \Theta_1\}$ for $\mathfrak{P}^S_{\theta_0}$ a.a. $y \in Y$, then $\varphi^*$ is a UMPU level $\alpha$-test for $\Theta_0 \times \Gamma : \Theta_1 \times \Gamma$.

b) If for $\mathfrak{P}^S$ a.a. $y \in Y$, $\mathfrak{P}^y = \{P^y_\theta;\ \theta \in \Theta\}$ has a monotone likelihood ratio in the real statistic T and $\varphi^*(x) = \varphi_{S(x)}(x)$, where $\varphi_y(x) = 1_{\{T > c(y)\}} + \gamma(y) 1_{\{T = c(y)\}}$ satisfies $\int \varphi_y dP^y_{\theta_0} =$ for all $y \in Y$, then $\varphi^*$ is a UMPU level $\alpha$-test for $\Theta_0 \times \Gamma$: $\Theta_1 \times \Gamma$.

Proof. a) If $\varphi \in \phi_{\alpha,u}(\mathfrak{P}) \subset \phi_{\alpha,s}(\mathfrak{P})$, then $\alpha = E_{\theta_0,\eta} \varphi = E_{\theta_0,\eta} E_{\theta_0}(\varphi | S)$, i.e. $E_{\theta_0}(\varphi | S) = \alpha$ $[\mathfrak{P}_{\theta_0}]$. For any $(\theta,\eta) \in \Theta_1 \times \Gamma$ holds:

$$E_{\theta,\eta}\varphi = E_{\theta,\eta}E_\theta(\varphi|S) = \int(\int\varphi dP_\theta^y)dP_{\theta,\eta}^S(y) \le \int(\int\varphi^* dP_\theta^y)dP_{\theta,\eta}^S(y) = E_{\theta,\eta}\varphi .$$

At the inequality we used the assumption on the null sets. Since $\varphi^* \in \phi_\alpha(\mathfrak{P})$, $\varphi^*$ is a UMPU level $\alpha$-test.

b) follows from a) since the UMP conditional test $\varphi^*$ lies in $\phi_{\alpha,u}(\mathfrak{P})$. □

The completeness assumption of Proposition 7 may be replaced by an ancillarity assumption in the following way implying that UMP (UMPU) tests are necessarily conditional tests.

<u>PROPOSITION</u> 8. Assume that S is weakly p-ancillary for $\theta$ and that $\mathfrak{P}^y$ has a monotone likelihood ratio in a real statistic T for $\mathfrak{P}^S$ a.a. $y \in Y$. Then:

a) For any $\varphi \in \phi_\alpha(\mathfrak{P})$ with $\alpha(y) = \int \varphi dP_{\theta_0}^y$ define $\psi(x) = \psi_{S(x)}(x)$, where

$$\psi_y = 1_{\{T > c(y)\}} + \gamma(y)\, 1_{\{T = c(y)\}} \text{ with } \int\psi_y dP_{\theta_0}^y = \alpha(y). \tag{15}$$

Then $\psi \in \phi_\alpha(\mathfrak{P})$ and for all $(\theta,\eta) \in \Theta_1 \times \Gamma$ holds

$$E_{\theta,\eta}\,\psi \ge E_{\theta,\eta} \quad . \tag{16}$$

b) If there exists a UMP (UMPU) level $\alpha$-test $\psi$ in the class of functions of (S,T), then $\psi$ is UMP (UMPU) level $\alpha$-test w.r.t. $\phi_\alpha(\mathfrak{P})$.

<u>Proof.</u> a) For $\varphi \in \phi_\alpha(\mathfrak{P})$ and $(\theta,\eta) \in \Theta_0 \times \Gamma$ holds

$$\begin{aligned} E_{\theta,\eta}\psi &= \int(\int\psi_y dP_\theta^y)\, dP_{\theta,\eta}^S(y) \le \int(\int\psi_y dP_{\theta_0}^y)\, dP_{\theta,\eta}^S(y) = \\ &= \int\alpha(y)\, dP_{\theta,\eta}^S(y) = \int\alpha(y)\, dP_{\theta_0,\eta(\theta_0)}^S(y) = \\ &= E_{\theta_0,\eta(\theta_0)}\varphi \le \alpha, \text{ where } P_{\theta_0,\eta(\theta_0)}^S = P_{\theta,\eta}^S, \text{ i.e. } \psi \in \phi_\alpha(\mathfrak{P}). \end{aligned}$$

Furthermore, for $(\theta,\eta) \in \Theta_1 \times \Gamma$ holds

$$E_{\theta,\eta} = \int(\int\varphi\, dP_\theta^y)\, dP_{\theta,\eta}^S(y) \le \int(\int\psi_y\, dP_\theta^y)\, dP_{\theta,\eta}^S(y) = E_{\theta,\eta}\psi ,$$

i.e. $\psi$ is better than $\varphi$ .

b) follows from a). □

Note that the weak ancillarity of S was used only for $\Theta_0 \times \Gamma$. Some examples for weakly p-ancillary models were discussed in section 2.

The problem of determining the MP level $\alpha_0$ test in $(\theta_1,\eta_1) \in \Theta_1 \times \Gamma_1$ leads to the following not easy variational problem (P):
Let $F_y$ be the df of $(P_{\theta_0}^y)^T$, $\gamma = \gamma(y, \alpha(y))$ be as defined in (15) and define

$$h(\alpha(y),y) = P_{\theta_1}^y\{T > F_y^{-1}(1-\alpha(y))\} + \gamma(y,\alpha(y))P_{\theta_1}^y\{T = F_y^{-1}(1-\alpha(y))\}. \tag{17}$$

$$\begin{aligned} &\text{(P): Determine } \sup_{\alpha\in\phi_{\alpha_0}} \int h(\alpha(y),y)dP_{\theta_1,\eta_1}^S(y), \text{ where } \phi_{\alpha_0} = \phi_{\alpha_0}(\mathfrak{P}^S) \\ &\text{is the set of all level } \alpha_0 \text{ tests w.r.t. } \{P_{\theta,\eta}^S;\ \eta\in\Gamma,\ \theta\in\Theta_0\}. \end{aligned} \tag{18}$$

Using now the close relation between unbiased tests and median unbiased estimates, as explained in Lehmann [15], p. 83, 176 - 177 and extended to the case of randomized tests (resp. estimators) as in Pfanzagl [17], one obtains for $\alpha = \frac{1}{2}$ the following result.

PROPOSITION 9. Under the assumptions of Proposition 6.b) or Proposition 7.b) or Proposition 8.b) (the unbiased case), there exists a uniformly optimal median unbiased (randomized) estimator $\delta^*$ for $\theta$ in the sense that for any other median unbiased estimator $\delta$ it holds that

$$E_{\theta,\eta} \int L(t,\theta)\delta^*(dt,\cdot) \leq E_{\theta,\eta} \int L(t,\theta)\delta(dt,\cdot) \tag{19}$$

for all loss functions $L(\cdot,\theta)$ decreasing on $(-\infty,\theta]$, increasing on $[\theta,\infty)$.

Proof. The proof is analogous to that given by Lehmann [15], p. 83, 176 resp. Pfanzagl [17]. Under the assumptions of Proposition 7.b), 8.b) we can argue as in Pfanzagl [17], relations (8), (9), (10), (13), while under the assumptions of Proposition 6.b) we argue as in Lehmann [15], p. 83, using the generalized randomized estimator (as in Pfanzagl [17], p. 191) if the distribution of T is not continuous. □

Proposition 9 was derived under the assumption of Proposition 7.b) by Pfanzagl [17] with the only difference that Pfanzagl assumes the whole class $\mathfrak{P}$ to be dominated, while our dominatedness assumptions concern each of the (smaller) conditional classes $\mathfrak{P}^y$, $y \in Y$. From the proof it is clear that the optimal estimator can be chosen nonrandomized if the (conditional) distribution of T is continuous.

## References


[1] Andersen, E. B.: On partial sufficiency and partial ancillarity. Skand Akturiet. 50 (1967), 137 - 152

[2] Andersen, E. B.: Asymptotic properties of conditional maximum-likelihood estimators. J. Roy. Stat. Soc., Ser. B, 32 (1970), 283 - 301

[3] Bahadur, R. R.: On unbiased estimates of uniformly minimum variance. Sankhya 18 (1957), 211 - 229

[4] Bahadur, R. R.: A note on UMV estimates and ancillary statistics. In: Memorial Volume dedicated to J. Hajek, Charles University, Prague, 1976

[5] Barndorff-Nielsen, O.: Information and Exponential Families in Statistical Theory. Wiley, 1978

[6] Basu, D.: On statistics independent of a complete sufficient statistic. Sankhya 15 (1955), 377 - 380

[7] Basu, D.: The family of ancillary statistics. Sankhya 21 (1959), 247 - 256

[8] Basu, D.: On the elimination of nuisance parameters. JASA 72 (1977), 355 - 366

[9] Bondesson, L.: On uniformly minimum variance unbiased estimation when no complete sufficient statistics exist. Metrika 30 (1983), 49 - 54

[10] Dawid, A. P.: On the concepts of sufficiency and ancillarity in the presence of nuisance parameters. J. Roy. Stat. Soc., Ser. B, 37 (1975), 248 - 258

[11] Fraser, D. A. S.: Sufficient statistics with nuisance parameters. Ann. Math. Statist. 27 (1956), 838 - 842

[12] Godambe, V. P.: On sufficiency and ancillarity in the presence of a nuisance parameter. Biometrika 67 (1980), 155 - 162

[13] Hajek, J.: On basic concepts of statistics. Proceedings of the Fifth Berkeley Symposium on Mathematical Statistics and Probability 1 (1965), 139 - 162

[14] Landers, D. and Rogge, L.: A note on completeness. Scand. J. Statist. 3 (1976), 139

[15] Lehmann, E. L.: Testing Statistical Hypotheses. Wiley, 1959

[16] Lehmann, E. L.: An interpretation of completeness and Basu's theorem. JASA 76 (1981), 335 - 340

[17] Pfanzagl, J.: On optimal median unbiased estimators in the presence of nuisance parameters. Ann. Statist. 7 (1979), 187 - 193

[18] Pfanzagl, J.: Contributions to a General Asymptotic Statistical Theory. Lecture Notes in Statistics 13, Springer, 1982

[19] Plachky, D.: A characterization of bounded completeness in the undominated case. Transactions of the Seventh Prague Conference 1974, 477 - 480

[20] Plachky, D. and Rüschendorf, L.: Conservation of the UMP and maximin property of tests under extension of probability measures. To appear in: Colloquium on Goodness of Fit, Debrecen, 1983

[21] Rao, C. R.: Minimum variance estimation in distributions admitting ancillary statistics. Sankhya 12 (1952), 53 - 56

[22] Sandved, E.: Ancillary statistics and estimation of the loss in estimation problems. Ann. Math. Statist. 39 (1968), 1755 - 1758

# Stability of Filtered Experiments

Helmut Strasser
Mathematisches Institut
Universität Bayreuth
D-8580 Bayreuth, FRG

**Summary:** A family of probability measures on a filtered probability space is called a filtered experiment. It is shown that sequences of filtered experiments, which are obtained by rescaling a fixed filtered experiment, can only have weak limits satisfying an invariance property called stability. This property allows a simplified approach to the problem of determining the sample size needed for separating parameter points by critical functions. In case of independent, identically distributed observations, the result covers previous assertions obtained by the author, [3]. In general, the result covers the case of dependent observations. It can be explained, how so-called mixed-normal situations arise in the limit. As a by-product we show, how an increasing family of experiments can be represented by a filtered experiment.

## 1. Introduction

Let us give a rough outline of what is done in this paper. We shall use the terminology of [4]. Assume that $E = (\Omega, \mathcal{A}, \{P_\theta: \theta \in \mathbb{R}\})$ is an experiment. Let $(\mathcal{A}_k)$ be a filtration of $\mathcal{A}$. If $\delta_n \downarrow 0$, then we might ask for weak limits of the sequence

$$E_n = (\Omega, \mathcal{A}_n, \{P_{\delta_n\theta}: \theta \in \mathbb{R}\}), \quad n \in \mathbb{N}.$$

This is the usual situation in case of independent, identically distributed observations, where $E$ is an infinite product of identical experiments and $(\mathcal{A}_k)$ is a filtration, consisting of cylindric $\sigma-$ fields.

In this paper we shall modify the problem slightly. For every $t \geq 0$ let

$$E_{nt} = (\Omega, \mathcal{A}_{[nt]}, \{P_{\delta_n\theta}: \theta \in \mathbb{R}\}), \quad n \in \mathbb{N}.$$

---

AMS 1980 *subject classifications.* Primary 62C99; Secondary 60F99.

*Key words and phrases.* Limits of experiments, dependent observations, stability of filtered experiments.

Contributions to Stochastics.
Ed. by W. Sendler
© Physica-Verlag Heidelberg 1987

Then, for fixed $n \in \mathbb{N}$ the family $t \longmapsto E_{nt}$ is increasing relative to deficiency. Assuming that for every $t \geq 0$ there is a weak limit $E_t$, the family $t \longmapsto E_t$ is increasing, too. In Section 3 of the paper, we prove that an increasing family typically is of the form $E_t = (\Omega_0, \mathcal{F}_t, \{Q_\theta\colon \theta \in \Theta\})$ where $(\mathcal{F}_t)_{t\geq 0}$ is a filtration on $\Omega_0$.

The main result of the paper is proved in Section 5, and it says, that the limiting family $(E_t)_{t\geq 0}$ necessarily satisfies the invariance property

$$(\Omega_0, \mathcal{F}_{\alpha t}, \{Q_\theta\colon \theta \in \Theta\}) \sim (\Omega_0, \mathcal{F}_t, \{Q_{\alpha^{1/p}\theta}\colon \theta \in \Theta\})$$

$\alpha \geq 0$, where $p > 0$ is a characteristic constant. This property is called stability and it is consistent with the notion of stability introduced in [3]: If $F$ is an infinitely divisible experiment, then it is stable in the sense of [3], iff the increasing family $t \longmapsto F^t$, $t \geq 0$, is stable in the sense of the present paper.

The practical implications of the notion of stability are discussed in Section 2.

In Section 4 we discuss several equivalent versions of stability. Moreover, for families whose likelihood ratios are of the form

$$\frac{dQ_\theta|\mathcal{F}_t}{dQ_0|\mathcal{F}_t} = \exp(X_\theta(t) - \frac{1}{2}[X_\theta, X_\theta]_t),$$

where $(X_\theta(t))_{t\geq 0}$ are continuous local martingales, stability can be expressed in terms of the processes $(X_\theta(t))_{t\geq 0}$, $\theta \in \Theta$. We discuss some special cases, showing that stability together with translation invariance imposes considerable restrictions to the class of possible limit experiments.

## 2. The practical significance of stability

The property of stability is of considerable practical importance. In order to explain this fact, let us consider the problem of determining the sample size which is necessary to separate two points $\theta_0 < \theta_1$ of the parameter space by means of tests with given level $\alpha$ and power $\beta > \alpha$.

**(2.1) Discussion.** First, let $E = (\Omega_0, (\mathcal{F}_t)_{t\geq 0}, \{Q_h\colon h \in \mathbb{R}\})$ be a fixed filtered experiment. Consider the parameter points $h_0 = 0$ and $h_1 > 0$. If

$$g(t,h) = \sup\{Q_h\,\varphi\colon \varphi \in \mathcal{F}_t,\ Q_0\,\varphi \leq \alpha\},$$

then the needed sample size is

$$t(h_1,\beta) = \inf\{t > 0\colon g(t,h_1) \geq \beta\}.$$

Next, consider a sequence of filtered experiments $E_n = (\Omega, (\mathcal{A}_{[nt]}), \{P_{\delta_n h}\colon h \in \mathbb{R}\}$, $n \in \mathbb{N}$, which converges weakly to $E$ in the sense that $E_n\,|\mathcal{A}_{[nt]} \longrightarrow E|\mathcal{F}_t$ weakly for every $t > 0$. Then we have

$$g_n([n\cdot t],\ \delta_n h) \longrightarrow g(t,h),$$

and provided that $t \longmapsto g(t,h)$ is continuous and strictly increasing, we obtain

$$t_n(h_1,\beta) \longrightarrow t(h_1,\beta).$$

This result shows us how to determine the sample size $k$ needed for $E_n$ to separate $\theta_0 = 0$ from $\theta_1 > 0$: Choose $h_1$ such that $\delta_n \cdot h_1 = \theta_1$. Then the quantity we are asking for is given by $k = [n \cdot t_n(h_1,\beta)]$. Since $t_n(h_1,\beta) \sim t(h_1,\beta)$, we get $k \sim [n \cdot t(\frac{\theta_1}{\delta_n}, \beta)]$.

We have seen that in general, for the correct asymptotic solution of our problem it is not sufficient to consider only a sequence of classical experiments. We have rather to consider the whole dynamical structure of each experiment, i.e. the filtration which describes the effect of drawing samples of different size. However, it is a matter of fact, that usually the problem is solved without considering experiments carrying a filtration. The usual device is as follows.

**(2.2) Discussion.** Consider a sequence of classical experiments

$$(\Omega, \mathcal{A}_n, \{P_{\delta_n h}: h \in \mathbb{R}\}), \quad n \in \mathbb{N},$$

converging weakly to

$$(\Omega_0, \mathcal{F}_1, \{Q_h: h \in \mathbb{R}\}).$$

Let $h_2 > 0$ be that parameter point which satisfies

$$h_2 = \inf\{h > 0: g(1,h) \geq \beta\}.$$

Then choose $k$ in such a way that $\theta_1 = \delta_k \cdot h_2$.

Obviously, the latter method is much simpler then the first one, but in general it solves a different problem. However, in those cases which usually are considered in asymptotic situations, the method works. Let us consider the most common example.

**(2.3) Example.** Let the limit experiment $(\Omega_0, (\mathcal{F}_t), \{Q_h: h \in \mathbb{R}\})$ be such that $\Omega_0 = \mathcal{C}([0,\infty))$, $Q_0$ is the Wiener measure, $(\mathcal{F}_t)$ is the natural history of the projection process $(W_t)$, and

$$\frac{dQ_h}{dQ_0} = \exp(h\,W_t - \frac{h^2}{2}t), \quad t > 0, \quad k \in \mathbb{R}.$$

For every fixed $t > 0$ the pertaining experiment is a simple Gaussian shift.
Assume that $\delta_n = n^{-1/2}$, $n \in \mathbb{N}$. Since

$$g(t,h) = \Phi(N_\alpha + \sqrt{t}\cdot h),$$

we get

$$t(h_1, \beta) = \left(\frac{N_\beta - N_\alpha}{h_1}\right)^2.$$

Hence, the correct solution is

$$k \sim \left[n \cdot \left(\frac{N_\beta - N_\alpha}{\sqrt{n}\,\theta_1}\right)^2\right] = \left(\frac{N_\beta - N_\alpha}{\theta_1}\right)^2.$$

The usual method of determining $k$ leads to

$$h_2 = \inf\{h > 0: \Phi(N_\alpha + h) \geq \beta\} = N_\beta - N_\alpha,$$

and the equation $\theta_1 = \frac{1}{\sqrt{k}} \cdot h_2$ then gives the same solution

$$k = \left(\frac{N_\beta - N_\alpha}{\theta_1}\right)^2.$$

What is the reason for this surprising coincidence we have observed in the preceding example? The reason is simply, that the limit experiment is a stable experiment.

In fact, if the limit experiment is stable of order $p > 0$, then we have

$$g(\gamma t, h) = g(t, \gamma^{1/p} \cdot h),$$

which leads to

$$t(h_1, \beta) = \left(\frac{1}{h_1} \cdot \inf\{h: g(1,h) \geq \beta\}\right)^p = \left(\frac{h_2}{h_1}\right)^p.$$

As we shall prove in the present paper, the sequence $(\delta_n)$ must be of the form $\delta_n = n^{-1/p}$ or $\delta_n = n^{-1/p} \cdot a_n$, where $(a_n)$ is of slow variation at infinity. Assuming the first case, we obtain by the correct method

$$k \sim \left[n \cdot \left(\frac{h_2}{n^{1/p}, \theta_1}\right)^p\right] = \left[\left(\frac{h_2}{\theta_1}\right)^p\right],$$

which is the same solution, which is obtained also by the usual device of solving the equation $\theta_1 = k^{1/p} \cdot h_2$. In the second case, things run similarly.

As a result, we have seen that the property of stability is the intrinsic reason why questions of sample size can be answered forgetting the dynamical structure of filtered experiments. The reason is a simple relation between the time parameter and the structural parameter. We shall prove in this paper that such a relation is typical for standard asymptotic situations.

## 3. Filtered experiments

Let $\Theta \neq \emptyset$ be an arbitrary parameter set.

**(3.1) Definition.** A family of experiments $E = (E_t: t \geq 0)$, $E_t \in \mathcal{E}(\Theta)$, $t \geq 0$, is a filtered experiment if there is an experiment $F = (\Omega, \mathcal{A}, \{P_\theta: \theta \in \Theta\})$ and a filtration $(\mathcal{F}_t)_{t \geq 0}$ of $\mathcal{A}$ such that every $E_t$ is the restriction of $F$ to $\mathcal{F}_t$, $t \geq 0$.

These are some examples of filtered experiments.

**(3.2) Examples.**

(1) Let $(\Omega, \mathcal{A}, \{Q_\theta: \theta \in \Theta\})$ be an experiment. Then the family of experiments $E_t = (\Omega^{\mathbb{N}}, \mathcal{A}^{[t]}, \{Q_\theta^{\mathbb{N}}: \theta \in \Theta\})$, $t \geq 0$, $(\mathcal{A}^0 := \{\emptyset, \Omega^{\mathbb{N}}\})$, defines a filtered experiment. Note, that $E_0$ is trivial, and $E_\infty := \lim_{t \to \infty} E_t$ is deterministic if $Q_\sigma \neq Q_\tau$ for $\sigma \neq \tau$.

(2) Let $(\Omega, \mathcal{A}, \{Q_\theta: \theta \in \Theta\})$ be an experiment and $(\mathcal{A}_k)_{k \geq 0}$ a countable filtration of $\mathcal{A}$. Then $E_t = (\Omega, \mathcal{A}_{[t]}, \{Q_\theta: \theta \in \Theta\})$, $t \geq 0$, defines a filtered experiment. This is the situation usually encountered in statistics with sequences of dependent observations.

(3) Let $(\Omega, \mathcal{A}, \{Q_\theta: \theta \in \Theta\})$ be an experiment and $(X_t)_{t \geq 0}$ a stochastic process. Supposing that $\mathcal{F}_t = \sigma\{X_s: s \leq t\}$, gives the experiment $E_t = (\Omega, \mathcal{F}_t, \{Q_\theta: \theta \in \Theta\})$. The family $(E_t: t \geq 0)$ is a filtered experiment.

For dealing with filtered experiments we need some well-known facts on likelihood ratios.

**(3.3) Remarks.**

(1) Let $P, Q$ be probability measures on a sample space $(\Omega, \mathcal{A})$ and let $\mathcal{B} \subseteq \mathcal{A}$ be a sub-$\sigma$-field. If $(X, M)$ is a Lebesgue-decomposition of $Q$ with respect to $P$ on $\mathcal{A}$, and $(Y, N)$ a decomposition on $\mathcal{B}$, then $P(X|\mathcal{B}) \geq Y$ $P$-a.e. and $M \subseteq N$ $Q$-a.e.

(2) Let $(\mathcal{A}_k)_{k \geq 0}$ be a sequence of sub-$\sigma$-fields and let $(X_k, N_k)$ be Lebesgue decompositions of $Q$ with respect to $P$ on $\mathcal{A}_k$, $k \geq 0$.

(a) If $\mathcal{A}_k \uparrow \mathcal{A}_\infty$, then $(X_k)$ is a positive supermartingale, and $(\lim X_k, \cup N_k)$ is a Lebesgue-decomposition of $Q$ with respect to $P$ on $\mathcal{A}_\infty$.

(b) If $\mathcal{A}_k \downarrow \mathcal{A}_\infty$, then $(X_k)$ is a reversed positive supermartingale, and $(\lim X_k, \cap N_k)$ is a Lebesgue-decomposition of $Q$ with respect to $P$ on $\mathcal{A}_\infty$.

(3) Let $E = (\Omega, \mathcal{A}, \{Q_\theta: \theta \in \Theta\})$ be an experiment and let $(\mathcal{A}_k)$ be a sequence of sub-$\sigma$-fields of $\mathcal{A}$. Denote $E_k = (\Omega, \mathcal{A}_k, \{Q_\theta: \theta \in \Theta\})$, $k \geq 0$. If $\mathcal{A}_k \uparrow \mathcal{A}_\infty$ or $\mathcal{A}_k \downarrow \mathcal{A}_\infty$ then

$$E_k \longrightarrow (\Omega, \mathcal{A}_\infty, \{Q_\theta: \theta \in \Theta\}), \quad \text{weakly.}$$

If $E = (E_t: t \geq 0)$ is a filtered experiment then $t \longmapsto E_t$ is increasing with respect to deficiency. Another example of an increasing family is $(E^t: t \geq 0)$, if $E$ is an infinitely divisible experiment. Let us mention some obvious properties of increasing families.

**(3.4) Remarks.** Let $(E_t: t \geq 0)$ be an increasing family.

(1) At every $t > 0$ there exist one-sided weak limits

$$E_{t+} = \lim_{s \downarrow t} E_s, \qquad E_{t-} = \lim_{s \uparrow t} E_s.$$

This is due to the fact, that the space of all experiment types is weakly compact (LeCam, [2]).

(2) Every increasing family can be regulated from the right. If the filtration of a filtered experiment is right continuous, then the corresponding increasing family is regulated from the right.

(3) Suppose that two increasing families $(E_t)_{t\geq 0}$ and $(F_t)_{t\geq 0}$ are equivalent on some countable, dense subset $D \subseteq [0, \infty)$. Then the corresponding regulated families are equivalent for every $t \geq 0$.

**(3.5) Definition.** Two increasing families are called equivalent, if the corresponding regulated families are equivalent for every $t \geq 0$.

This notion of equivalence is a preliminary one. It is sufficient for the purposes of the present paper. For more refined problems dealing with the dynamical structure of increasing families and filtered experiments a finer equivalence relation is needed. We shall discuss this problem elsewhere.

We shall prove that every increasing family of dominated experiments is equivalent to a filtered experiment, if for $s < t$ the experiment $E_t$ is not only more informative, but even exhaustive for $E_s$. Conditions for exhaustivity are given in [4], Corollary 55.11.

**(3.6) Remark.** For the proof of the next theorem we shall need the following fact:

Let $E = (\Omega_1, \mathcal{A}_1, \{P_\theta: \theta \in \Theta\})$ be a dominated experiment. Then there exists an experiment $F = (\Omega_2, \mathcal{A}_2, \{Q_\theta: \theta \in \Theta\})$ satisfying the conditions:

(i) The $\sigma$-field $\mathcal{A}_2$ is minimal-sufficient.

(ii) There is a semi-compact system $\mathcal{C} \subseteq \mathcal{A}_2$, having the approximation property for every $Q_\theta$, $\theta \in \Theta$.

(iii) $E$ and $F$ are mutually exhaustive.

E.g., take a probability measure $P_0 \sim \{P_\theta: \theta \in \Theta\}$, $\Omega_2 = [0, \infty)^\Theta$, $\mathcal{A}_2 = \mathcal{B}([0, \infty))^\Theta$, $Q_0 = \mathcal{L}((\frac{dP_\theta}{dP_0})_{\theta \in \Theta} \mid P_0)$, $Q_\theta = p_\theta \cdot Q_0$, $\theta \in \Theta$. For the proof of properties (i) and (iii) confer [4], Lemma 20.6 and Lemma 23.5.

**(3.7) Theorem.** Let $T = [0,\infty)$ and let $E_t = (\Omega_t, \mathcal{A}_t, \{P_{t,\theta}: \theta \in \Theta\})$, $t \in T$, be an increasing family of dominated experiments. If for $s < t$ the experiment $E_t$ is exhaustive for $E_s$, then $(E_t)_{t\in T}$ is equivalent to a filtered experiment.

*Proof.* W.l.g. we may assume that every $E_t$, $t \in T$, satisfies conditions (i) and (ii) of Remark (3.6). For $s < t$ let $K_{s,t}: \Omega_t \times \mathcal{A}_s \longrightarrow [0,1]$ be a stochastic kernel such that

$$P_{s,\theta} = K_{s,t} \circ P_{t,\theta}, \quad \theta \in \Theta.$$

For reasons of minimal sufficiency, these kernels are uniquely determined ([4], Theorem 55.16) and hence satisfy the semigroup relation

$$K_{t_1,t_3} = K_{t_1,t_2} \circ K_{t_2,t_3} \quad \text{if} \quad t_1 < t_2 < t_3.$$

Choose a countable and dense subset $D \subseteq T$ and let $\Omega = \prod_{t\in D} \Omega_t$ and $\mathcal{A} = \bigotimes_{t\in D} \mathcal{A}_t$. Let $\alpha = \{t_1,\cdots,t_k\} \in A(D)$ be such that $0 \le t_1 < t_2 < \cdots < t_k$. For every $\theta \in \Theta$ define a probability measure $R_{\theta,\alpha}$ on $\mathcal{B}^\alpha \times \Omega^{D\setminus\alpha}$ by

$$\begin{aligned} &R_{\theta,\alpha}(A_{t_1} \times \cdots \times A_{t_k} \times \Omega^{D\setminus\alpha}) \\ &:= \int_{A_{t_2}} \cdots \int_{A_{t_k}} K_{t_1,t_2}(\omega_{t_2}, A_{t_1}) \cdots \\ &\cdots K_{t_{k-1},t_k}(\omega_{t_k}, d\omega_{t_{k-1}})\, P_{t_k,\theta}(d\omega_{t_k}), \end{aligned}$$

if $A_{t_i} \in \mathcal{B}_{t_i}$, $1 \le i \le k$. The semigroup property implies that $(R_{\theta,\alpha})_{\alpha\in A(D)}$ is a consistent system of probability measures and by property (3.6), (ii), there exist projective limits $R_\theta \,|\, \mathcal{A}$, $\theta \in \Theta$.

Let $F = (\Omega, \mathcal{A}, \{R_\theta: \theta \in \Theta\})$ and $\mathcal{F}_t = \bigotimes_{s\le t, s\in D} \mathcal{A}_s \times \prod_{s>t, s\in D} \Omega_s$, $t \in D$. We claim that $F \,|\, \mathcal{F}_t \sim E_t$, $t \in D$. First, it is clear that $F \,|\, \mathcal{F}_t \supseteq E_t$ since

$$R_\theta(A_t \times \Omega^{T\setminus\{t\}}) = P_{t,\theta}(A_t), \quad A_t \in \mathcal{A}_t.$$

For the proof of $F \,|\, \mathcal{F}_t \subseteq E_t$, consider $\alpha = \{t_1,\cdots,t_k\} \subseteq D$ such that $t_1 < t_2 < \cdots < t_k = t$. Then

$$\begin{aligned} &(\omega_{t_k}, A_{t_1} \times \cdots \times A_{t_k}) \\ &\longmapsto 1_{A_{t_k}}(\omega_{t_k}) \int_{A_{t_2}} \cdots \int_{A_{t_{k-1}}} K_{t_1,t_2}(\omega_{t_2}, A_{t_1}) \cdots \\ &\cdots K_{t_{k-1},t_k}(\omega_{t_k}, d\omega_{t_{k-1}}) \end{aligned}$$

defines a randomization from $E_{t_k}$ to $F \,|\, \bigotimes_{s\in\alpha} \mathcal{A}_s \times \Omega^{T\setminus\alpha}$. This implies that

$$F \,|\, \bigotimes_{s\in\alpha} \mathcal{A}_s \times \Omega^{T\setminus\alpha} \subseteq E_t.$$

If $(\alpha_n)$ is a sequence in $A(D)$ satisfying $\alpha_n \uparrow \{s \in D: s \le t\}$, the left hand side converges weakly to $F|\mathcal{F}_t$, which proves that $F|\mathcal{F}_t \subseteq E_t$. Thus, we have shown that $F|\mathcal{F}_t \sim E_t$, $t \in D$. For every $t \in T$ let $\mathcal{F}_{t+} = \bigcap_{s>t, s\in D} \mathcal{F}_t$. Then $(F \,|\, \mathcal{F}_{t+})_{t\ge 0}$ is equivalent to $(E_{t+})_{t\ge 0}$, which proves the assertion. ∎

## 4. Stability of increasing families

In this section we assume that $\Theta$ is a convex cone of vertex $0$ in a linear space. If $E = (\Omega, \mathcal{A}, \{P_\theta: \theta \in \Theta\})$ then let $U_\alpha E = (\Omega, \mathcal{A}, \{P_{\alpha\theta}: \theta \in \Theta\})$, $\alpha \geq 0$.

**(4.1) Remark.** As is shown in [3], Lemma (2.3), $(U_\alpha)_{\alpha \geq 0}$ is a semigroup of weakly continuous transformations satisfying:

(1) $U_\alpha U_\beta = U_{\alpha\beta}, \quad \alpha \geq 0, \quad \beta \geq 0$.

(2) $U_\alpha E \otimes U_\alpha F = U_\alpha(E \otimes F), \quad \alpha \geq 0$.

(3) $E \sim F \implies U_\alpha E \sim U_\alpha F, \quad \alpha \geq 0$.

(4) If $E$ is continuous at $\theta = 0$, then $\lim\limits_{\alpha \to 0} U_\alpha E$ is trivial.

(5) If $E$ is continuous at $\theta = 0$ and not trivial, then $U_\alpha E \sim U_\beta E$ implies $\alpha = \beta$.

Let us introduce a family of transformations concerning the time parameter. If $(E_t)_{t \geq 0}$ is an increasing family, then let $V_\alpha E_t = E_{\alpha t}$, $\alpha \geq 0$, $t \geq 0$. The following assertion describes the relations between the transformations $(U_\alpha)$ and $(V_\alpha)$.

**(4.2) Lemma.** Let $(E_t)_{t \geq 0}$ and $(F_t)_{t \geq 0}$ be increasing families which are regulated from the right. Then the following assertions are valid:

(1) $V_\alpha V_\beta = V_{\alpha\beta}, \quad \alpha \geq 0, \quad \beta \geq 0$.

(2) $(E_t)_{t \geq 0} \sim (F_t)_{t \geq 0} \implies V_\alpha E_t \sim V_\alpha F_t, \quad t \geq 0$.

(3) $\alpha_n \downarrow \alpha \implies V_{\alpha_n} E_t \longrightarrow V_\alpha E_t$ weakly, $t \geq 0$.

(4) $\beta_n \uparrow \beta \implies V_{\beta_n} E_t \longrightarrow V_\beta E_{t-}$ weakly, $t \geq 0$.

(5) $V_\alpha U_\beta = U_\beta V_\alpha, \quad \alpha \geq 0, \quad \beta \geq 0$.

*Proof.* Obvious. ∎

The following theorem is basic for the notion of stability.

**(4.3) Theorem.** Let $(E_t)_{t \geq 0}$ be an increasing family which is regulated from the right. If $t \longmapsto E_t$ is not constant and $E_t$ is a continuous experiment for every $t \geq 0$, then the following assertions are equivalent:

(1) For every $\alpha > 0$ there is $\gamma > 0$ such that $V_\alpha E_t \sim U_\gamma E_t$, $t \geq 0$.

(2) For every $n \in \mathbb{N}$ there is $c_n > 0$ such that $V_n E_t \sim U_{c_n} E_t$, $t \geq 0$.

(3) There exists $p > 0$ such that $V_n E_t \sim U_{n^{1/p}} E_t$, $t \geq 0$, $n \in \mathbb{N}$.

(4) There exists $p > 0$ such that $V_\alpha E_t \sim U_{\alpha^{1/p}} E_t$, $t \geq 0$, $\alpha > 0$.

*Proof.* (1) $\implies$ (2): Obvious.

(2) $\implies$ (3): Let $t_0 \geq 0$ be such that $E_0 \not\sim E_{t_0}$. Then it follows easily that $U_{c_{m \cdot n}} E_{t_0} \sim U_{c_m \cdot c_n} E_{t_0}$ which implies by (4.2), (5), that $c_{m \cdot n} = c_m \cdot c_n$.

Let us show that $(c_n)$ is increasing. We have for every $t \geq 0$

$$V_{\frac{n}{m}} E_t = V_n E_{\frac{1}{m}} \sim U_{c_n} E_{\frac{1}{m}} = U_{\frac{c_n}{c_m}} U_{c_m} E_{\frac{1}{m}} \sim U_{\frac{c_n}{c_m}} E_t.$$

This implies by induction that

$$V_{(\frac{n}{m})^k} E_{t_0} \sim U_{(\frac{c_n}{c_m})^k} E_{t_0}, \quad k \in \mathbb{N}.$$

If $n \geq m$, but $c_n < c_m$, then it follows that

$$E_{t_0} \subseteq \lim_{n\to\infty} V_{(\frac{n}{m})^k} E_{t_0} \sim U_0 E_{t_0}$$

which is a contradiction, since $E_{t_0}$ cannot be trivial. The sequence $(c_n)$ is even strictly increasing since if it were constant we would obtain for $n < m$

$$E_0 \sim \lim_{k\to\infty} V_{(\frac{n}{m})^k} E_{t_0} \sim E_{t_0}.$$

(3) $\Longrightarrow$ (4): From the preceding arguments we see that for $m, n \in \mathbb{N}$

$$V_{(\frac{n}{m})} E_t \sim U_{(\frac{n}{m})^{1/p}} E_t, \quad t \geq 0.$$

Let $\alpha \geq 0$ and choose $(k(n)) \subseteq \mathbb{N}$ in such a way that

$$\frac{k(n)}{n} \downarrow \alpha \quad \text{for} \quad n \to \infty.$$

Then

$$V_\alpha E_t \sim \lim_{n\to\infty} V_{\frac{k(n)}{n}} E_t \sim \lim_{n\to\infty} U_{(\frac{k(n)}{n})^{1/p}} E_t \sim U_{\alpha^{1/p}} E_t, \quad t \geq 0.$$

(4) $\Longrightarrow$ (1): Obvious. ■

**(4.4) Definition.** Let $(E_t)_{t\geq 0}$ be an increasing family of continuous experiments which is not constant and is regulated from the right. The family is called stable if any of the assertions (1)–(4) of Theorem (4.3) is satisfied.

The relation of the preceding definition with the concept of stability introduced in [3] is as follows.

**(4.5) Corollary.** Let $E$ be a non-trivial, continuous experiment. The following assertions are equivalent:

(1) $E$ is stable (in the sense of [3], Definition (2.5)).

(2) $E$ is infinitely divisible and $(E^t)_{t\geq 0}$ is stable (in the sense of Definition (4.4)).

*Proof.* Easy. ■

Let us discuss some consequences of stability at hand of particular examples.

**(4.6) Definition.** Let $E_t = (\Omega, \mathcal{F}_t, \{P_\theta: \theta \in \Theta\})$, $t \geq 0$, be a filtered, homogeneous, experiment with a right-continuous filtration. It is of continuous type if

$$\frac{dP_\theta|\mathcal{F}_t}{dP_0|\mathcal{F}_t} = \exp(X_\theta(t) - \frac{1}{2}[X_\theta, X_\theta]_t), \quad t \geq 0,$$

where $(X_\theta(t))_{t\geq 0}$ is a continuous local martingale for each $\theta \in \Theta$.

This definition deserves some comments.

**(4.7) Remarks.**

(1) Experiments of continuous type are the typical limit experiments, if the jumps of the log-likelihood processes vanish as $n \to \infty$. This is proved by Swensen, [6], and by a different method by Strasser, [5].

(2) A Gaussian shift experiment can always be embedded into a filtered experiment. Let $\Theta$ be a finite-dimensional linear space and $B\colon \Theta \times \Theta \longrightarrow \mathbb{R}$ a positive definite bilinear function. Let $(W_\theta(t))_{t\geq 0}$ be Brownian motions on a probability space $(\Omega, \mathcal{A}, P_0)$, adapted to a filtration $(\mathcal{F}_t)_{t\geq 0}$, such that for $\sigma$, $\tau \in \Theta$,

$$P_0(W_\sigma(t), W_\tau(t)) = t \cdot B(\sigma,\tau), \quad t \geq 0.$$

Then

$$\frac{dP_\theta|\mathcal{F}_t}{dP_0|\mathcal{F}_t} = \exp(W_\theta(t) - \frac{t}{2} \cdot B(\theta,\theta))$$

defines experiments $(\Omega, \mathcal{F}_t, \{P_\theta\colon \theta \in \Theta\})$ which are Gaussian shifts with covariance structure $t \cdot B$, $t \geq 0$.

**(4.8) Theorem.** Assume that $E_t = (\Omega, \mathcal{F}_t, \{P_\theta\colon \theta \in \Theta\})$ is of continuous type. Then $(E_t)_{t\geq 0}$ is stable of exponent $p > 0$ iff

$$\mathcal{L}((X_\theta(\alpha t))_\theta \,|\, P_0) = \mathcal{L}((X_{\alpha^{1/p}\theta}(t))_\theta \,|\, P_0)$$

for all $\alpha \geq 0$, $t \geq 0$.

*Proof.* It is clear that stability is equivalent with

$$\mathcal{L}\left(\left(\frac{dP_\theta|\mathcal{F}_{\alpha t}}{dP_0|\mathcal{F}_{\alpha t}}\right)_\theta |\, P_0\right) = \mathcal{L}\left(\left(\frac{dP_{\alpha^{1/p}\theta}|\mathcal{F}_t}{dP_0|\mathcal{F}_t}\right)_\theta |\, P_0\right), \quad \alpha \geq 0, \quad t \geq 0.$$

Hence the condition is sufficient since $\mathcal{L}((X_\theta(t))_\theta|P_0)$, $t \geq 0$, determines the distributions of the likelihood processes. But also the converse is easy. Denoting $S_\theta(t) = X_\theta(t) - \frac{1}{2}\,[X_\theta, X_\theta]_t$, we have by stability

$$\mathcal{L}((S_\theta(\alpha t))_\theta \,|\, P_0) = \mathcal{L}((S_{\alpha^{1/p}\theta}(t))_\theta \,|\, P_0), \quad \alpha \geq 0, \quad t \geq 0.$$

Since $[S_\theta, S_\theta]_t = [X_\theta, X_\theta]_t$, it follows

$$X_\theta(t) = S_\theta(t) + \frac{1}{2}\,[S_\theta, S_\theta]_t, \quad t \geq 0, \quad \theta \in \Theta,$$

which proves the assertion. ∎

Let us discuss the problem of characterizing stability in terms of the increasing processes $([X_\sigma, X_\tau]_t)$, $\sigma, \tau \in \Theta$.

It is clear that stability implies a property of the quadratic variations $([X_\sigma, X_\tau]_t)$, $\sigma, \tau \in \Theta$, which corresponds to the condition given in Theorem (4.8). But stability cannot be characterized in terms of this property, if the quadratic variation does not determine the process $(X_\theta(t))$ completely. Only in cases, where the quadratic variation determines the martingale we can check stability by considering only the quadratic variation term. E.g., this is the case if the quadratic variation is deterministic.

**(4.9) Example.** Assume that $(E_t)_{t\geq 0}$ is of continuous type. If the quadratic variation

$$[X_\sigma, X_\tau]_t = K(\sigma,\tau;\, t), \quad \sigma, \tau \in \Theta, \quad t \geq 0,$$

is deterministic, then for every $t \geq 0$ the experiment $E_t$ is a Gaussian experiment with covariance kernel $(\sigma,\tau) \longmapsto K(\sigma,\tau;\, t)$, $\sigma, \tau \in \Theta$. If $(E_t)_{t\geq 0}$ is stable of exponent $p > 0$ and consists of translation invariant experiements, then

$$K(\sigma,\tau;\, t) \;=\; f(t\cdot|\sigma|^p) \,+\, f(t\cdot|\tau|^p) \,-\, f(t|\sigma-\tau|^p),$$

where

$$f(t) \;=\; -4\log\left(1-d^2(P_0\,|\,\mathcal{F}_t,\; P_1\,|\,\mathcal{F}_t)\right), \quad t \geq 0.$$

Let us consider a special case:

Sometimes it is known that a factorization $K(\sigma,\tau;\, t) = g(t)\,A(\sigma,\tau)$ is possible. Then $(E_{g^{-1}(t)})_{t\geq 0}$ is an infinitely divisible semi-group of Gaussian experiments. In the stable and translation invariant case we have even $f(t\cdot|\sigma|^p) = f(t)\cdot f(|\sigma|^p)$, which implies $f(t) = t^b$, $t \geq 0$, for some $b > 0$. This is a Gaussian shift situation iff $p \cdot b = 2$.

A slightly more general situation than the deterministic one is a mixed situation, which sometimes occurs in the limit when one is dealing with dependent observations.

**(4.10) Example.** Assume that $(E_t)_{t\geq 0}$ is of continuous type and there is a $\sigma$-field $\mathcal{T} \subseteq \mathcal{F}_\infty$ satisfying:

(i) $[X_\sigma, X_\tau]_t$ is $\mathcal{T}$-measurable for all $\sigma, \tau \in \Theta$, $t \geq 0$.

(ii) The conditional $P_0$-distribution of $(X_\theta(t))_\theta$ given $\mathcal{T}$ is Gaussian with covariance $(\sigma,\tau) \longmapsto [X_\sigma, X_\tau]_t$, $\sigma, \tau \in \Theta$.

(Confer Basawa-Scott, [1])

It is obvious, how the remarks of Example (4.9) have to be put to cover this situation.

## 5. Limits of increasing families

**(5.1) Definition.** A sequence $(E_{nt})_{t\geq 0}$ of increasing families converges weakly to an increasing family $(E_t)_{t\geq 0}$ if $E_{nt} \longmapsto E_t$ weakly for all $t \in D$, where $D \subseteq [0,\infty)$ is a countable dense subset of $[0,\infty)$.

This is a preliminary concept. Confer the remarks following Definition (3.5).

The following assertions are almost trivial.

**(5.2) Remarks.**

(1) The weak limit of a sequence of increasing families is uniquely determined up to equivalence (in the sense of Definition (3.5)).

(2) If a sequence $(E_{nt})_{t\geq 0}$ of increasing families converges weakly to an increasing family $(E_t)_{t\geq 0}$ of experiments then $E_{nt} \longrightarrow E_t$ for all continuity points $t \geq 0$ of $(E_t)_{t\geq 0}$.

(3) If the limit $(E_t)_{t\geq 0}$ is continuous, $E_0$ is trivial and $E_\infty$ is orthogonal, then the convergence $E_{nt} \longrightarrow E_t$ is uniform for all $t \geq 0$, with respect to the deficiencies based on finite parameter sets.

(4) Let $(E_{nt})_{t\geq 0}$, $n \in \mathbb{N}$, be a sequence of increasing families. If $(E_{nt})_{n\in\mathbb{N}}$ is weakly sequentially compact for $t \in D$, where $D \subseteq [0,\infty)$ is countable and dense, then there is a subsequence $\mathbb{N}_0 \subseteq \mathbb{N}$, such that $(E_{nt})_{t\geq 0}$, $n \in \mathbb{N}_0$, converges weakly to an increasing family.

We are interested in limits of increasing families of a special kind.

Assume that $\Theta$ is a convex cone of vertex 0 in a linear space. Let $E = (\Omega, \mathcal{A}, \{P_\theta: \theta \in \Theta\})$ be an experiment, $(\mathcal{A}_k)_{k\geq 0}$ a filtration of $\mathcal{A}$, and $\delta_n \downarrow 0$ a sequence of numbers. Define

$$E_{nt} = (\Omega, \mathcal{A}_{[nt]}, \{P_{\delta_n \theta}: \theta \in \Theta\}), \quad t \geq 0, \quad n \in \mathbb{N}.$$

We shall prove, that weak limits of $(E_{nt})_{t\geq 0}$, $n \in \mathbb{N}$, typically are stable in the sense of Definition (4.4).

In the statement of the next theorem we keep the notation just introduced.

**(5.3) Theorem.** Assume that $(E_{nt})_{t\geq 0}$, $n \in \mathbb{N}$, converges weakly to an increasing family $(E_t)_{t\geq 0}$ of separable experiments, which is regulated from the right. If $(E_{nt})_{n\in\mathbb{N}}$ is equicontinuous for every $t \geq 0$, then $(E_t)_{t\geq 0}$ is either constant or stable.

*Proof.* Let $m \in \mathbb{N}$. There is a subsequence $\mathbb{N}_m \subseteq \mathbb{N}$ such that ${}^{\delta_{m\cdot n}}/_{\delta_n} \longrightarrow a_m$ if $n \in \mathbb{N}_m$. Let $D \subseteq [0,\infty)$ be countable and dense such that $E_{nt} \longrightarrow E_t$ weakly for $t \in D$. Let $t_1 \leq t_2$ be such that $t_1 \in D$ and $m \cdot t_2 \in D$. Then we have

$$E_{mn,t_1} \longrightarrow E_{t_1} \quad \text{weakly for} \quad n \in \mathbb{N}_m,$$

and by equicontinuity

$$E_{mn,t_2} \longrightarrow V_m\, U_{a_m}\, E_{t_2} \quad \text{weakly for} \quad n \in \mathbb{N}_m.$$

This implies that

$$E_{t_1} \subseteq V_m\, U_{a_m}\, E_{t_2}.$$

Since $(E_t)_{t\geq 0}$ is regulated from the right, it follows that

$$E_t \subseteq V_m\, U_{a_m}\, E_t, \quad t \geq 0.$$

A similar argument yields

$$V_m\, U_{a_m}\, E_t \subseteq E_t, \quad t \geq 0.$$

Now, the assertion follows from Theorem (4.3). ■

**(5.4) Remark.** If $p > 0$ is the exponent of stability of the limit family in Theorem (5.3), then it follows that

$$\frac{\delta_{m\cdot n}}{\delta_n} \longrightarrow (\frac{1}{m})^{1/p} \quad \text{if} \quad n \in \mathbb{N},$$

which implies

$$\delta_n = n^{-1/p} \cdot a_n, \quad n \in \mathbb{N},$$

where $(a_n)$ is a sequence of slow variation at infinity.

It is clear that in the statement of Theorem (5.3) the filtered experiments $(E_{nt})_{t\geq 0}$, $n \in \mathbb{N}$, may be replaced by increasing families.

**(5.5) Corollary.** If a separable, filtered experiment $(E_t)_{t\geq 0}$ is stable of exponent $p > 0$ then there exists an experiment $E = (\Omega, \mathcal{A}, \{P_\theta: \theta \in \Theta\})$, and a filtration $(\mathcal{A}_k)$ of $\mathcal{A}$, such that the sequences of increasing families

$$E_{nt} = (\Omega, \mathcal{A}_{[nt]}, \{P_{n^{-1/p}\theta}: \theta \in \Theta\}), \quad t \geq 0, \quad n \in \mathbb{N},$$

converges weakly to $(E_t)_{t\geq 0}$.

*Proof.* Suppose that $E_t = (\Omega, \mathcal{F}_t, \{P_\theta : \theta \in \Theta\})$. Define $\mathcal{A}_k := \mathcal{F}_k$. Then

$$E_{nt} = (\Omega, \mathcal{F}_{[nt]+1}, \{P_{n^{-1/p}\theta} : \theta \in \Theta\}) \sim (\Omega, \mathcal{F}_{\frac{1}{n}([nt]+1)}, \{P_\theta : \theta \in \Theta\}).$$

Since $\mathcal{F}_{\frac{1}{n}([nt]+1)} \downarrow \mathcal{F}_t$ if $n \to \infty$, $t \geq 0$, the assertion follows. ∎

To put it slightly more generally, every stable increasing family of separable experiments may be obtained as limit of increasing families, which are obtained be rescaling a fixed increasing family.

Let us consider a special case.

**(5.6) Example.** In many situations (confer Basawa-Scott, [1]) we have

$$\frac{dP_{\delta_n \theta} | \mathcal{A}_{[nt]}}{dP_0 | \mathcal{A}_{[nt]}} = \exp\left( \frac{1}{a_n} \cdot \sum_{k=1}^{[nt]} X_k(\theta) - \frac{1}{2} \cdot \frac{1}{a_n^2} \cdot \sum_{k=1}^{[nt]} X_k^2(\theta) + o_{P_0}(1) \right),$$

where $a_n \uparrow \infty$, and $(X_k(\theta))_{k \in \mathbb{N}}$ are centered sequences. If

$$\frac{1}{a_n^2} \sum_{k=1}^{n} X_k(\sigma) \cdot X_k(\tau) \xrightarrow{P_0} Z(\sigma, \tau),$$

where $Z(\sigma,\tau)$ may be random variables, then in the limit we have the situation of Example (4.10). If the limit experiment is of continuous type, then the quadratic variation is $[X_\sigma, X_\tau]_t = t^b \cdot Z(\sigma, \tau)$, where

$$t^b = \lim_{n \to \infty} \left(\frac{a_{[nt]}}{a_n}\right)^2, \quad t \geq 0.$$

The limit experiment defines a stable, increasing family $(E_t)_{t \geq 0}$. In case $\Theta = \mathbb{R}$, it consists of translation invariant experiments iff

$$[X_\sigma, X_\tau]_t = (t|\sigma - \tau|^p)^b \cdot Z(0,1).$$

Every $E_t$ is a mixture of Gaussian shifts iff $p \cdot b = 2$, i.e. iff $a_n \cdot \delta_n \longrightarrow 1$ for $n \to \infty$.

## References

[1] Basawa, I.V. and D.J. Scott: Asymptotic optimal inference for non-ergodic models. Springer, New York, 1983.

[2] Lecam, L.: Limits of experiments. Proc. 6th Berkeley Symp. Math. Stat. Prob., Vol. 1, 245-261, 1972.

[3] Strasser, H.: Scale invariance of statistical experiments. Probability and Mathematical Statistics, Vol. 5, 1-20, 1985.

[4] Strasser, H.: Mathematical Theory of Statistics: Statistical Experiments and Asymptotic Decision Theory. De Gruyter Studies in Mathematics 7, de Gruyter, Berlin, 1985.

[5] Strasser, H.: Martingale difference arrays and stochastic integrals. Probab. Th. Rel. Fields 72, 83-98 (1986).

[6] Swensen, A.R.: Conditions for contiguity of probability measures under an asymptotic negligibility condition. Ph. D., Berkeley, 1980.

# Nonparametric Selection Procedures in Complete Factorial Experiments

Baldeo K. Taneja
Department of Mathematics
and Statistics
Case Western Reserve University
Cleveland, Ohio 44106, USA

SUMMARY: The behavior of many real-world systems depends on two or more factors which can be set at various levels. In such systems factorial experiments are usually conducted so that one can select or rank factor-level combinations, and study the performance of the system at those selected factor-level combinations. For the goal of selecting the best factor-level combination, all the existing theory in ranking and selection assumes normality of the observations. In this paper, we consider selection procedures for the above goal, in two-factor factorial experiments without relying on the assumption of normality. These procedures are then campared under no-interaction and interaction cases, and adaptive procedures are formulated.

## 1. INTRODUCTION

In his pioneering paper on selection, Bechhofer [2] considered a factorial procedure for the goal of selecting the largest mean in complete factorial experiments without interaction and with known variance. Bawa [1] compared Bechhofer's procedure with one-at-a-time procedure under the same setting and found Bechhofer's procedure to be superior. In this connection, an interaction procedure was given by Dudewicz [5] and Bechhofer [3], independently and simultaneously. Dudewicz and Taneja [6] compared these three existing procedures under no-interaction and interaction cases, and found interaction procedure to be fully robust to the presence of interaction whereas factorial and one-at-a-time procedures to be fully vitiated by interaction. Then, Taneja and Dudewicz ([12], [13]) formulated and studied a procedure which is valid even in the

AMS 1980 subject Classification: 62F07
Key words and phrases: Nonparametric selection, complete factorial experiments, indifference zone approach, asymptotic efficiency.

Contributions to Stochastics.
Ed. by W. Sendler
© Physica-Verlag Heidelberg 1987

presence of unbounded interaction, yet adapts to the presence/absence of interaction. Recently, Taneja [11] studied two-stage selection procedures when the common variance is unknown. All the above procedures require the assumption of normality of observations. The aim of this paper is to develop selection procedures for the goal of selecting the largest location parameter in complete factorial experiments without relying on the assumption of normality.

## 2. NONPARAMETRIC SELECTION IN COMPLETE FACTORIAL EXPERIMENTS WITHOUT INTERACTION

Suppose we have a two-factor factorial experiment with $k_1$ levels of factor 1 and $k_2$ levels of factor 2. Each factor-level combination is called a population. So we have $K = k_1 k_2$ populations $\pi_{ij}$ $(i=1, 2, \ldots, k_1;\ j=1, 2, \ldots, k_2)$. Let us assume that $\pi_{ij}$ has a continuous distribution function $F(x-\mu_{ij})$ where

$$\mu_{ij} = \mu + \alpha_i + \beta_j$$

along with the side-conditions

$$\sum_{i=1}^{k_1} \alpha_i = 0 \quad \text{and} \quad \sum_{j=1}^{k_2} \beta_j = 0.$$

Here $\mu$ is the overall general effect, $\alpha_i$ is the $i^{th}$ level effect of factor 1, and $\beta_j$ is the $j^{th}$ level effect of factor 2. Let the ordered $\alpha_i$'s be

$$\alpha_{[1]} \leq \alpha_{[2]} \leq \cdots \leq \alpha_{[k_1]}; \quad \text{and}$$

the ordered $\beta_j$'s be

$$\beta_{[1]} \leq \beta_{[2]} \leq \cdots \leq \beta_{[k_2]}.$$

Let $\ell$ be a subscript in one-to-one correspondence with $(i,j)$ such that $\ell$ takes values from 1 through K if and only if $(i,j)$ takes values from $(1,1)$ through $(k_1,k_2)$ and the ordered $\mu_{ij}$'s or $\mu_\ell$'s be

$$\mu_{[1]} \leq \mu_{[2]} \leq \cdots \leq \mu_{[K]}.$$

A population associated with $\mu_{[K]}$ is termed as the <u>best population</u>. Define

$$\delta_1 = \alpha_{[k_1]} - \alpha_{[k_1-1]},$$

$$\delta_2 = \beta_{[k_2]} - \beta_{[k_2-1]},$$

$$\delta = \mu_{[K]} - \mu_{[K-1]},$$

and let $\underset{\sim}{\delta} = (\delta_1, \delta_2)$. Our <u>goal</u> is to select a population (that is, a factor-level combination) associated with the largest location parameter $\mu_{[K]}$ (that is, the best population). Achieving this goal is called making a <u>Correct</u> <u>Selection</u> (abbreviated as CS). We seek procedures $P$ which satisfy the <u>probability</u> <u>requirement</u>

$$P\ (CS|P) \geq P^*$$

when $\underset{\sim}{\delta}$ is subject to the condition that

$$\underset{\sim}{\delta} \geq \underset{\sim}{\delta}^* \quad \text{(componentwise)}$$

where $P^*$ and $\underset{\sim}{\delta}^* = (\delta_1^*, \delta_2^*)$ are set prior to the experimentation such that

$$1/K < P^* < 1, \quad 0 < \delta_1^* < \infty, \quad 0 < \delta_2^* < \infty.$$

Note that

$$\begin{aligned} \delta &= \mu_{[K]} - \mu_{[K-1]} \\ &= \{\mu + \alpha_{[k_1]} + \beta_{[k_2]}\} - \max\{\mu + \alpha_{[k_1]} + \beta_{[k_2-1]}, \\ &\qquad \mu + \alpha_{[k_1-1]} + \beta_{[k_2]}\} \\ &= \min\{\delta_1, \delta_2\}. \end{aligned}$$

Hence, $\underset{\sim}{\delta} \geq \underset{\sim}{\delta}^*$ (componentwise) and $\delta \geq \delta^*$ with $\delta^* = \min\{\delta_1^*, \delta_2^*\}$ furnish comparable probability requirements for the following two procedures.

## 2.1 Procedures Based on Means

### Selection Procedure $P_1$:

When $F$ is normal, Bechhofer [2] has proposed the following procedure: Take $N$ independent observations $\{X_{ijk} : k=1, 2, \ldots, N\}$ independently from each of the $K=k_1k_2$ populations $\pi_{ij}$ $(i=1, 2, \ldots, k_1;\ j=1, 2, \ldots, k_2)$. Compute

$$\overline{X}_{ij\cdot} = \frac{1}{N} \sum_{k=1}^{N} X_{ijk},$$

$$\overline{X}_{i\cdot\cdot} = \frac{1}{k_2} \sum_{j=1}^{k_2} \overline{X}_{ij\cdot} = \frac{1}{Nk_2} \sum_{j=1}^{k_2} \sum_{k=1}^{N} X_{ijk}, \quad \text{and}$$

$$\overline{X}_{\cdot j\cdot} = \frac{1}{k_1} \sum_{i=1}^{k_1} \overline{X}_{ij\cdot} = \frac{1}{Nk_1} \sum_{i=1}^{k_1} \sum_{k=1}^{N} X_{ijk};$$

where $i=1, 2, \ldots, k_1$ and $j=1, 2, \ldots, k_2$. Let the ordered $\overline{X}_{i\cdot\cdot}$'s be

$$\overline{X}_{[1]\cdot\cdot} \leq \overline{X}_{[2]\cdot\cdot} \leq \cdot \cdot \cdot \leq \overline{X}_{[k_1]\cdot\cdot},$$

and the ordered $\overline{X}_{\cdot j\cdot}$'s be

$$\overline{X}_{\cdot[1]\cdot} \leq \overline{X}_{\cdot[2]\cdot} \leq \cdot \cdot \cdot \leq \overline{X}_{\cdot[k_2]\cdot}.$$

Select the levels associated with the largest marginal sample means $\overline{X}_{[k_1]\cdot\cdot}$ and $\overline{X}_{\cdot[k_2]\cdot}$; and assert that the combination of these two levels is best. Note that NK observations are used. There exists the smallest sample size per population $n$ for which this procedure satisfies the probability requirement:

$$P\ (CS|P_1) \geq P^* \tag{2.1.1}$$

where $\underset{\sim}{\delta}$ is subject to the restriction

$$\underset{\sim}{\delta} \geq \underset{\sim}{\delta}^*. \tag{2.1.2}$$

If the assumption of normality of $F$ is considered unreasonable, we can find a large-sample solution (for smallest sample size per population) which depends only on $\sigma^2$ (the variance of $F$). Consider a sequence of situations for increasing $n$ and define the restriction (2.1.2) as

$$\underset{\sim}{\delta} \geq \underset{\sim}{\delta}^{(n)} \tag{2.1.3}$$

where $\underset{\sim}{\delta}^{(n)} = (\delta_1^{(n)}, \delta_2^{(n)})$ is given by (2.1.6). Note that this definition is only a mathematical device for approximating the actual situation. If in a real-life situation the definition (2.1.2) applies with a given value of $\underset{\sim}{\delta}^*$, then $\underset{\sim}{\delta}^{(n)}$ will be identified with $\underset{\sim}{\delta}^*$. Since the family of distributions $F\ (x-\mu_{ij})$ with $\mu_{ij} = \mu + \alpha_i + \beta_j$

has the stochastic increasing property, $P(CS|P_1)$ takes on its minimum value for all $F$, when the following configuration of parameters hold:

$$\alpha_{[1]} = \alpha_{[2]} = \cdots = \alpha_{[k_1-1]} = \alpha_{[k_1]} - \delta_1^{(n)},$$
$$\beta_{[1]} = \beta_{[2]} = \cdots = \beta_{[k_2-1]} = \beta_{[k_2]} - \delta_2^{(n)}. \tag{2.1.4}$$

This configuration of parameters is referred to as the Least Favorable Configuration (LFC). We wish to find the smallest sample size per population $n$ for which

$$\inf_{\underset{\sim}{\delta} \geq \underset{\sim}{\delta}^{(n)}} P(CS|P_1) = P^* \tag{2.1.5}$$

Since there is no interaction between the levels of two factors, it is meaningful to select the "best" level of factor 1 (with probability of correct selection at least $P_1^*$ where $1/k_1 < P_1^* < 1$) and the "best" level of factor 2 (with probability of correct selection at least $P_2^*$ where $1/k_2 < P_2^* < 1$) independently so that there is an added requirement on $P^*$, namely, $P^* = P_1^* \cdot P_2^*$.

Theorem 2.1.1

For fixed $P^* = P_1^* \cdot P_2^*$ and under the condition (2.1.3), let $n$ be the smallest sample size per population for which (2.1.5) holds. Then as $n \to \infty$,

$$\delta_1^{(n)} = \Delta_1 \sigma k_2^{-\frac{1}{2}} n^{-\frac{1}{2}} + o(n^{-\frac{1}{2}})$$
$$\delta_2^{(n)} = \Delta_2 \sigma k_1^{-\frac{1}{2}} n^{-\frac{1}{2}} + o(n^{-\frac{1}{2}}) \tag{2.1.6}$$

where $\Delta_1$ and $\Delta_2$ are determined by the conditions

$$Q_{k_1-1}(\underbrace{\Delta_1 2^{-\frac{1}{2}}, \Delta_1 2^{-\frac{1}{2}}, \ldots, \Delta_1 2^{-\frac{1}{2}}}_{(k_1-1)\ \text{times}}) = P_1^*,$$

and

$$Q_{k_2-1}(\underbrace{\Delta_2 2^{-\frac{1}{2}}, \Delta_2 2^{-\frac{1}{2}}, \ldots, \Delta_2 2^{-\frac{1}{2}}}_{(k_2-1)\ \text{times}}) = P_2^* \tag{2.1.7}$$

where $Q_{k_t-1}$ is the distribution function of a normally distributed random vector $(Z_{t1}, Z_{t2}, \ldots, Z_{t,k_t-1})$ with $E(Z_{ti}) = 0$, $Var(Z_{ti}) = 1$, $(i = 1,2,\ldots,k_t-1;\ t = 1,2)$.

Proof:

Let $\overline{X}_{(i)\cdot\cdot}$ be the average of sample means from populations having location parameters $\mu + \alpha_{[i]} + \beta_j$, where average is being taken on subscript $j$; and $\overline{X}_{\cdot(j)\cdot}$ be the average of sample means from populations having location parameters $\mu + \alpha_i + \beta_{[j]}$, where average is being taken on the subscript $i$. Then the probability of correct selection using the procedure $P_1$ is

$$P(CS|P_1) = P(\overline{X}_{(k_1)\cdot\cdot} = \overline{X}_{[k_1]\cdot\cdot} \text{ and } \overline{X}_{\cdot(k_2)\cdot} = \overline{X}_{\cdot[k_2]\cdot})$$
$$= P(\overline{X}_{(i)\cdot\cdot} < \overline{X}_{(k_1)\cdot\cdot},\ i=1, 2, \ldots, k_1-1;$$
$$\overline{X}_{\cdot(j)\cdot} < \overline{X}_{\cdot(k_2)\cdot},\ j=1, 2, \ldots, k_2-1).$$

Let $W_{1i} = (\overline{X}_{(i)\cdot\cdot} - \overline{X}_{(k_1)\cdot\cdot}) \Big/ \sqrt{\frac{2\sigma^2}{nk_2}}$, $i=1, 2, \ldots, k_1-1$; and

$W_{2j} = (\overline{X}_{\cdot(j)\cdot} - \overline{X}_{\cdot(k_2)\cdot}) \Big/ \sqrt{\frac{2\sigma^2}{nk_1}}$, $j=1, 2, \ldots, k_2-1$; and let $(\mu^{(n)}, \alpha_1^{(n)}, \ldots, \alpha_{k_1}^{(n)}; \beta_1^{(n)}, \ldots, \beta_{k_2}^{(n)})$ be a sequence of parameter points satisfying (2.1.4). Then

$$P(CS|P_1) = P(W_{1i} < 0,\ i=1,2,\ldots,k_1-1;\ W_{2j} < 0,\ j=1,2,\ldots,k_2-1).$$

By the central limit theorem and the fact that the convergence is uniform in the arguments of the distribution function, we have

$$P(CS|P_1) = \lim_{n\to\infty} P\Bigg(Z_{1i} < \frac{\alpha_{[k_1]}^{(n)} - \alpha_{[i]}^{(n)}}{\sqrt{\frac{2\sigma^2}{nk_2}}},\ i=1, 2, \ldots, k_1 - 1;$$

$$Z_{2j} < \frac{\beta_{[k_2]}^{(n)} - \beta_{[j]}^{(n)}}{\sqrt{\frac{2\sigma^2}{nk_1}}},\ j=1, 2, \ldots, k_2-1\Bigg)$$

where the random vector $(Z_{11}, \ldots, Z_{1,k_1-1}; Z_{21}, \ldots, Z_{2,k_2-1})$ is normally distributed with

$$E(Z_{1i}) = 0, \text{ Var } (Z_{1i}) = 1 \quad (i=1, 2, \ldots, k_1-1)$$

$$E(Z_{2j}) = 0, \text{ Var } (Z_{2j}) = 1 \quad (j=1, 2, \ldots, k_2-1)$$

$$\text{Cov}(Z_{1i}, Z_{1i'}) = \frac{1}{2} \quad (i \neq i')$$

$$\text{Cov}(Z_{2j}, Z_{2j'}) = \frac{1}{2} \quad (j \neq j')$$

and

$$\text{Cov}(Z_{1i}, Z_{2j}) = 0 \quad (i=1, 2, \ldots, k_1-1;\ j=1, 2, \ldots, k_2-1).$$

Since $\text{Cov}(Z_{1i}, Z_{2j}) = 0$ $(i=1, 2, \ldots, k_1-1;\ j=1, 2, \ldots, k_2-1)$, the vectors $(Z_{11}, Z_{12}, \ldots, Z_{1,k_1-1})$ and $(Z_{21}, Z_{22}, \ldots, Z_{2,k_2-1})$ are independent, and under the LFC of parameters given by (2.1.4), we have

$$\inf_{\underset{\sim}{\delta} \geq \underset{\sim}{\delta}^{(n)}} P(CS|P_1) = \lim_{n\to\infty} P\left(Z_{1i} < \delta_1^{(n)} \sqrt{\frac{nk_2}{2\sigma^2}},\ i=1, 2, \ldots, k_1-1\right)$$

$$\cdot \lim_{n\to\infty} P\left(Z_{2j} < \delta_2^{(n)} \sqrt{\frac{nk_1}{2\sigma^2}},\ j=1, 2, \ldots, k_2-1\right).$$

If $\underset{\sim}{\Delta} = (\Delta_1, \Delta_2)$ is defined by (2.1.7), then

$$\inf_{\underset{\sim}{\delta} \geq \underset{\sim}{\delta}^{(n)}} P(CS|P_1) = P_1^* \cdot P_2^* = P^*$$

if and only if (2.1.6) holds. This completes the proof.

Now if $P^*$ and $\underset{\sim}{\delta}^*$ are specified, then it follows from Theorem 2.1.1 that a large-sample solution for the smallest sample size per population is

$$n = \left[\text{the smallest integer} \geq \max\left\{\frac{\Delta_1^2\ \sigma^2}{k_2\ \delta_1^{*2}},\ \frac{\Delta_2^2\ \sigma^2}{k_1\ \delta_2^{*2}}\right\}\right] \qquad (2.1.8)$$

for which the probability requirement is satisfied.

<u>Selection Procedure $P_2$:</u>

When $F$ is normal, Bechhofer [3] and Dudewicz [5] simultaneously but independently proposed the following procedure: Take $N$ independent observations $\{X_{ijk}: k=1, 2, \ldots, N\}$ independently from each of the $K=k_1k_2$ populations $\pi_{ij}$ $(i=1, 2, \ldots, k_1;\ j=1, 2, \ldots, k_2)$. Compute

$$\overline{X}_{ij\cdot} = \frac{1}{N} \sum_{k=1}^{N} X_{ijk}$$

where $i=1, 2, \ldots, k_1$ and $j=1, 2, \ldots, k_2$. Let $\ell$ be a subscript in one-to-one correspondence with $(i,j)$ such that $\ell$ takes values from 1 through $K$ if and only if $(i,j)$ takes values from (1,1) through $(k_1,k_2)$. Let the ordered $\bar{X}_{ij\cdot}$'s or $\bar{X}_\ell$'s be

$$\bar{X}_{[1]} \leq \bar{X}_{[2]} \leq \cdots \leq \bar{X}_{[K]}.$$

Select the population associated with $\bar{X}_{[K]}$ as best. Note that NK observations are used. There exists the smallest sample size per population $n$ for which this procedure satisfies the probability requirement

$$P\ (CS|P_2) \geq P^* \tag{2.1.9}$$

where $\delta \equiv \mu_{[K]} - \mu_{[K-1]}$ is subject to the condition that

$$\delta \geq \delta^* \tag{2.1.10}$$

where $\delta^* = \min \{\delta_1^*, \delta_2^*\}$.

If we do not wish to rely on the assumption of normality of $F$, we can find a large-sample solution for the smallest sample size per population which depends on $\sigma^2$ (the variance of $F$). For this, consider a sequence of situations for increasing $n$, and define (2.1.10) as

$$\delta \geq \delta^{(n)} \tag{2.1.11}$$

where $\delta^{(n)}$ is given by (2.1.14). If for a real-life situation (2.1.10) applies, then $\delta^{(n)}$ in (2.1.11) will be identified with $\delta^*$. Because of stochastic increasing property of $F(x-\mu_{ij})$ or $F(x-\mu_\ell)$, the LFC of $\mu_{ij}$'s or $\mu_\ell$'s is given by

$$\mu_{[1]} = \mu_{[2]} = \cdots = \mu_{[K-1]} = \mu_{[K]} - \delta^{(n)}. \tag{2.1.12}$$

We wish to find the smallest sample size per population $n$ for which

$$\inf_{\delta \geq \delta^{(n)}} P\ (CS|P_2) = P^*. \tag{2.1.13}$$

Theorem 2.1.2

For fixed $P^*$ and under the condition (2.1.11), let $n$ be the smallest sample size per population for which (2.1.13) holds. Then as $n \to \infty$,

$$\delta^{(n)} = \Delta\sigma n^{-\frac{1}{2}} + o\ (n^{-\frac{1}{2}}) \tag{2.1.14}$$

where $\Delta$ is determined by the condition

$$Q_{K-1}\underbrace{(\Delta 2^{-\frac{1}{2}}, \Delta 2^{-\frac{1}{2}}, \ldots, \Delta 2^{-\frac{1}{2}})}_{K-1 \text{ times}} = P^* \tag{2.1.15}$$

where $Q_{K-1}$ is the distribution function of a normally distributed random vector $(Z_1, Z_2, \ldots, Z_{K-1})$ with

$$E(Z_\ell) = 0, \ \text{Var}(Z_\ell) = 1 \qquad (\ell=1, 2, \ldots, K-1)$$

$$\text{Cov}(Z_\ell, Z_{\ell'}) = \frac{1}{2} \quad \text{for} \quad \ell \neq \ell'.$$

<u>Proof</u>:

Let $\overline{X}_{(\ell)}$ be the sample mean from the population having location parameter $\mu_{[\ell]}$, $\ell=1, 2, \ldots, K$. Then the probability of correct selection using procedure $P_2$ is

$$P(CS|P_2) = P(\overline{X}_{(K)} = \overline{X}_{[K]})$$

$$= P(\overline{X}_{(\ell)} < \overline{X}_{(K)}, \ \ell=1, 2, \ldots, K-1).$$

Let $W_\ell = (\overline{X}_{(\ell)} - \overline{X}_{(K)})/\sqrt{\frac{2\sigma^2}{n}}$ and let $\underset{\sim}{\mu}^{(n)} = (\mu_1^{(n)}, \mu_2^{(n)}, \ldots, \mu_K^{(n)})$ be a sequence of parameter points satisfying (2.1.12). Then

$$P(CS|P_2) = P(W_\ell < 0, \ \ell=1, 2, \ldots, K-1).$$

By the central limit theorem and the fact that the convergence is uniform in the arguments of the distribution function, we have

$$P(CS|P_2) = \lim_{n \to \infty} P\left(Z_\ell < \frac{\mu_{[K]}^{(n)} - \mu_{[\ell]}^{(n)}}{\sqrt{\frac{2\sigma^2}{n}}}, \ \ell=1, 2, \ldots, K-1\right).$$

Under the LFC of $\mu_{ij}$'s or $\mu_\ell$'s given by (2.1.12),

$$\inf_{\delta \geq \delta^{(n)}} P(CS|P_2) = \lim_{n \to \infty} P\left(Z_\ell < \delta^{(n)}\sqrt{\frac{n}{2\sigma^2}}, \ \ell=1, 2, \ldots, K-1\right).$$

If $\Delta$ is defined by the condition (2.1.15), then

$$\inf_{\delta \geq \delta^{(n)}} P(CS|P_2) = P^*$$

if and only if (2.1.14) holds. This completes the proof.

Now if $P^*$ and $\delta^*$ are specified, it follows from Theorem 2.1.2

that a large-sample solution for the smallest sample size per population is

$$n = \left[ \text{the smallest integer} \geq \frac{\Delta^2 \sigma^2}{\delta^{*2}} \right] \qquad (2.1.16)$$

for which the probability requirement is satisfied.

Comparison of $P_1$ and $P_2$

In order to compare the two selection procedures $P_1$ and $P_2$, let $n_1$ and $n_2$ denote the respective smallest number of observations each procedure needs in order to guarantee the probability requirement. The relative efficiency (RE) and asymptotic relative efficiency (ARE) of $P_2$ relative to $P_1$ are defined as

$$RE\ (P_1,\ P_2) = \frac{n_1}{n_2}, \quad \text{and} \quad ARE_{P^*}\ (P_1, P_2) = \lim_{P^* \to 1} \frac{n_1}{n_2}.$$

Similar definitions of RE $(R_1,\ R_2)$ and $ARE_{P^*}$ $(R_1,\ R_2)$ will be used in the next section for the two procedures $R_1$ and $R_2$. Although in practice the sample sizes are integers, we treat them as nonnegative continuous variables for computing the ARE. Using standard asymptotics as in Dudewicz and Taneja [6], we obtain the following

Theorem 2.1.3

$$ARE_{P^*}\ (P_1,\ P_2) = \frac{\delta^{*2}}{K} \max\left\{ \frac{k_1}{\delta_1^{*2}} , \frac{k_2}{\delta_2^{*2}} \right\}. \qquad (2.1.17)$$

It follows from Theorem 2.1.3 that $ARE_{P^*}$ $(P_1, P_2) \leq 1/2$ for large $P^*$. Hence, when there is no interaction, procedure $P_1$ is better than $P_2$ in terms of asymptotic relative efficiency.

## 2.2 Procedures Based on Ranks

Selection Procedure $R_1$

Suppose that the observations $X_{ijk}$ $(i=1, 2, \ldots, k_1;\ j=1, 2, \ldots, k_2;\ k=1, 2, \ldots, N)$ are ranked from smallest to largest, and the rank of $X_{ijk}$ is denoted by $R_{ijk}$. Let

$$\psi(i,j,M,r) = \begin{cases} 1 & \text{if the } r^{th} \text{ smallest of } M\ (=Nk_1k_2) \text{ observations is from the population } \pi_{ij} \\ 0 & \text{Otherwise.} \end{cases}$$

Define

$$T_{ij} = \frac{1}{N} \sum_{r=1}^{M} E\,(V^{(r)})\ \psi(i,j,M,r)$$

$$T_{i\cdot} = \frac{1}{k_2} \sum_{j=1}^{k_2} T_{ij}, \quad T_{\cdot j} = \frac{1}{k_1} \sum_{i=1}^{k_1} T_{ij} \tag{2.2.1}$$

where $i=1, 2, \ldots, k_1$ and $j=1, 2, \ldots, k_2$ and $V^{(1)} \leq V^{(2)} \leq \ldots \leq V^{(M)}$ is an ordered sample from a given continuous distribution $F_o$. Let the ordered $T_{i\cdot}$'s be

$$T_{[1]\cdot} \leq T_{[2]\cdot} \leq \cdot\cdot\cdot \leq T_{[k_1]\cdot} \tag{2.2.2}$$

and the ordered $T_{\cdot j}$'s be

$$T_{\cdot[1]} \leq T_{\cdot[2]} \leq \cdot\cdot\cdot \leq T_{\cdot[k_2]}. \tag{2.2.3}$$

Select the levels associated with $T_{[k_1]\cdot}$ and $T_{\cdot[k_2]}$ and assert that the combination of these two levels is best.

We wish to find the smallest sample size per population $m$ for which $R_1$ guarantees the probability requirement

$$P\,(CS|R_1) \geq P^* \quad \text{with} \quad P^* = P_1^* \cdot P_2^* \tag{2.2.4}$$

whenever a sequence $\underset{\sim}{\theta}^{(m)} = (\mu^{(m)};\ \alpha_1^{(m)}, \ldots, \alpha_{k_1}^{(m)};\ \beta_1^{(m)}, \ldots, \beta_{k_2}^{(m)})$ of parameter points satisfying

$$\alpha_{[k_1]} - \alpha_{[i]} \equiv \delta_{1i} = \Delta_{1i}\ m^{-\frac{1}{2}} + o(m^{-\frac{1}{2}}),\ i=1, 2, \ldots, k_1-1$$

$$\beta_{[k_2]} - \beta_{[j]} \equiv \delta_{2j} = \Delta_{2j}\ m^{-\frac{1}{2}} + o(m^{-\frac{1}{2}}),\ j=1, 2, \ldots, k_2-1 \tag{2.2.5}$$

is subject to the conditions

$$\alpha_{[k_1]} - \alpha_{[k_1-1]} \geq \tilde{\delta}_1^{(m)}$$

$$\beta_{[k_2]} - \beta_{[k_2-1]} \geq \tilde{\delta}_2^{(m)} \tag{2.2.6}$$

where $\Delta$'s are positive constants and $\underset{\sim}{\tilde{\delta}}^{(m)} = (\tilde{\delta}_1^{(m)}, \tilde{\delta}_2^{(m)})$ is given by (2.2.11).

In order to find LFC of the parameters we use the following Lemma. Proof of the lemma follows from Puri and Puri [10] and hence is omitted.

Lemma 2.2.1

For $m=1, 2, \ldots,$ let $\{X_{ijk}: k=1, 2, \ldots, m\}$ be independently distributed according to $F_{ij}(x) = F(x - \mu^{(m)} - \alpha_i^{(m)} - \beta_j^{(m)})$ with the sequence of parameter points $\underset{\sim}{\theta}^{(m)} = (\mu^{(m)}; \alpha_1^{(m)}, \ldots, \alpha_{k_1}^{(m)}; \beta_1^{(m)}, \ldots, \beta_{k_2}^{(m)})$ and suppose that the assumptions of Theorem 6.1 and Lemma 7.2 of Puri [9] are satisfied. Let $T_{(i)\cdot}$ be the statistic associated with populations having locations $\mu + \alpha_{[i]} + \beta_j$; and $T_{\cdot(j)}$ be the statistic associated with populations having locations $\mu + \alpha_i + \beta_{[j]}$. Then the limiting $(m \to \infty)$ distribution of the random vector $(U_{11}, U_{12}, \ldots, U_{1,k_1-1}; U_{21}, U_{22}, \ldots, U_{2,k_2-1})$ with

$$U_{1i} = (k_2 m/2A^2)^{\frac{1}{2}} (T_{(i)\cdot} - T_{(k_1)\cdot} - \alpha_{[i]}(\underset{\sim}{\theta}^{(m)}) + \alpha_{[k_1]}(\underset{\sim}{\theta}^{(m)}))$$

$$U_{2j} = (k_1 m/2A^2)^{\frac{1}{2}} (T_{\cdot(j)} - T_{\cdot(k_2)} - \beta_{[j]}(\underset{\sim}{\theta}^{(m)}) + \beta_{[k_2]}(\underset{\sim}{\theta}^{(m)}))$$

(2.2.7)

where

$$\alpha_i(\underset{\sim}{\theta}^{(m)}) = \sum_{j=1}^{k_2} \{\int J(H(x))\, d F_{ij}(x)\}/k_2, \quad i=1, 2, \ldots, k_1$$

$$\beta_j(\underset{\sim}{\theta}^{(m)}) = \sum_{i=1}^{k_1} \{\int J(H(x))\, d F_{ij}(x)\}/k_1, \quad j=1, 2, \ldots, k_2$$

$$H(x) = \sum_{i=1}^{k_1} \sum_{j=1}^{k_2} F_{ij}(x)/k_1 k_2, \qquad J = F_0^{-1}$$

$$A^2 = \int J^2(x)\, dx - (\int J(x)\, dx)^2 \qquad (2.2.8)$$

is the distribution of $(k_1 + k_2 - 2)$-variate normal vector $(Z_{11}, Z_{12}, \ldots, Z_{1,k_1-1}; Z_{21}, Z_{22}, \ldots, Z_{2,k_2-1})$ with

$$E(Z_{1i}) = 0, \quad Var(Z_{1i}) = 1 \qquad (i=1, 2, \ldots, k_1-1)$$

$$E(Z_{2j}) = 0, \quad Var(Z_{2j}) = 1 \qquad (j=1, 2, \ldots, k_2-1)$$

$$Cov(Z_{1i}, Z_{1i'}) = \frac{1}{2} \qquad (i \neq i')$$

$$Cov(Z_{2j}, Z_{2j'}) = \frac{1}{2} \qquad (j \neq j')$$

and

$$\text{Cov}\,(Z_{1i}, Z_{2j}) = 0 \qquad (i=1, 2, \ldots, k_1-1;\ j=1, 2, \ldots, k_2-1).$$

Using the Lemma 2.2.1, the LFC of the parameters is

$$\alpha_{[1]} = \alpha_{[2]} = \ldots = \alpha_{[k_1-1]} = \alpha_{[k_1]} - \tilde{\delta}_1^{(m)}$$

$$\beta_{[1]} = \beta_{[2]} = \ldots = \beta_{[k_2-1]} = \beta_{[k_2]} - \tilde{\delta}_2^{(m)}. \qquad (2.2.9)$$

Theorem 2.2.1

For fixed $P^* = P_1^* P_2^*$ and under the conditions (2.2.5) and (2.2.6), let $m$ be the smallest sample size per population for which

$$\inf_{\underset{\sim}{\delta} \geq \underset{\sim}{\tilde{\delta}}^{(m)}} P\,(CS|R_1) = P^* \qquad (2.2.10)$$

and suppose that $F$ and $J = F_0^{-1}$ satisfy the regularity conditions of Theorem 6.1 and Lemma 7.2 of Puri [9]. Then as $m \to \infty$

$$\tilde{\delta}_1^{(m)} = \Delta_1 m^{-\frac{1}{2}} k_2^{-\frac{1}{2}} A / \int \frac{d}{dx}\,(J(F(x))dF(x) + o(m^{-\frac{1}{2}})$$

$$\tilde{\delta}_2^{(m)} = \Delta_2 m^{-\frac{1}{2}} k_1^{-\frac{1}{2}} A / \int \frac{d}{dx}\,(J(F(x))dF(x) + o(m^{-\frac{1}{2}}) \qquad (2.2.11)$$

where $\underset{\sim}{\Delta} = (\Delta_1, \Delta_2)$ is determined by (2.1.7).

Proof:

Let $T_{(i)\cdot}$ be the statistics associated with the populations having location parameters $\mu + \alpha_{[i]} + \beta_j$, and $T_{\cdot(j)}$ be the statistics associated with the populations having location parameters $\mu + \alpha_i + \beta_{[j]}$, $i=1, 2, \ldots, k_1$ and $j=1, 2, \ldots, k_2$. Then

$$\begin{aligned}
P(CS|R_1) &= P(T_{(k_1)\cdot} = T_{[k_1]\cdot} \text{ and } T_{\cdot(k_2)} = T_{\cdot[k_2]}) \\
&= P(T_{(i)\cdot} - T_{(k_1)\cdot} < 0, \quad i=1, 2, \ldots, k_1-1; \\
&\qquad T_{\cdot(j)} - T_{\cdot(k_2)} < 0, \quad j=1, 2, \ldots, k_2-1) \\
&= P(U_{1i} < (k_2 m/2A^2)^{\frac{1}{2}}(\alpha_{[k_1]}(\underset{\sim}{\theta}^{(m)}) - \alpha_{[i]}(\underset{\sim}{\theta}^{(m)})), \\
&\qquad i=1, 2, \ldots, k_1-1; \\
&\qquad U_{2j} < (k_1 m/2A^2)^{\frac{1}{2}}(\beta_{[k_2]}(\underset{\sim}{\theta}^{(m)}) - \beta_{[j]}(\underset{\sim}{\theta}^{(m)})), \\
&\qquad j=1, 2, \ldots, k_2-1).
\end{aligned}$$

Using the LFC (2.2.9) of parameters, techniques of Puri and Puri [10], and independence of vectors $(Z_{11}, Z_{12}, \ldots, Z_{1,k_1-1})$ and $(Z_{21}, Z_{22}, \ldots, Z_{2,k_2-1})$, we have

$$\inf_{\underset{\sim}{\hat{\delta}} \geq \underset{\sim}{\delta}^{(m)}} P(CS|P_1) = \lim_{m\to\infty} P(Z_{1i} < (k_1 m/2A^2)^{\frac{1}{2}} \delta_{1i}^{(m)}, \quad i=1, \ldots, k_1-1)$$
$$\cdot \lim_{m\to\infty} P(Z_{2j} < (k_2 m/2A^2)^{\frac{1}{2}} \delta_{2j}^{(m)}, \quad j=1, \ldots, k_2-1).$$

If $\underset{\sim}{\Delta} = (\Delta_1, \Delta_2)$ is obtained from (2.1.7), then

$$\inf_{\underset{\sim}{\hat{\delta}} \geq \underset{\sim}{\delta}^{(m)}} P(CS|R_1) = P^*$$

if and only if (2.2.11) holds.

Now if $P^*$ and $\underset{\sim}{\delta}^*$ are specified, identifying $\hat{\tilde{\delta}}^{(m)}$ with $\underset{\sim}{\delta}^*$, it follows from the Theorem 2.2.1 that a large-sample solution for procedure $R_1$ is

$$m = \left[ \text{the smallest integer} \geq \max \left\{ \frac{A^2\Delta_1^2}{k_2 \delta_1^{*2} \left( \int \frac{d}{dx}(J(F(x))dF(x)\right)^2}, \frac{A^2\Delta_2^2}{k_1 \delta_2^{*2} \left( \int \frac{d}{dx}(J(F(x)))dF(x)\right)^2} \right\} \right] \tag{2.2.12}$$

for which the probability requirement is satisfied.

Selection Procedure $R_2$

Proceed as in Selection Procedure $R_1$ and define $T_{ij}(i=1, 2, \ldots, k_1;\ j=1, 2, \ldots, k_2)$. Let $\ell$ be a subscript in one-to-one correspondence with $(i,j)$. Let the ordered $T_{ij}$'s or $T_\ell$'s be

$$T_{[1]} \leq T_{[2]} \leq \cdot \cdot \cdot \leq T_{[k]}. \tag{2.2.13}$$

Select the population associated with $T_{[K]}$ as best. Note that NK observations are used.

We wish to find the smallest sample size per population m for which $R_2$ guarantees the probability requirement

$$P(CS|R_2) \geq P^* \tag{2.2.14}$$

whenever a sequence $\underset{\sim}{\mu}^{(m)} = (\mu_1^{(m)}, \mu_2^{(m)}, \ldots, \mu_K^{(m)})$ of parameter points satisfying

$$\mu_{[K]} - \mu_{[\ell]} \equiv \delta_\ell = \Delta_\ell m^{-\frac{1}{2}} + o(m^{-\frac{1}{2}}), \quad \ell=1, 2, \ldots, K-1 \tag{2.2.15}$$

is subject to the condition

$$\mu_{[K]} - \mu_{[K-1]} \geq \tilde{\delta}^{(m)} \tag{2.2.16}$$

where the $\Delta_\ell$'s are positive constants and $\tilde{\delta}^{(m)}$ is given by (2.2.21). It can be shown as in Puri and Puri [10] that the LFC of $\mu_{ij}$'s or $\mu_\ell$'s is given by

$$\mu_{[1]} = \mu_{[2]} = \cdots = \mu_{[K-1]} = \mu_{[K]} - \tilde{\delta}^{(m)}. \tag{2.2.17}$$

This requires the following lemma. Proof follows from Puri and Puri [10] and hence is omitted.

<u>Lemma 2.2.2</u>

For $m=1, 2, \ldots,$ let $X_{ijk}(i=1, 2, \ldots, k_1;\ j=1, 2, \ldots, k_2;\ k=1, 2, \ldots, m)$ be independently distributed according to $F_\ell(x) = F(x-\mu_\ell^{(m)})$ with the sequence of parameter points $\underset{\sim}{\mu}^{(m)} = (\mu_1^{(m)}, \mu_2^{(m)}, \ldots, \mu_K^{(m)})$ and suppose that the assumptions of Theorem 6.1 and Lemma 7.2 of Puri [9] are satisfied. Let $T_{(\ell)}$ be the statistic associated with the population having location parameter $\mu_{[\ell]}$, $\ell=1, 2, \ldots, K$. Then the limiting $(m \to \infty)$ distribution of the random vector $(U_1, U_2, \ldots, U_{K-1})$ with

$$U_\ell = (m/2A^2)^{\frac{1}{2}}(T_{(\ell)} - T_{(K)} - \mu_{[\ell]}(\underset{\sim}{\mu}^{(m)}) + \mu_{[K]}(\underset{\sim}{\mu}^{(m)})) \tag{2.2.18}$$

where

$$\mu_\ell(\underset{\sim}{\mu}^{(m)}) = \int J(H(x))dF_\ell(x), \qquad \ell=1, 2, \ldots, K$$

$$H(x) = \sum_{\ell=1}^{K} F_\ell(x)/K, \quad J = F_0^{-1}$$

$$A^2 = \int J^2(x)dx - (\int J(x)dx)^2 \tag{2.2.19}$$

is the distribution of (K-1)-variate normal vector $(Z_1, Z_2, \ldots, Z_{K-1})$ with

$$E(Z_\ell) = 0, \quad \text{Var}(Z_\ell) = 1 \qquad (\ell = 1, 2, \ldots, K-1)$$

$$\text{Cov}(Z_\ell, Z_{\ell'}) = \frac{1}{2} \qquad (\ell \neq \ell').$$

Theorem 2.2.2

For fixed $P^*$ and under the conditions (2.2.15) and (2.2.16), let $m$ be the smallest sample size per population for which

$$\inf_{\delta \geq \tilde{\delta}^{(m)}} P(CS|R_2) = P^* \tag{2.2.20}$$

and suppose that $F$ and $J = F_0^{-1}$ satisfy the regularity conditions of Theorem 6.1 and Lemma 7.2 of Puri [9]. Then as $m \to \infty$,

$$\tilde{\delta}^{(m)} = \Delta m^{-\frac{1}{2}} A / \int \frac{d}{dx}(J(F(x)))dF(x) + o(m^{-\frac{1}{2}}) \tag{2.2.21}$$

where $\Delta$ is determined by the condition (2.1.15).

Proof:

Let $T_{(\ell)}$ be the statistic associated with the population having location parameter $\mu_{[\ell]}$, $\ell=1, 2, \ldots, K$. Then

$$\begin{aligned} P(CS|R_2) &= P(T_{(K)} = \max\{T_{(1)}, T_{(2)}, \ldots, T_{(K)}\}) \\ &= P(T_{(\ell)} - T_{(K)} < 0, \quad \ell=1, 2, \ldots, K-1) \\ &= P(U_\ell < (m/2A^2)^{\frac{1}{2}} \cdot (\mu_{[K]}(\underset{\sim}{\mu}^{(m)}) - \mu_{[\ell]}(\underset{\sim}{\mu}^{(m)})), \\ &\qquad \ell=1, 2, \ldots, K-1). \end{aligned}$$

Under the LFC of $\mu_{ij}$'s or $\mu_\ell$'s, and using techniques of Puri and Puri [10], we have

$$\inf_{\delta \geq \tilde{\delta}^{(m)}} P(CS|R_2) = \lim_{m\to\infty} P(Z_\ell < (m/2A^2)^{\frac{1}{2}} \delta_\ell^{(m)}, \quad \ell=1, \ldots, K-1).$$

If $\Delta$ is obtained from (2.1.15), then

$$\inf_{\delta \geq \tilde{\delta}^{(m)}} P(CS|R_2) = P^*$$

if and only if (2.2.21) holds. This completes the proof.

Now if $P^*$ and $\delta^*$ are specified, identifying $\tilde{\delta}^{(m)}$ with $\delta^*$, it follows from the Theorem 2.2.2 that a large-sample solution for procedure $R_2$ is

$$m = \left[\text{the smallest integer} \geq \frac{A^2\Delta^2}{\delta^{*2}\left(\int \frac{d}{dx}(J(F(x)))dF(x)\right)^2}\right] \tag{2.2.22}$$

for which the probability requirement is satisfied.

Comparison of $R_1$ and $R_2$

Using standard asymptotics as in Dudewicz and Taneja [6], we obtain the following

Theorem 2.2.3

$$\text{ARE}_{P^*}(R_1,R_2) = \frac{\delta^{*2}}{K}\max\left\{\frac{k_1}{\delta_1^{*2}},\frac{k_2}{\delta_2^{*2}}\right\}. \qquad (2.2.23)$$

It follows that for large $P^*$, $\text{ARE}_{P^*}(R_1,R_2) \leq 1/2$. Hence, when there is no interaction, procedure $R_1$ performs better than $R_2$ in terms of asymptotic relative efficiency.

2.3 Asymptotic Efficiencies of Ranks-Procedures Relative to Means-Procedures

Let $n$ and $m = g(n)$ be the smallest sample sizes for means-procedure $P$ and ranks-procedure $R$ respectively to satisfy the same probability requirements under the same conditions in both the cases. Then Lehmann [7] definition of the asymptotic efficiency of $R$ relative to $P$ is

$$e_{R,P}(F) = \lim_{n\to\infty} \frac{n}{m}. \qquad (2.3.1)$$

The proof of the following theorem follows by equating $\underset{\sim}{\delta}^{(n)}$ and $\underset{\sim}{\tilde{\delta}}^{(m)}$ to $\underset{\sim}{\delta}^*$ for $P_1$ and $R_1$; and by equating $\delta^{(n)}$ and $\tilde{\delta}^{(m)}$ to $\delta^*$ for $P_2$ and $R_2$.

Theorem 2.3.1

$$e_{R_1,P_1}(F) = e_{R_2,P_2}(F) = \sigma^2 A^{-2}\left(\int \frac{d}{dx}(J(F(x)))dF(x)\right)^2. \qquad (2.3.2)$$

These efficiencies are given below for the ease of comparison for three F's and two $F_o$'s.

| $F_o$ | F | | |
|---|---|---|---|
| | Normal | Uniform | Double Exponential |
| Normal N(0,1) | 1 | $\infty$ | $4/\pi \sim 1.273$ |
| Uniform U(0,1) | 0.955 | 1 | 0.927 |

More generally, if $F_o$ is Uniform U(0,1), these efficiencies are $\geq$ 0.864 for all F; and if $F_o$ is Normal N(0,1), these efficiences are $>$ 1 for all nonnormal F. Thus, from the efficiency point of view ranks- and normal-scores procedures appear to be advantageous when compared with the means-procedures, unless one can be reasonably sure of the absence of gross errors and other departures of normality.

Remark 2.3.1

These asymptotic relative efficiences $e_{R_1,P_1}(F)$ and $e_{R_2,P_2}(F)$ are the same as that of Lehmann [7] for selecting the best population among $s$ populations, and as that of Puri [9] for corresponding tests in the c-sample problem.

## 3. NONPARAMETRIC SELECTION IN COMPLETE FACTORIAL EXPERIMENTS WITH INTERACTION

Now we assume that we have a complete factorial experiment with $k_1$ levels of factor 1 and $k_2$ levels of factor 2, and $\{X_{ijk}: k=1, 2, \ldots, N\}$ are independent random samples from populations $\pi_{ij}$ $(i=1, 2, \ldots, k_1; j=1, 2, \ldots, k_2)$ with continuous distribution functions $F(x-\mu_{ij})$ where

$$\mu_{ij} = \mu + \alpha_i + \beta_j + \gamma_{ij}$$

with side-conditions

$$\sum_{i=1}^{k_1} \alpha_i = 0.$$

$$\sum_{j=1}^{k_2} \beta_j = 0,$$

$$\sum_{i=1}^{k_1} \gamma_{ij} = 0 \quad \text{for all } j, \text{ and}$$

$$\sum_{j=1}^{k_2} \gamma_{ij} = 0 \quad \text{for all } i.$$

Here $\mu$ is the overall general effect, $\alpha_i$ is the $i^{th}$ level effect of factor 1, $\beta_j$ is the $j^{th}$ level effect of factor 2, and $\gamma_{ij}$ is the interaction effect of $i^{th}$ level of factor 1 and $j^{th}$ level of factor 2. Let the ordered $\mu_{ij}$'s be

$$\mu_{[1]} \leq \mu_{[2]} \leq \cdots \leq \mu_{[K]}, \; K = k_1 k_2.$$

Let us further assume that variance of $F$ exists and is given by $\sigma^2$.

Our <u>goal</u> is to select a population associated with $\mu_{[K]}$. We seek procedures $P$ which satisfy the probability requirement

$$P(CS|P) \geq P^*$$

where $\delta \equiv \mu_{[K]} - \mu_{[K-1]}$ is subject to the restriction

$$\delta \geq \delta^*$$

where $P^*$ and $\delta^*$ are set prior to the experiment such that

$$1/K < P^* < 1 \quad \text{and} \quad 0 < \delta^* < \infty.$$

Recall that for $\delta^* = \min\{\delta_1^*, \delta_2^*\}$, we have comparable probability requirements under $\delta \geq \delta^*$ and $\underset{\sim}{\delta} \geq \underset{\sim}{\delta}^*$ (componentwise).

We now wish to show that the procedure $P_1$ is vitiated by the presence of interaction (that is, it does not satisfy the probability requirement) whereas procedure $P_2$ remains valid (that is, it does satisfy the probability requirement). An interaction scheme described in Dudewicz and Taneja [6] can be utilized to prove the following theorem.

Theorem 3.1

Let n be the smallest sample size per population for which the procedure $P_1$ guarantees the probability requirement

$$P(CS|P_1) \geq P^*$$

whenever $\underset{\sim}{\delta}$ is restricted to the condition

$$\underset{\sim}{\delta} \geq \underset{\sim}{\delta}^* \quad \text{(componentwise)}$$

in a complete factorial experiment with zero interactions. Then in a general non-zero interaction model

$$\inf_{\underset{\sim}{\delta} \geq \underset{\sim}{\delta}^*} P(CS|P_1) \leq \frac{1}{K}.$$

Now, in order to support the claim that the procedure $P_2$ remains valid in the presence of interaction, we prove the following

Theorem 3.2

Let n be the smallest sample size per population for which procedure $P_2$ guarantees the probability requirement

$$P(CS|P_2) \geq P^*$$

whenever $\delta \equiv \mu_{[K]} - \mu_{[K-1]}$ is restricted to the condition

$$\delta \geq \delta^*, \quad \delta^* = \min\{\delta_1^*, \delta_2^*\}$$

in a complete factorial experiment with zero interactions. Then in a general non-zero interaction model

$$P(CS|P_2) \geq P^* \quad \text{whenever} \quad \delta \geq \delta^*, \quad \delta^* = \min\{\delta_1^*, \delta_2^*\}.$$

Proof:

In a model with zero interactions, if $n$ observations per population are taken, then the probability of correct selection using the procedure $P_2$ is

$$P(CS|P_2) = P(\overline{X}_{(\ell)} < \overline{X}_{(K)}, \ \ell=1, 2, \ldots, K-1)$$

$$\geq \lim_{n\to\infty} P\left(Z_\ell < \frac{\min\{\delta_1^{(n)}, \delta_2^{(n)}\}}{\sqrt{\frac{2\sigma^2}{n}}}, \ \ell=1, 2, \ldots, K-1\right)$$

whenever condition (2.1.3) along with (2.1.6) holds.

Now in a general non-zero interaction model, if we choose

$$\delta^{(n)} = \min\{\delta_1^{(n)}, \delta_2^{(n)}\},$$

then from Theorem 2.1.2 it follows that

$$\inf_{\delta \geq \delta^*} P(CS|P_2) = P^*$$

This completes the proof.

So, the procedure $P_1$ is vitiated by the presence of interaction whereas procedure $P_2$ remains valid for selecting a population with the largest location parameter in complete factorial experiments with interactions where the assumption of normality of the underlying distribution function $F$ is unreasonable.

Similarly, it can also be shown that the procedure $R_1$ is vitiated by the presence of interactions whereas the procedure $R_2$ remains valid.

## 4. FORMULATION OF ADAPTIVE PROCEDURES

### 4.1 Procedure Based on Means

We are now in a position to formulate a new adaptive procedure $P_3$. The procedure $P_3$ first tests the hypothesis that interactions are all zero, and then proceeds as does $P_1$ if the hypothesis is accepted, while proceeding as does $P_2$ if it is rejected. Work is in progress on this procedure $P_3$,

4.2 Procedure Based on Ranks

We have shown that procedure $R_1$ is not valid in the presence of interactions whereas procedure $R_2$ is; and that $R_1$ performs better than $R_2$ when there are no interactions (in which case both are valid). One may therefore wish to first test the presence of interactions (see Mehra and Smith [8], Bradley [4]) and then use accordingly between $R_1$ and $R_2$. Further work is in progress in developing a procedure $R_3$ which acts as does $R_1$ when interactions are insignificant and as does $R_2$ when interactions are significant.

## REFERENCES


[1] Bawa, V.S. (1972). Asymptotic efficiency of one R-factor experiment relative to R one-factor experiments for selecting the best normal population. J. Amer. Statist. Assoc., 67, 660-661.

[2] Bechhofer, R.E. (1954). A single-sample multiple decision procedure for ranking means of normal populations with known variances. Ann. Math. Statist., 25, 16-39.

[3] Bechhofer, R.E. (1977). Selection in factorial experiments. Proeedings of the 1977 Winter Simulation Conference, 65-70.

[4] Bradley, J.V. (1979). A nonparametric test for interactions of any order. J. Quality Tech., 11, 177-184.

[5] Dudewicz, E.J. (1977). New procedures for selection among (simulated) alternatives. Proceedings of the 1977 Winter Simulation Conference, 59-62.

[6] Dudewicz, E.J. and Taneja, B.K. (1982). Ranking and selection in designed experiments: complete factorial experiments. J. Japan Statist. Soc., 12, 51-62.

[7] Lehmann, E.L. (1963). A class of selection procedures based on ranks. Math. Annalen., 150, 268-275.

[8] Mehra, K.L. and Smith, G.E.J. (1970). On nonparametric estimation and testing for interactions in factorial experiments. J. Amer. Statist. Assoc., 65, 1283-1296.

[9] Puri, M.L. (1964). Asymptotic efficiency of a class of c-sample tests. Ann. Math. Statist., 35, 102-121

[10] Puri, M.L. and Puri, P.S. (1969). Multiple decision procedures based on ranks for certain problems in analysis of variance. Ann. Math. Statist., 40, 619-632.

[11] Taneja, B.K. (1986). Selection of best normal mean in complete factorial experiments with interaction and with common unknown variance. J. Japan Statist. Soc., 16, 53-65.

[12] Taneja, B.K. and Dudewicz, E.J. (1984). Selection of the best cell in $2 \times 2$ factorial experiments with interaction. Transactions of the Annual Quality Control Conference of the Rochester Section, American Society for Quality Control, 40, 305-350.

[13] Taneja, B.K. and Dudewicz, E.J. (1984). Selection in factorial experiments with interaction, especially the $2 \times 2$ case. Preprint #84-2, Department of Mathematics and Statistics, Case Western Reserve University, Cleveland, Ohio 44106. Submitted for publication.

# Bayesian Statistics in the Regional and Information Sciences

Reinhard Viertl
Institut für Statistik und
Wahrscheinlichkeitstheorie
Technische Universität Wien
A-1040 Wien, Austria

SUMMARY: The development of Statistics shows growing importance of Bayesian Inference. Especially in applications where all available information has to be used the Bayesian paradigm is superior by the possibility of using expert information in the measurable form of a-priori distributions. Different inference techniques for "Regional- and Information Statistics" are pointed out where Bayesian methods are of advantage compared with classical statistical inference.

## 1. ELEMENTS OF BAYESIAN STATISTICS

The fundamental ideas of Bayesian statistics are the following.

(i) All unknown quantities have to be described by stochastic quantities (also called random quantities).

(ii) The description of the uncertainty of these quantities have to be expressed by probabilities, i.e. coherent assessments by the analyst.

(iii) Probabilities don´t exist except in the mind but are assessments of experts which have to follow the rules of probability and are depending on the available information.

(iv) The transformation of the probability assessments if new information from data is available, is by using Bayes´theorem.

Contributions to Stochastics.
Ed. by W. Sendler
© Physica-Verlag Heidelberg 1987

## 1.1 Bayes´Theorem

If a stochastic Model $X \sim f(x|\theta)$ depends on the parameter $\theta \in \Theta$ this parameter is described by a stochastic quantity $\tilde{\theta}$. The a-priori uncertainty on $\theta$ is described by the so called *a-priori distribution* $\pi(\theta)$. If no special information is available a uniform distribution can be used.

If data $D = x_1, \ldots, x_n$ from X are available the a-priori information in the a-priori distribution is updated to the so called *a-posteriori distribution* $\pi(\theta|D)$ using the data D by Bayes´theorem

$$\pi(\theta|D) \quad \propto \quad \pi(\theta)l(\theta;D)$$

where $l(\theta;D)$ is the likelihood function. The symbol $\propto$ stands for proportionality, since the right hand side of the above equation is a density up to a constant.

## 1.2 Decision Analysis, Estimation, and Prediction

The formal elements in *decision analysis* are the following

$\tilde{\theta}$ parameter

$\pi(\theta|H_t)$ distribution of the parameter $\tilde{\theta}$ under information $H_t$ at time t

$L(\theta,d)$ loss for decision d if $\tilde{\theta} = \theta$

$r(\pi,d) = E_\pi L(\tilde{\theta},d) = \int_\Theta L(\theta,d)\,\pi(\theta|H_t)d\theta$ Bayes risk

A Bayesian decision $d^*$ in the class $\mathfrak{D}$ of possible decisions $d \in \mathfrak{D}$ is a decision which minimizes the Bayes risk $r(\pi,d)$, i.e.

$$r(\pi,d^*) = \min_{d \in \mathfrak{D}} r(\pi,d)$$

where $\pi = \pi(\theta|H_t)$ depends on the problem.

| | |
|---|---|
| $H_o$ | a-priori knowledge |
| $\pi(\theta\|H_o) = \pi(\theta)$ | a-priori distribution of $\tilde{\theta}$ |
| $H_1 = H_o,D$ | a-posteriori knowledge |
| $\pi(\theta\|H_1) = \pi(\theta\|D)$ | a-posteriori distribution of $\tilde{\theta}$. |

The decision analysis can equivalently be formulated using utility functions instead of loss functions.

The basis for *estimations* is formed by the a-posteriori distribution of the parameter (vector) $\tilde{\theta}$. Different characteristics of the a-posteriori distribution can be used for estimates. In connection with decision analysis these estimates have interpretations as optimal decisions relative to different loss functions. For example the mean of the a-posteriori distribution of a one-dimensional parameter $\theta$ is the optimal decision relative to quadratic loss. The median is the optimal estimate for the same problem under absolute value loss function.

*Bayesian confidence regions* $\Theta^*$, so called *HPD-regions* are obtained by solving the equation

$$\int_{\Theta^*} \pi(\theta|D)\,d\theta = \gamma$$

under the condition $\pi(\theta|D) \geq c$ for all $\theta \in \Theta^*$ and c is the largest possible value.

*Predictive distributions* for stochastic quantities X are obtained using the a-posteriori distribution $\pi(\theta|D)$ and the stochastic model $X \sim f(x|\theta)$ by

$$f(x|D) = \int_{\Theta} f(x|\theta)\,\pi(\theta|D)\,d\theta \quad .$$

## 2. BAYESIAN INFERENCE IN REGIONAL- AND INFORMATION SCIENCE

It is not easy to define what information is but in the Bayesian context a definition can be given. *Information is everything which changes the assessment of uncertainty*. This assessments can be used to treat different problems in applied statistics related to regional science and information science.

### 2.1 Allocation Problems

The allocation of funds by formula is a major use of data collected by statistical agencies. Allocation formulas $\underline{f}(\underline{z})$ are functions mapping data points $\underline{z}$ into n-dimensional Enclidean space. The data point $\underline{z}$ is a long vector containing all values used to determine allocations to all recipients in a particular allocation period. The number n is the number of recipients and the i-th coordinate of $\underline{f}(\underline{z})$ is the allocation to the i-th recipient. The vector of allocations intended by the policy-makers will be denoted by $\underline{\theta}$ and is called the optimal allocation.

Usually analyses begin from the assumption that the legislated formula is correct. This means that if $\underline{z}$ were observed without error then $\underline{\theta}$ would be allocated by the formula. To find an optimal allocation loss functions are used and as optimality principle the minimization of expected losses. This leads immediately to Bayesian decision analysis which is described in section 1.2.

## 2.2 Cluster Analysis

If one wishes to establish not only the identity but the affinity of observations the aim might be to group the data into homogeneous classes or *clusters*. Let $\underline{x}_1,\ldots,\underline{x}_n$ be measurements of p variables on each object which are believed to be heterogeneous. Then the measurements should be grouped into g homogeneous classes where g is unknown.

The stochastic formulation of the problem assumes $\underline{x}_1,\ldots,\underline{x}_n$ as independent. Further each may arise from any one of g possible sub-populations with probability density $f(\underline{x};\theta_k)$ for $k = 1,\ldots,g$, where g is assumed to be known.

Let $\gamma = (\gamma_1,\ldots,\gamma_n)^t$ be a set of identifying labels so that $\gamma_i = k \Rightarrow \underline{x}_i$ comes from the k-th sub-population. Denote by $C_k$ the set of $\underline{x}_i$ assigned to the k-th group by $\gamma$. Then the likelihood function is

$$l(\gamma;\theta_1,\ldots,\theta_g) = \prod_{\underline{x}\in C_1} f(\underline{x};\theta_1)\ldots\prod_{\underline{x}\in C_g} f(x;\theta_g)$$

and the likelihood method can be used to estimate $\gamma$ and the $\theta_k$'s respectively. The allocation rule yields the partition $\hat{C}_1,\ldots,\hat{C}_g$ under the m.l. estimate $\hat{\gamma}$.

Now a-priori information on $\underline{\theta} = (\theta_1,\ldots,\theta_g)^t$ in form of an a-priori distribution $\pi(\underline{\theta})$ can be used to construct Bayesian allocation rules using the a-posteriori distribution $\pi(\underline{\theta}|D)$ and suitable loss functions $L(.)$ .

## 2.3 Discriminant Analysis

If g populations or groups $G_1,\dots,G_g$ , $g \geq 2$ are considered and to each group $G_j$ there exists a probability density $f_j(\underline{x})$ on $\mathbb{R}^p$, so that if an individual comes from group $G_j$ it has p.d.f. $f_j(\underline{x})$. The subject of discriminant analysis is to allocate an individual to one of these g Groups on the basis of its measurements $\underline{x}$. It is desirable to make as few "mistakes" as possible in a sense to be made precise.

A *discriminant rule* d corresponds to a division of $\mathbb{R}^p$ into disjoint regions $R_1,\dots,R_g$ whose union is $\mathbb{R}^p$. The rule d is defined by

allocate $\underline{x}$ to $G_j$ if $\underline{x} \in R_j$ for $j = 1,\dots,g$ .

If a-priori information on the population an individual is likely to come from is available, it can be incorporated into a Bayesian approach.

*Discrimination when Populations are Known*

Writing $l_j(\underline{x})$ for the density $f_j(\underline{x})$ the maximum likelihood discriminant rule says one should allocate $\underline{x}$ to $G_j$, where

$$l_j(\underline{x}) = \max_i l_i(\underline{x}) .$$

Using a-priori probabilities of the various populations this information can be used to define Bayesian discriminant rules. Let us assume all a-priori probabilities $\pi_j$ are strictly positive $j = 1,\dots,g$ .

Definition: If the populations $G_1,\dots,G_g$ have a-priori probability vector $\underline{\pi} = (\pi_1,\dots,\pi_g)^t$, then the Bayes discriminant rule with respect to $\underline{\pi}$ allocates the observation $\underline{x}$ to the population $G_j$ for which

$$\pi_j \, l_j(\underline{x}) = \max_i \pi_i \, l_i(\underline{x}) .$$

The ML rule is a special case of the Bayes rule.

The probability of allocating an individual to group $G_i$, when in fact it comes from $G_j$, is given by

$$p_{ij} = \int_{R_i} l_j(\underline{x})\, d\underline{x} .$$

In particular, if an individual is from $G_i$, the probability of correct allocation is $p_{ii}$. The performance of the discriminant rule can be summarized by the numbers $p_{11},\ldots,p_{gg}$ .

Definition: A discriminant rule d with probabilities of correct allocation $(p_{ii})$ is as good as another discriminant rule $d'$, with probabilities $(p'_{ii})$ if

$$p_{ii} \geq p'_{ii} \quad \text{for all } i = 1,\ldots,g \ .$$

The discriminant rule d is better than $d'$ if at least one of the inequalities is strict. If d is a rule for which there exists no better rule, then d is called admissible.

Theorem 1: All Bayes discriminant rules are admissible.

For the proof see [ 6 ] p. 306.

Theorem 2: If populations $G_1,\ldots,G_g$ have a-priori probabilities $\pi_1,\ldots,\pi_g$ then no discriminant rule has a larger a-posteriori probability of correct allocation than the Bayes rule with respect to this a-priori distribution.

For the proof compare [ 6 ].

If costs are considered the discrimination problem can be stated in the language of decision theory. Let the cost be described by

$$L(i,j) = \begin{cases} 0 & \text{for } i = j \\ c_{ij} & \text{for } i \neq j \end{cases}$$

representing a *loss function* which represents the lost or loss incurred when an observation is allocated to $G_i$ when in fact it comes from $G_j$. Suppose $c_{ij} > 0$ for all $i \neq j$. If d is a discriminant rule with allocation functions $\varphi_j(\underline{x})$ defined by

$$\varphi_j(\underline{x}) = \begin{cases} 1 & \text{if } \pi_j \, l_j(\underline{x}) = \max\limits_i \pi_i \, l_i(\underline{x}) \\ 0 & \text{otherwise} \ , \end{cases}$$

then the *risk function* is given by

$$\begin{aligned} R(d,j) &= E(L(d(\underline{x}),j) \mid G_j) \\ &= \sum_i L(i,j) \int \varphi_i(\underline{x}) \, l_j(\underline{x}) \, d\underline{x} = \sum_{i \neq j} c_{ij} \, p_{ij} \ . \end{aligned}$$

For a-priori probabilities $\underline{\pi} = (\pi_1, \ldots, \pi_g)^t$ the *Bayes risk* is

$$r(\underline{\pi}, d) = \sum_j \pi_j R(d,j)$$

which is the a-posteriori expected loss. A decision rule d is *admissible* if there exists no other rule d' such that $R(d',j) \leq R(d,j)$ for all j, with at least one strict inequality.

Definition: The *Bayes rule* with respect to the a-priori probabilities $\pi_1, \ldots, \pi_g$ is as follows

allocate $\underline{x}$ to $G_j$ if $\sum_{k \neq j} c_{jk} \pi_k l_k(\underline{x}) = \min_i \sum_{k \neq i} c_{ik} \pi_k l_k(\underline{x})$

Theorem 3: If the populations $G_1, \ldots, G_g$ have a-priori probabilities $\pi_1, \ldots, \pi_g$ , then no discriminant rule has smaller Bayes risk for the risk function R than the Bayes rule with respect to $\underline{\pi}$ .

*Discrimination under Estimation*

If the forms of the distributions are known, but their parameters must be estimated from a data matrix $\underline{X}(n \times p)$ the m.l. method can be used.

A Bayesian method is to use an a-priori distribution $\pi(\theta)$ for the parameter $\tilde{\theta}$. Then the likelihood of an observation $\underline{x}$ given the data $\underline{X}$, on the assumption that $\underline{x}$ comes from the j-th population, is given by averaging the probability density of $\underline{x}$ given $\tilde{\theta}=\theta$ with respect to the a-posteriori density of $\theta$ given $\underline{X}$. That is

$$l_j(\underline{x}|\underline{X}) = \int f_j(\underline{x}|\theta)\,\pi(\theta|\underline{X})\,d\theta$$

with

$$\pi(\theta|\underline{X}) \propto f(\underline{X}|\theta)\,\pi(\theta) .$$

## 2.4 Sample Surveys

In sampling the Bayesian approach allows an immediate description of the uncertainty concerning proportions in a population.

For sampling with replacement the Multinomial distribution $M(n,p)$ with discrete density

$$f(x|p) = \frac{n!}{\prod_{i=1}^{k} (x_i!)} \prod_{i=1}^{k} p_i^{x_i} \quad \text{for} \quad \sum_{i=1}^{k} x_i = n$$

where $x = (x_1,\ldots,x_k)^t$ and $x_i$ are integers between 0 and n, $p = (p_1,\ldots,p_k)^t$ with $\sum_{i=1}^{k} p_k = 1$ and $0 \leq p_i \leq 1$ for all $i = 1,\ldots,k$ with its natural conjugate Dirichlet family $\mathfrak{D}(\alpha)$ with density

$$f(x|\alpha) = \frac{\Gamma(\alpha_1 + \ldots + \alpha_k)}{\prod_{i=1}^{k} \Gamma(\alpha_i)} \prod_{i=1}^{k} x_i^{\alpha_i - 1} \quad \text{for} \quad \sum_{i=1}^{k} x_i = 1$$

where $x = (x_1,\ldots,x_k)^t$ and $0 \leq x_i \leq 1$ for all $i = 1,\ldots,k$ , $\alpha = (\alpha_1,\ldots,\alpha_k)^t$ and $\alpha_i > 0$, is an effective model for the analysis and interpretation.

For inference on one proportion this model specializes to the Binomial distribution and the conjugate family of Beta distributions.

For the estimation problem if three possible outcomes for each unit exist the uncertainty can be expressed graphically by the a-posteriori density $\pi(x_1,x_2,x_3|D)$ using $x_1 + x_2 + x_3 = 1$ .

*Figure 1: A-posteriori density of pro and contra votings*

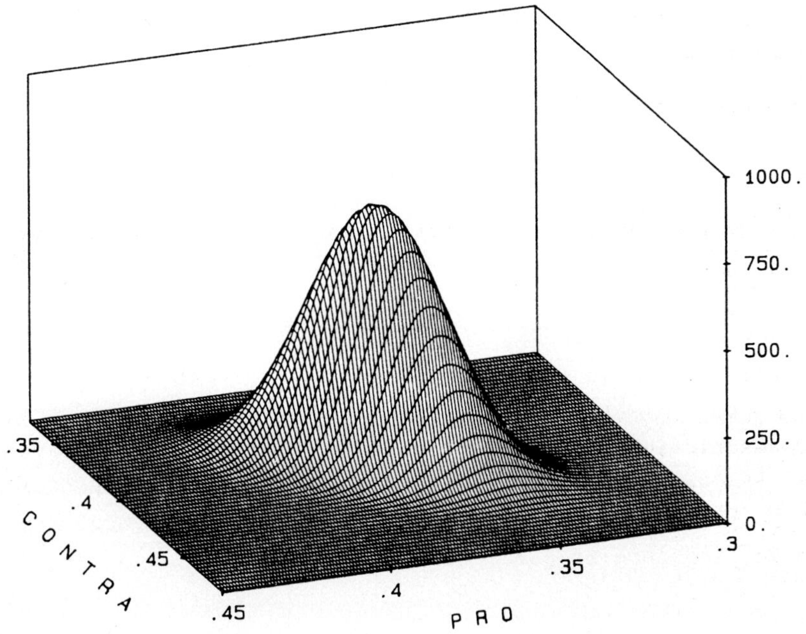

Figure 1 on the foregoing page is related to a sample survey taken in Vienna which concerns a power plant in the Hainburg region.

The a-posteriori distribution is the basis for decisions related to the survey problem.

One of the most important advantages of Bayesian Analysis is the possibility of a coherent answer to the question: What is the probability that the pro votings in the population are less than the contra votings.

$$\Pr\{\tilde{p} < \tilde{c}\} = \int_0^1 \int_0^c \pi(p,c,1-(p+c)\,|\,D)\,dp\,dc \ .$$

## 2.5 Regression Analysis

Bayesian regression analysis considers the regression parameter (vector) $\theta = (\theta_1,\ldots,\theta_k)'$ in the model

$$y = x'\theta + \varepsilon$$

as a stochastic quantity $\tilde{\theta}$. Depending on the distribution on the error term $\varepsilon$ with parameters $\alpha_1,\ldots,\alpha_m$ the likelihood function

$$l(\theta,\alpha_1,\ldots,\alpha_m;y)$$

is used to obtain the a-posteriori distribution

$$\pi(\theta,\alpha_1,\ldots,\alpha_m|y)$$

of the parameters and the a-posteriori marginal distribution of $\tilde{\theta}$

$$\pi(\theta|y)$$

which contains all relevant information for further analysis and decisions.

## 2.6 Time Series Analysis and Prediction

A dynamic description of time series is very natural in the Bayesian context by the Kalman filter. The general model is a sequence of observations $\ldots,X_t,X_{t+1},\ldots$ whose distributions are governed by the corresponding sequence of parameters $\ldots,\theta_t,\theta_{t+1},\ldots$ .

This results in a sequence

$$\ldots,\theta_t,X_t,\theta_{t+1},X_{t+1},\ldots$$

and the corresponding sequence of distributions

$$\ldots,\pi(\theta_t),f(x_t|\theta_t),\pi(\theta_{t+1}),f(x_{t+1}|\theta_{t+1}),\ldots \ .$$

The transition from $\pi(\theta_t)$ to $\pi(\theta_{t+1})$ is by using Bayes'theorem and the so called *system equation*

$$\tilde{\theta}_{t+1} = \psi(\tilde{\theta}_t,\tilde{u}_{t+1})$$

which is the model for the dynamic evolution of the sequence.

The transition is a two step procedure. The first step is using Bayes' theorem to obtain $\pi(\theta_t|x_t)$

$$\pi(\theta_t|x_t) \propto \pi(\theta_t) \cdot f(x_t|\theta_t) \ .$$

The second step is using the transformation theorem for probability densities of functions of stochastic quantities. Here the function is the system equation

$$\tilde{\theta}_{t+1} = \psi(\tilde{\theta}_t,\tilde{u}_t)$$

and by the transformation theorem the density $\pi(\theta_{t+1})$ of $\tilde{\theta}_{t+1}$ is obtained.

From this distribution the *predictive distribution* $f(x_{t+1})$ of $X_{t+1}$ can be calculated by

$$f(x_{t+1}) = \int_{\Theta} f(x_{t+1}|\theta_{t+1})\pi(\theta_{t+1})d\theta_{t+1} \ .$$

For normal distributed quantities in the *dynamic linear model* with *observation equation*

$$X_t = F_t\theta_t + \tilde{v}_t$$

where $F_t$ is a known matrix and the *observation error* $\tilde{v}_t \sim N(\underline{O},V_t)$, known variance matrix $V_t$, under the assumption of a *system equation*

$$\tilde{\theta}_{t+1} = G_{t+1}\,\tilde{\theta}_t + \tilde{w}_{t+1},$$

$G_t$ being a known matrix and the *system equation error* $\tilde{w}_t \sim N(\underline{O},W_t)$, with $W_t$ known also, the corresponding distributions can be calculated.

$$\tilde{\theta}_t | x_{t-1} \sim N(G_t \hat{\theta}_{t-1}, R_t)$$

with $R_t = G_t \Sigma_{t-1} G_t' + W_t$ where $\hat{\theta}_{t-1}$ is the expectation and $\Sigma_{t-1}$ the variance of $\tilde{\theta}_{t-1} | x_{t-1}$. After the observation of $x_t$ the distribution of $\tilde{\theta}_t$ is

$$N(G_t \hat{\theta}_{t-1} + R_t F_t' (V_t + F_t R_t F_t')^{-1} (Y_t - F_t G_t \hat{\theta}_{t-1}) \ ,$$
$$R_t - R_t F_t' (V_t + F_t R_t F_t')^{-1} F_t R_t) \ .$$

This is the desired a-posteriori distribution. For details of the calculation compare [7].

## 2.7 Further Fields of Application

Besides the above mentioned applications of Bayesian analysis in regional and information science further fields where Bayesian inference is useful are for example in connection with environment statistics

Analysis of Variance
Entropy Models
Factor Analysis
Pattern Recognition .

# 3. FUTURE DIRECTIONS OF DEVELOPMENT

One of the main problems in regional statistics is the situation that usually no identical repetitions of experiments are possible. Therefore a theory of similar observations would be necessary. Another problem is the fuzzyness in many data. To model these phenomena fuzzy numbers and fuzzy probabilities could be used. In order to use a-priori information a modification of Bayesian analysis for fuzzy data is necessary. An extension of Bayesian inference is on the evolutionary path deterministic models - true stochastic models - Bayesian models - Fuzzy Bayesian modelling.

REFERENCES

[1] J.P. Ancot: Bayesian Estimation of a Simple Model with a Spatially Distributed Explanatory Variable, Contributed paper at the 23rd R.S.A. European Meeting, 1983

[2] J.O. Berger: *Statistical Decision Theory and Bayesian Analysis*, 2nd edition, Springer Series in Statistics, New York, 1985

[3] L. Bosse, W. Eberl: *Stochastische Verfahren in den technischen Wissenschaften und in der amtlichen Statistik*, Schriftenreihe der Technischen Universität Wien, Bd. 18, Wien, 1981

[4] G.E.P. Box, G.C. Tiao: *Bayesian Inference in Statistical Analysis*, Addison - Wesley Publ., Reading, Massachusetts, 1973

[5] W. Eberl, O. Moeschlin: *Mathematische Statistik*, de Gruyter, Berlin, 1982

[6] K.V. Mardia, J.T. Kent, J.M. Bibby: *Multivariate Analysis*, Academic Press, London, 1979

[7] R.J. Meinhold, N.D. Singpurwalla: Understanding the Kalman Filter, *The American Statistician*, Vol. 37, No. 2 (1983)

[8] G. Paaß: *Prognose und Asymptotik Bayes'scher Modelle*, Oldenbourg, München, 1984

[9] J. Pilz: *Bayesian Estimation and Experimental Design in Linear Regression Models*, Teubner-Texte zur Mathematik, Bd. 55, Leipzig, 1983

[10] P.M. Schulze: *Region und Informationssystem*, Minerva-Publikation, München, 1980

[11] B.D. Spencer: *Benefit-Cost Analysis of Data Used to Allocate Funds*, Lecture Notes in Statistics, Vol. 3, Springer-Verlag, New York, 1980

[12] R. Viertl: Bayesian Statistics and Applications in Regional- and Information Statistics, *Proceedings 15th IARUS Conference*, Innsbruck, 1986

[13] R. Viertl, K. Bousek: Regional Statistics for Social and Economic Comparison: An Austrian Example, *Proceedings 44th Session of the International Statistical Institute*, Vol. 2, Madrid, 1983

# Inequalities for Convex Functions

István Vincze
Mathematical Institute of the
Hungarian Academy of Sciences
H-1053 Budapest, Hungary

SUMMARY: In the present paper a completed form of Jensen's inequality along with some of its consequences is given. As an application a stronger form of an inequality by G. J. MINTY [1] is obtained, which claims that the "gradient" of a convex function defined in a (certain kind of) linear space is also "monotone", like the derivative in the case of a real, convex function.

## §.1. A completed form of Jensen's inequality

1.1. Let $f(x)$ be a convex function in an open interval $I=(a,b)$ of the real line. Let further $X$ be a random variable taken on values in $I$ ; the expected value and the variance will be denoted by $E$ and $D^2$, respectively. In the sequel $f'(x)$ will mean any value between $f'(x-0)$ and $f'(x+0)$ when these values differ, which may occur at countably many points only. From the convexity it follows that for any pair $u,x \in I$ the relation

$$f(x) \geq f(u) + (x-u)f'(u) \tag{1.1}$$

holds, i.e. the graph of $f(x)$ lies above all of its supporting lines. Replacing $u$ by $EX$ and taking $x=X$, the expected value of the relation gives the inequality of Jensen

$$0 \leq Ef(X) - f(EX) . \tag{1.2}$$

Key words: Jensen's inequality, convexity, inequalities for mean values, "monotonicity" of gradient
AMS subject classification: 26A48, 26D15

Contributions to Stochastics.
Ed. by W. Sendler
© Physica-Verlag Heidelberg 1987

But carrying through this procedure conversely, i.e. $x=EX$ and $u=X$, applying also (1.2), we obtain a completed form of the inequality of Jensen:

Theorem 1.1. *Using our above notations and assumptions the relation*

$$0 \leq Ef(X) - f(EX) \leq E[(X-EX)f'(X)] \tag{1.3}$$

*is true; the equality signs hold if and only if* $f$ *is linear, or if* $X$ *is concentrated in a single point.*

Theorem 1.1 contains the statements that the quantity on the right hand side of (1.3) is nonnegative, moreover that the value of it has the lower bound $Ef(X)-f(EX)$ which can be called the convexity or Jensen excess of $f(x)$ with respect to the random variable $X$.

We also have

Corollary 1.1. *If* $f(x)$ *is convex in* $I$, *then*

$$0 \leq Ef(X) - f(EX) \leq D(X)D[f'(X)]$$

*holds.* -

1.2. Here we turn to two special cases:

a) Let $X$ be uniformly distributed in $I = (a,b)$; then $EX = \frac{a+b}{2}$,

$$E\left[(X - \tfrac{a+b}{2})f'(X)\right] = \frac{f(a)+f(b)}{2} - \frac{1}{b-a}\int_a^b f(t)dt .$$

Consequently we have

$$0 \leq \frac{1}{b-a}\int_a^b f(t)dt - f(\tfrac{a+b}{2}) \leq \frac{f(a)+f(b)}{2} - \frac{1}{b-a}\int_a^b f(t)dt . \tag{1.4}$$

From this relation follows the known inequality of Hadamard: From the convexity of $f(x)$ in $(a,b)$ we have the following relation

$$f(\tfrac{a+b}{2}) \leq \frac{1}{b-a}\int_a^b f(t)dt \leq \frac{f(a)+f(b)}{2} \tag{1.5}$$

b) Here we turn to the case of discrete distributions. Suppose $X$ takes values $x_1 < x_2 < \ldots < x_n$ $(a < x_1, b > x_n)$ with corresponding probabilities $\alpha_1, \alpha_2, \ldots, \alpha_n$ $\alpha_i \geq 0$, $i = 1,2,\ldots,n$, $\sum_{i=1}^n \alpha_i = 1$. Then with the notation $\bar{x}_\alpha = \sum_{i=1}^n \alpha_i x_i$ our inequality (1.3) has the form

$$0 \leq \sum_{i=1}^n \alpha_i f(x_i) - f\left(\sum_{i=1}^n \alpha_i x_i\right) \leq \sum_{i=1}^n \alpha_i (x_i - \bar{x}_\alpha) f'(x_i) . \tag{1.6}$$

For obtaining a best upper bound, the value of $f(x_i)$ has to be taken equal to $f(x_i+0)$, when $x_i < x_\alpha$ and to $f(x_i-0)$ when $x_i > x_\alpha$.

In the special case $n=2$ - making use of the notation $\alpha_2=\alpha$ and being $x_1-\bar{x}_\alpha = -\alpha(x_2-x_1)$, $x_2-\bar{x}_\alpha = (1-\alpha)(x_2-x_1)$ - our relation has the form

$$0 \le (1-\alpha)f(x_1)+\alpha f(x_2) - f((1-\alpha)x_1+\alpha x_2) \le$$
$$\le \alpha(1-\alpha)(x_2-x_1)(f'(x_1)-f'(x_0)) \qquad (1.7)$$

proving also the simple statement that *the derivative of a convex function is monotonously increasing.*

## §.2. Inequalities for two point distributions

2.1. The same conditions and notations will be used as above. We remark that in our inequalities for $\alpha\to 0$ or $\alpha\to 1$ the equality sign holds in the first and in the second place both.

Proposition 2.1. *For the convex function* $f(x)$, $x\in I$ *the following inequality is true*

$$0 \le (1-\alpha)f(x_1)+\alpha f(x_2) - \frac{1-\alpha}{x_\alpha-x_1}\int_{x_1}^{x_\alpha} f(t)\,dt - \frac{\alpha}{x_2-x_\alpha}\int_{x_\alpha}^{x_2} f(t)\,dt \le$$
$$\le 2\alpha(1-\alpha)\left\{\frac{x_2-x_1}{4}(f'(x_2)-f'(x_1)) + \frac{f(x_1)+f(x_2)}{2} - \right.$$
$$\left. - \frac{1}{x_2-x_1}\int_{x_1}^{x_2} f(t)\,dt\right\}, \qquad (2.1)$$

*where* $x_\alpha = (1-\alpha)x_0 + \alpha x_1$, $0\le\alpha\le 1$.

Proof: Let us write $x_2=t$ in (1.7); integrating both sides with respect to $t$ from $x_1$ to $x_2$, we obtain - using a slight calculation -

$$0 \le (1-\alpha)f(x_1)(x_2-x_1) + \int_{x_1}^{x_2} f(t)\,dt - \frac{1}{\alpha}\int_{x_1}^{x_\alpha} f(t)\,dt \le$$
$$\le \alpha(1-\alpha)\left\{f(x_2)(x_2-x_1) - \int_{x_1}^{x_2} f(t)\,dt - f'(x_1)\frac{(x_2-x_1)^2}{2}\right\}. \qquad (2.2)$$

Turning now in (1.7) to $x_1=t$ and integrating according to $t$ from $x_1$ to $x_2$ we obtain

$$0 \le (1-\alpha)\int_{x_1}^{x_2} f(t)\,dt + \alpha f(x_2)(x_2-x_1) - \frac{1}{1-\alpha}\int_{x_\alpha}^{x_2} f(t)\,dt \le$$

$$\le \alpha(1-\alpha)\left\{ f(x_1)(x_1-x_0) - \int_{x_1}^{x_2} f(t)\,dt + f'(x_2)\frac{(x_2-x_1)^2}{2}\right\} . \qquad (2.3)$$

Adding (2.2) and (2.3) after a simple alteration we obtain our statement in (2.1). -

We remark that the left hand part of (2.1) can be written in the form

$$0 \le (1-\alpha)f(x_1)+\alpha f(x_2) + \frac{1}{x_2-x_1}\int_{x_1}^{x_2} f(t)dt - \frac{1}{x_\alpha - x_1}\int_{x_1}^{x_\alpha} f(t)dt - \frac{1}{x_2-x_\alpha}\int_{x_\alpha}^{x_2} f(t)dt ,$$

holding equality for $\alpha=0$ and $\alpha=1$.

The next relation is an extension of the inequality by Hadamard:

Proposition 2.2. *Whenever* $f(x)$ *is convex in* $(a,b)$ *the following relation holds* — $a < x_1 < x_2 < b$ , $0 \le \alpha \le 1$ :

$$f((1-\alpha)x_1+\alpha x_2) \le \frac{1-\alpha}{x_\alpha - x_1}\int_{x_1}^{x_\alpha} f(t)\,dt + \frac{\alpha}{x_2-x_\alpha}\int_{x_\alpha}^{x_2} f(t)\,dt \le$$

$$\le (1-\alpha)f(x_1)+\alpha f(x_2) \quad . \text{ —} \qquad (2.4)$$

Proof: Let us consider a supporting line (tangent) to the graph of $f(x)$ at the point $X = X_\alpha$. This straight line cuts the ordinates belonging to $x_1$ and $x_2$ in heights (say) $z_1$ and $z_2$ respectively. Trivially

$$f(X_\alpha) = (1-\alpha)z_1+\alpha z_2 = \frac{1}{2}\Big[(1-\alpha)(z_1+f(x_\alpha))+\alpha(z_2+f(x_\alpha))\Big] .$$

This way we can write

$$f(x_\alpha) = \frac{1-\alpha}{x_\alpha - x_1}\,\frac{(z_1+f(x_\alpha))(x_\alpha-x_1)}{2} + \frac{\alpha}{z_2-z_\alpha}\,\frac{(z_2+f(x_\alpha))(x_2-x_\alpha)}{2} \le$$

$$\le (1-\alpha)\frac{1}{x_\alpha - x_1}\int_{x_1}^{x_\alpha} f(t)\,dt + \alpha\frac{1}{x_2-x_\alpha}\int_{x_\alpha}^{x_2} f(t)\,dt ,$$

justifying the left hand side of inequality (2.4).

The statement on the right hand side of (2.4) is contained in our Proposition 2.1. —

## §.3. A stronger version of an inequality by G. J. MINTY

In his paper [1] G. J. MINTY proved that in a linear space $Y$ the "gradient" of a convex function $\Phi$ defined in a convex domain $G \subset Y$ is monotone in the following sense: for any two points $x_1, x_2 \in G$ the inner product

$$\langle x_2 - x_1,\ \text{grad}\ \Phi(x_2) - \text{grad}\ \Phi(x_1)\rangle \tag{3.1}$$

is nonnegative. In the mentioned paper the conditions for $Y$, along with the definition of a generalized gradient, under which the statement is true, are given. We intend to strengthen this inequality showing that *the quantity in* (3.1) *is even larger than (or equal to) the nonnegative Jensen excess*

$$(1-\alpha)\Phi(x_1) + \alpha\Phi(x_2) - \Phi((1-\alpha)x_1 + \alpha x_2) \tag{3.2}$$

*divided by* $\alpha(1-\alpha)$ .

For proving our statement we do not turn to the question in the generality used by MINTY; we restrict ourselves to a convex function $\Phi$ defined in a convex domain $G$ of the Euclidean $k$-space $Y = R_k$ $(k > 1)$. The corresponding part of the proof of MINTY can be changed by applying our procedure below.

We assume that the normal vectors of $\Phi(x)$ exist in $G$, i.e. the gradient of $\Phi(x)$ is well defined $x \in G$ ; considering the supporting hyperplane of $\Phi$ in a point $u \in G$, from the convexity follows the relation for all $x \in G$

$$\Phi(x) \ge \Phi(u) + \langle x-u,\ \text{grad}\ \Phi(u)\rangle\ . \tag{3.3}$$

Taking $u=x_i$ , $x_i \in G$ , $i=1,2,\ldots,n$ , and $x = \bar{x}_\alpha = \sum_{i=1}^{n} \alpha_i x_i$ , $\alpha_i \ge 0$ , $\sum_{i=1}^{n} \alpha_i = 1$ , multiplying by $\alpha_i$ and summarizing with respect to $i$ , we have

$$0 \le \sum_{i=1}^{n} \alpha_i \Phi(x_i) - \Phi\left(\sum_{i=1}^{n} \alpha_i x_i\right) \le \sum_{i=1}^{n} \alpha_i \langle x_i - \bar{x}_\alpha,\ \text{grad}\ \Phi(x_i)\rangle \tag{3.4}$$

obtaining the $k$-variate form of (1.6). Turning to the case $n=2$, we obtain the relation

$$0 \le \sup_\alpha \frac{1}{\alpha(1-\alpha)}\left\{(1-\alpha)\Phi(x_1) + \alpha\Phi(x_2) - \Phi((1-\alpha)x_1 + \alpha x_2\right\} \le$$

$$\le \langle x_2 - x_1,\ \text{grad}\ \Phi(x_2) - \text{grad}\ \Phi(x_1)\rangle$$

giving a stronger form of the inequality by MINTY.

REFERENCES

[1] G. J. Minty: On the monotonicity of the gradient of a convex function. *Pacific Journal of Mathematics* 14 (1964), 243-247.

# Em. o. Univ. Prof. Dr. Walther Eberl

Walther Eberl was born on 11 April 1912 in Wien. Here he obtained his elementary and high school education. Then he joined the University of Vienna as a student of mathematics. In 1936 he finished his studies with a Ph.D. degree in mathematics. His doctoral thesis was in the field of algebra. In the same year he started studying law and finished the first part of the studies with the so called "first state examination". But in 1939 he had to join the army and was there up to the end of the second world war in 1945.

From 1945 to 1948 Dr. Eberl got first experience with industry in a pharmacentical company. In 1949 he was the leader of a precision nivellement in the Austrian regions Salzburg and Tirol in charge of the national office for survey measurement. In the same year he became assistent in the mathematics department at the Technische Hochschule Wien. The conditions for research in these times were hard. Therefore he accepted a stipend by the ministry of transportation to study the applications of statistical methods in industry during half a year in England. There he worked as statistician in the division of quality control of Lucas Ltd. at Birmingham. Also he took courses at the department for industrial engineering at Birmingham University.

Back at Vienna in 1956 Dr. Eberl obtained the "venia legendi" for "Mathematical Statistics and their Analytic Foundations" and became Dozent at the University of Technology Vienna (TH Wien).

In the same year he worked as a visitor at the department of management of the ETH Zürich. The fellowships at Birmingham and Zürich had essential influence on the curriculum of the studies in "computing" and for the work of the "Mathematical Laboratory" at the TH Wien. During the winter semester 1965/66 Dozent Eberl was visiting professor at the University Göttingen and during the academic year 1966/67 he was holding the position of a visiting professor at the University Karlsruhe.

In 1967 he accepted the position of a full professor at the new founded "Institut für Statistik" at the University of Technology Vienna. Here he had to organize a number of different courses ranging from measure theory trough stochastic processes to applied and mathematical statistics. As a result of professor Eberl´s influence the curriculum of mathematics at the TU Wien contains different courses in modern stochastics. To cover all these fields the department was rather understaffed and professor Eberl was trying hard to solve this shortage. Short time before his retirement he was succesful and the minister of science and research founded one additional chair for applied statistics with special emphasis on regional and information sciences in 1980 and another chair for probability theory and theory of stochastic processes in 1981. The dedication of the chair held by professor Eberl was changed to engineering statistics and he worked with emphasis on applications until his retirement in 1982.

During his work at the Technische Universität Wien professor Eberl was successful as academic teacher. A number of his assistents obtained the "venia docendi" and three of his students are holding chairs at different universities. He also initiated a postgraduate curriculum in management, law and economics at the TU Wien which was realized after his retirement.

The publications of professor Eberl are ranging from abstract mathematics to applied statistics and articles for the general reader to support the availability of stochastic modelling for the applied sciences.

Besides his academic work professor Eberl also gave courses in continuing engineering education for practicing engineers and he was the initiator for founding an Austrian Society of Quality Control. By his numerous altruistic activities which provided the basis for appropriate research and teaching in stochastics at the TU Wien and his work outside the university professor Walther Eberl is an outstanding promotor of statistics for technolgy in Austria.

# Publications by Professor Walther Eberl


[1] Zur Multiplikation reeller Zahlen, *Zeitschrift für Angewandte Mathematik und Physik*, No. 1 (1950).

[2] Zusammenstellung von Bezeichnungsweisen der Vektor- und Tensorrechnung, *Abhandlungen des Dokumentationszentrums der Technik*, No. 2, Wien, 1951.

[3] Ein Zufallsweg in einer Markoffschen Kette von Alternativen, *Monatshefte für Mathematik*, Vol. 58, No. 3 (1954).

[4] *Technische Statistik und Planungsforschung als Hilfsmittel der Produktionstechnik* (Operational Research in Engineering Production), Bericht über einen Studienaufenthalt in England, 1954.

[5] Planungsforschung, *Statistische Vierteljahresschrift* 8 (1955).

[6] Die Summenverteilung verketteter Alternativen, *Österreichisches Ingenieur-Archiv* 9 (1955).

[7] Die Kontrolle der Druckfestigkeit von Beton durch Stichproben, *Österr. Ingenieur-Archiv*, Vol. 11, No. 3 (1957) (with G. Schneeweiß).

[8] *Die Mathematik als Hilfsmittel der Betriebswissenschaften*, Bericht über einen Studienaufenthalt an der ETH Zürich, 1957.

[9] Quellen und Schwinden des Holzes, *Holzforschung und Holzverwertung*, Vol. 11, No. 3 (1959) (with A. Gratzl).

[10] Über die statistische Kontrolle der Härte von Zigaretten, *Fachliche Mitteilungen der Österr. Tabakregie*, Sonderheft, 1959, (with W. Sedlatschek).

[11] Die Ausgleichung vermittelnder Beobachtungen im Rahmen der mathematischen Statistik, *Österr. Zeitschrift für Vermessungswesen*, Vol. 47, No. 3 (1959).

[12] Statistische Kontrollen in der industriellen Produktion, *Österr. Ingenieur-Zeitschrift*, Vol. 3, No. 3 (1960).

[13] Bemerkung über ein Theorem für additive Mengenfunktionen, *Metrika* 3 (1960).

[14] Die Anwendung der Probentheorie auf die Prüfung wirtschaftlicher Sachverhalte und Vorgänge, *Mathematik-Technik-Wirtschaft*, No. 8 (1961) (with P. Swoboda).

[15] Verbandstheoretische Verallgemeinerung der Poincaréschen Formel der Wahrscheinlichkeitsrechnung, *Archiv der Mathematik*, Vol. 12 (1961).

[16] Statistische Kontrollen im Betrieb, *Zeitschrift des österr. Vereines zur Förderung der betriebswissenschaftlichen Forschung und Ausbildung*, No. 1 (1962).

[17] Lineare Planungsrechnung (Linear Programming), *Mathematik-Technik-Wirtschaft*, No. 10 (1963).

[18] Zur wahrscheinlichkeitstheoretischen Deutung gewisser Mannschaftswettkämpfe, *Österr. Ingenieur-Archiv*, Vol. 10 (1965).

[19] Lineare Programme, *Unternehmensforschung* 1 (1965).

[20] Statistische Verfahren in der Regelungstheorie, *Operations Research Verfahren* 2 (1966).

[21] Ein Zusammenhang der Zufallsteilung von Strecken mit dem Rencontreproblem, *Zeitschrift für Angewandte Mathematik und Mechanik*, Vol. 46, No. 3/4 (1966).

[22] Die Ermittlung von Konfidenzintervallen für eindimensionale stetige Parameter diskreter Verteilungen, *Operations Research Verfahren* 3 (1967).

[23] Statistik - Brücke zwischen den Wissenschaften, *Antrittsvorlesungen an der Technischen Hochschule in Wien*, No. 8 (1968).

[24] Die Bedeutung der Statistik für die Allgemeinbildung, *Die österreichische Höhere Schule*, No. 21 (1969).

[25] Das Testen von Fraktilen normalverteilter Merkmale, *Operations Research Verfahren* 5 (1969).

[26] Reine und Angewandte Mathematik, Arbeitsgruppen und Gruppenarbeit, *Österreichische Hochschulzeitung*, Juni 1970.

[27] Die Entwicklung der Wirtschaftswissenschaften an den Technischen Hochschulen, *Informationen der TH Wien*, No. 2 (1971).

[28] Die asymptotische Verteilung von Koinzidenzen, *Zeitschrift für Wahrscheinlichkeitstheorie und verwandte Gebiete* 18, (1971) (with R. Hafner).

[29] Die Bedeutung des Satzes von Ionescu Tulcea für die Entscheidungstheorie, *Operations Research Verfahren* 14 (1972) (with H. Stadler).

[30] Die Stochastik als Wissenschaft, *Österr. Hochschulzeitung*, September 1972.

[31] Die Stochastik an Technischen Hochschulen, *Österr. Hochschulzeitung*, Oktober 1972.

[32] Die Bedeutung der Stochastik an Technischen Hochschulen, *Informationen der TH Wien*, No. 3 (1972).

[33] Mathematische Labors, *Die Industrie*, No. 44 (1972).

[34] Angewandte Statistik und EDV, *Informationen der TH Wien*, No. 1 (1974).

[35] Aufgaben und Stellung der Stochastik an Technischen Hochschulen, *Die Industrie*, No. 39 (1974).

[36] Statistische Qualitätskontrolle (SQK), *Die Industrie*, No. 46 (1974).

[37] Probleme der Anwendung stochastischer Methoden in Technik und Wirtschaft, *Der Privatangestellte*, No. 9 (1974).

[38] Stochastics and All-Round-Education, in: *Statistics at the School Level*, Almquist & Wicksell, Stockholm, 1975.

[39] Mathematik an Technischen Hochschulen, *Die Industrie*, No. 17 (1975).

[40] Mathematik an Technischen Universitäten, *Informationen der TU Wien*, No. 2 (1975).

[41] Statistische Qualitätskontrolle (SQK) als Führungsaufgabe, *Die Industrie*, No. 36 (1975).

[42] Kongreß der European Organization for Quality Control, *Die Industrie*, No. 27 (1976).

[43] Reichsbrücke und Qualitätskontrolle, *Die Industrie*, No. 41 (1976).

[44] Statistische Qualitätskontrolle (SQK), in: Heft 29 der Schriftenreihe der Bundeswirtschaftskammer, Wien, 1977.

[45] Einige Bemerkungen zur Terminologie der Statistischen Qualitätssicherung, *Qualität und Zuverlässigkeit*, No. 7 (1977).

[46] Stochastikinstitute an technischen Universitäten, *Österr. Hochschulzeitung*, Juli 1979.

[47] *Bericht über die Situation der Stochastikinstitute an technischen Universitäten im deutschen Sprachraum*, Wien, 1979.

[48] Betonung technischer und wirtschaftlicher Notwendigkeiten, *Österr. Hochschulzeitung*, Juli 1981.

[49] Ein Institut im Aufbau, *Österr. Hochschulzeitung*, Mai 1982.

[50] Reine und Angewandte Mathematik an technischen Universitäten ("Rettet die Mathematik"), *Die Industrie*, No. 9 (1983).

Books and Monographs

[51] *Einführung in die Stochastik*, part 1: Wahrscheinlichkeitstheorie, part 2: Mathematische Statistik, Schriftenreihe der Wiener Schwachstromwerke, No. 19, 1965.

[52] *Produktion und statistische Qualitätskontrolle* (ed.), Schriftenreihe der Bundeswirtschaftskammer, No. 29, Wien, 1977 (with J. Kühne).

[53] *Stochastische Verfahren in den technischen Wissenschaften und in der amtlichen Statistik* (ed.), Schriftenreihe der Technischen Universität Wien, Wien, 1981 (with L. Bosse).